MW01643973

AUSTRALIA'S OIL EXPLORERS ACROSS TWO CENTURIES

WHERE GOD NEVER TROD

AUSTRALIA'S OIL EXPLORERS
ACROSS TWO CENTURIES

WHERE GOD NEVER TROD

Rick Wilkinson

David Ell Press Pty Ltd

First published in 1991 by
David Ell Press Pty Ltd
95 Beattie Street, Balmain NSW 2041

Design by Christie & Eckermann Art and Design Studio, Sydney
Typeset by Post Typesetters, Brisbane
Printed in Australia by Southwood Press Pty Ltd, Sydney

National Library of Australia
Cataloguing-in-Publication Data

Wilkinson, Rick, 1947–
Where God never trod.

Bibliography.
Includes index.
ISBN 0 908197 12 8.

1. Petroleum industry and trade — Australia — History.
2. Petroleum — Australia — Prospecting — History.
3. Petroleum industry and trade — Papua New Guinea — history. 4. Petroleum — Papua New Guinea — Prospecting — History. I. Title.

338.272820994

AUTHOR'S NOTE

I have always found exploration to be the most colourful part of the oil industry. When trying to chronicle Australia's oil story in *A Thirst For Burning* during the early 1980s, it was not possible to devote more than a few chapters to the country's search for petroleum. Consequently a great deal of fascinating material had to be culled. I thought then that to present the subject of exploration fully would require a separate book, but the idea didn't really progress until the official luncheon of the 1986 Australian Petroleum Exploration Association (APEA) conference in Adelaide. Several veterans of the industry kept our table entertained throughout the meal with a series of humorous stories and banter about 'the old days'. Afterwards, the person next to me lent over and said that I really should capture these things on paper before it was too late. The pioneers, he pointed out, were not getting any younger.

The comment spurred me into action. Advised to go beyond just a collection of funny stories, the aim was to portray Australia's oil exploration as seen and lived by the people who were, and still are, involved. The project became a sort of eyewitness account of the challenges thrown up by the isolation, difficult terrain and changing politics, as well as a story of the innovation, perseverence and humour of the teams which made the effort. Also, I was keen to set this picture against the slow rise in understanding of the country's geology and the gradual improvements in oil search technology over the years. Work in eastern New Guinea was included because of the Australian Government's long administration there and the continued interest of Australian companies in the region since Independence.

The list of willing helpers is almost as long as a roll call of the industry itself. My thanks go to all the 'co-authors' of this book. Some appear in person in the numerous quotations, but there are many others who provided invaluable background, lent hard-to-find references and pointed me in the direction of numerous sources I would never have known existed. A special thanks is due to librarian Fiona Finlayson and editor of publications Doug Rose at the Australian Institute of Petroleum in Melbourne, Evelyn Young and Charlie Modrack at the

BMR in Canberra, and the staff of the APEA secretariat in Sydney for their help and patience with my continual requests for maps, files and statistical data. I am also grateful to Emeritus Professor of Geology at Adelaide University, Eric Rudd, still an active explorer after 60 years in Australia's oil and mining industry; to APEA's executive director, Keith Orchison; and to geological consultant Colin Glazebrook — all three of whom read the manuscript and gave valuable advice. I appreciate, too, the help of those who lent and supplied historical photos. Last, but not least, thanks to my wife Sharron, who not only put up with my obsession at the computer keyboard, but also gave her own time and skill to photographically revive many of the fading images for reproduction in this book.

Finally, with David Ell Press, I gratefully acknowledge the following companies and organizations for their support: Australian Institute of Petroleum Ltd, Australian Petroleum Exploration Association Ltd, BHP Petroleum Pty Ltd, BP Australia Ltd, Caltex Oil (Australia) Pty Ltd, Esso Australia Ltd, Mobil Oil Australia Ltd, Potts West Trumbull and Company, The Shell Company of Australia Ltd, West Australian Petroleum Pty Ltd and Woodside Offshore Petroleum Pty Ltd.

Rick Wilkinson

March 1991

CONTENTS

An important lesson taught by the history of geology as a science is that little real progress was made until geologists began to combine patient observation in the field with work in the laboratory. The world would have been spared much useless controversy . . . had all geologists followed the simple advice of Desmarest: 'Go and See'.[1]

Edgeworth David, 1892

What we have to do now is go back to the courage of the 1950s and 1960s. We have to go back out to the regions where God never trod and take more risks, instead of drilling right next to the service station.[2]

Colin Glazebrook, 1989

Introduction

THE CROSSROADS

In the 1990s Australia's oil explorers stand uncertainly on the threshold of a new century — the third in which they will play a part. Some fundamental decisions must be made. Bass Strait oil production began to decline in 1986 and, even though several new fields have been found and brought on stream in the region, they are simply making the slide less steep. The same can be said for recent discoveries in the nation's other offshore and onshore basins. Despite having all the modern, sophisticated tools of the trade, Australia is not finding enough oil within its boundaries to keep pace with indigenous demand. Some argue that government taxation regimes place too onerous a burden on the explorers, which in turn dissuades them from giving full rein to new surveys and drilling in areas of continuing geological promise like the North West Shelf and the Timor Sea. Companies do have the alternative of moving their search to overseas locations where taxation regimes appear less harsh for the same geological risk, and better infrastructure means swifter cash return for any success. But there are hidden traps for the unwary and Australian explorers cannot blithely expect to master the rules in foreign climes without a great deal of effort. Nor will success overseas bolster the declining oil reserves at home. To do that, companies and individuals will have to blaze a trail ahead into the unknown. Fighting for a better taxation regime cannot be dismissed, but it is vital that explorers also have the courage to invest time and effort to ensure that creative geological thinking is instilled into the search ahead. There is no future in simply drilling holes using the same philosophy as in the past.

That is not to say the past should be forgotten. What amazes most about Australia's upstream oil industry is that commercial production of indigenous oil reserves is such a recent occurrence. So swiftly and completely did success come during the 1960s that the toil of 150 years prior to that decade has been erased from many memories, if indeed it was heeded at all. Some facts of history, perhaps, were 'buried' deliberately because the early search was studded with hoaxers, cranks and companies with a misguided idea of geology. All of them brought the industry into disrepute during the late 1800s and early 1900s when Australia was struggling to weld its far-flung group of colonies into a nation and find its own oil instead of relying on imports from the Americas. Looking back today, the best that can be said of those shady characters is that they did prompt early interest in, and discussion about, exploration, even if that meant putting a number of unwarranted twists in the road to the country's eventual oil success.

High on the list was a Russian 'visitor' named Eugene de Hautpick MRES, MIRE, who arrived in Australia in the 1920s claiming to be a geologist and using the self-styled title of 'Captain'. The newspaper *Smith's Weekly* sarcastically noted that he was immediately established as an authority on oil by reason of the letters after his name, the meaning of which no one understood. He came with references from the London paper the *Mining Standard,* although there were suspicions that his learned articles were cribbed verbatim from text books on oil geology. De Hautpick plunged straight into the Coorongite controversy — a debate centred on a naturally occurring petroliferous substance found in the swampy Coorong district of South Australia which was eventually proved to be an algal growth. The Russian 'Captain', however, was sure it was the escape to surface of a vast underground reservoir of oil and he formed a company with public funds to mount a drilling programme. The Coorong yielded no petroleum. Instead, the repeated failures and references to the subject antagonized the public to such an extent that de Hautpick felt it wise to return to Europe.

Another hoaxer was an American named Max Steinbuchel who appeared in southeast Australia during the 1930s armed with a device he called his 'Doodlebug' and claiming to be able to accurately detect the presence of oil fields. Amazingly, his apparatus appeared to be an old torch shell with a bit of wire and a marble, but that did not deter the flood of gullible subscribers. However when the money was not used to drill the promised wells, there was official government concern in Victoria. Steinbuchel apparently saw his ruse coming to an end and quit the country leaving a number of unpaid debts.

One blatant fraud of this period was even written up in the London *Mining Journal.* An 'oil' sample allegedly taken from a bore near Southern Cross in the gold fields of Western Australia turned out to be a mixture of Vaseline, soap and lubricating oil. Detectives following up the case found a tank of the concoction under the culprit's bed connected to pipes which led into the nearby 'discovery' bore. Yet another West Australian boom was cut short when the mother lode turned out to be a dead pig wedged in a creek bed and oozing its juices downstream in an oily scum.

Leading the bunch of more genuine, if over-enthusiastic promoters, was Bendigo-born banker and broker, George Dick Meudell, who based himself in Melbourne but travelled widely round the world including visits to the Californian oil fields. Seized by a desire to find oil in Australia, he began several companies and took up exploration permit interests in Tasmania, Victoria and South Australia. Noting the first confirmed Australian oil finds round the town of Lakes Entrance in eastern Victoria during the 1920s, and no doubt influenced by American operations along the Californian coast, he even contemplated building piers out from the beach into Bass Strait so that his rigs could drill in the shallows. Although the plan did not come to fruition, he had anticipated the country's major hydrocarbon province 40 years ahead of its discovery.

Unfortunately, the enthusiastic but unscientific promotions of men like Meudell, combined with outright fraud from the likes of de Hautpick and Steinbuchel, served only to alienate the geological giants of the early twentieth century. Sir Edgeworth David, Professor of Geology at Sydney University and arguably Australia's greatest geologist, was never enthusiastic about the country's chances of finding oil in commercial quantities. Keith Ward, a student of David's and then Director of Mines in South Australia for most of the first half of the 1900s, was more blunt. He declared there was as much chance of finding oil in his State as there was of finding the letter 'W' in the French language. Their antagonism towards amateur explorers prevailed despite the accidental discovery of gas in a water well at Roma, Queensland, in 1900 and the subsequent natural gas lighting of the town, albeit briefly, six years later. Even the sale of condensate from wells in the Roma region during the 1920s was marred by acts of sabotage and accusations that companies were manipulating the flows and the share markets to their own ends.

During the mid-1920s there was more hope for large-scale success in Papua and New Guinea, a region that came under Australian Government jurisdiction and where oil and gas seeps had been noted in a number of locations. But the explorers had to overcome enormous difficulties in merely visiting the surface hydrocarbon occurrences and mapping the geology, let alone transporting drilling rigs and supplies. In many cases it was the stuff of adventure stories, as the parties hacked their way through dense jungle, across crocodile-inhabited rivers and over mountain ranges. They battled humidity, constant rain, malaria and isolation, and were often out of contact with headquarters for weeks at a time.

A number of characters emerge from the pages of the New Guinea search, but perhaps the best known adventurer was Australian geologist G.A.V. (George) Stanley. He arrived in the country as a cadet geologist in 1928 to take part in a two-year survey programme being run by the British company Anglo Persian (now BP) for the Australian Government. Teaming up with a surveyor named Henry 'Dinkum' Eve, Stanley did the preliminary work in the old Mandated Territory (northern New Guinea). He quickly became an expert bushman and took a great interest in the indigenous people, eventually marrying a local girl and settling in Port Moresby. Stanley stayed in the Sepik region as a coastwatcher during the Japanese occupation of New Guinea in the Second World

War, a feat which earned him decorations from the Australian and American armies. He continued as a geologist when oil exploration resumed in the 1950s and became something of a legend. Adored by the natives, he was also well liked by his colleagues, although less so by visiting senior dignitaries whom he delighted in 'losing' in the bush. He was also a little eccentric, and known to stride trouserless through villages or collect old beer bottles round Port Moresby towing a trailer full of mischievous, grinning black bois. But he was nonetheless a sound geologist and botanist who penned a two-volume report on exploration in Papua which is still required reading for modern explorers.

Australia had its geological 'heroes' too. Federal Government concerns about the country's lack of indigenous oil supplies even for defence purposes after World War I led to the appointment of George Walter Woolnough as Commonwealth Geological Advisor. Like Ward, Woolnough was a student of Professor David but, unlike both other men, he soon became an enthusiast in the search for Australian oil. Ably assisted by government paleontologists Frederick Chapman and Irene Crespin, Woolnough set out to encourage exploration and the proper collation of all geological information from wells and surveys. He was also a pioneer in the use of air photography as a geological survey tool, often flying great distances and long hours hunched over a camera in the cramped, draughty aeroplanes of the day. But perhaps the real guiding light on the country's road to eventual success was Woolnough's successor, Sir Harold Raggatt.

A geologist who spent his early years in the New South Wales Geological Survey, Raggatt has been likened in talent and demeanour to Sir Edgeworth David. At the end of World War II, he saw clearly that Australia had to have a comprehensive inventory of its mineral (including oil) potential. As head of the newly formed Bureau of Mineral Resources, Geology and Geophysics, he organized the first coordinated mapping of Australia's onshore sedimentary basins. Geological and geophysical surveys carried out during the late 1940s and throughout the 1950s were combined with drilling results to provide detailed cross sections as well as surface outcrop charts. In some instances the basin boundaries were outlined for the first time. Raggatt's contribution went even further, for in 1946 he was instrumental in encouraging Sir William Walkley, managing director of the (then) small Australian petrol retail company Ampol, to set up a major commercial exploration effort in a barren, isolated stretch of country on the Exmouth Peninsula of Western Australia.

Walkley himself was a courageous man. A New Zealander who came to Australia in the mid-1930s to set up Ampol in the face of enormous competition from the world's oil majors, he conceived the idea that the company would have a firmer base if it had its own oil field and refinery. Oil discovery was a first priority and Raggatt's advice prompted him to take up the Western Australia permits, initially along the central coast, but quickly followed up by application for the State's entire onshore sedimentary basin area. Realizing Ampol was not strong enough technically or financially to carry out the mammoth exploration task he had set, Walkley canvassed a number of the big companies before Chevron and Texaco joined the venture in 1951. Drilling began on the Exmouth Peninsula in some low hills known as Rough Range in September 1953. The scene three months later, described by Ampol employee Jesse Haddock, signalled the beginning of Australia's modern oil search:

> It was one of those hot days when the drill crews stripped to the waist. But I couldn't tell if they had clothes on or not because they were covered in a green waxy stuff that smelt like kerosene. As each stand of pipe was broken out, an oily liquid came out and flowed over the rig floor and everything on it. In spite of the heat, the liquid set quickly on contact with the air. It was impossible to get a firm footing on the floor, but nobody seemed to care, they were so excited. It was the first time most of us had seen crude oil flowing from the earth.[1]

Rough Range turned out to be noncommercial, but in a sense that was immaterial because news of the discovery electrified the nation. It was proof that Australia did have the potential to produce its own oil. Suddenly there was a mad scramble as other explorers rushed into the field, most of them with more enthusiasm than science. The 1950s became a rollicking time, with new companies springing up and publishing prospectuses declaring they were sure to be the ones to make Australia's first commercial strike. The decade, perhaps more than any other, was punctuated with humour as the explorers fanned out into all States like a boisterous, but good-natured football crowd leaving the ground after a win. Stories abound, like the Englishman in a bush camp for the first time dousing the ground round his stretcher with fly-spray in an effort to keep the snakes away. On a similar note, one geologist in Western Australia did his best to lower the bushfly population by allowing them time to settle on his back before jumping into a vehicle and subjecting his captives to a dose of insecticide from a high-powered spray gun. There is also the tale of the Gippsland farmer who sent his ferret on a journey without return when he mistook a vertical shot hole on a seismic survey line for a rabbit burrow. At the same time, oil field improvisation reached its height in central Australia when a pile of empty beer cans was rammed down a well as a last resort to stop an uncontrolled gas flow.

But humour aside, when no discoveries were made in the years immediately following Rough Range, enthusiasm waned and the pessimists returned. As the 1960s approached, the Federal Government, once again advised by Sir Harold Raggatt, created incentives in the form of an exploration subsidy scheme. Around the same time the explorers formed themselves into the industry association known as APEA, spiritedly led by South Australian geologist Reg Sprigg. Although rivalries flared between the local and foreign explorers, the new decade saw a far more professional approach to the oil search. It also saw the first surveys in offshore regions, and onto the stage to direct the first operations off southeastern Australia stepped the 67-year-old American geological consultant Lewis Weeks. Although he is now seen as having been the right man in the right place at the right time when he advised steel company BHP to take up oil exploration leases in Bass Strait, Weeks was no stranger to the region. He had not visited previously but, through his connections with Esso and other contacts in the Australian industry, he was familiar with the geology. The young Tertiary-age sediments in the offshore Gippsland Basin in particular fitted neatly into his theories of oil generation and entrapment. When asked, he had no hesitation in pointing the Australian explorers in that direction.

The result was an amazing run of discoveries which, virtually at one stroke, unveiled the country's potential for self-sufficiency in oil before the 1960s had ended. All the dreams of previous years suddenly came true. There was support from a number of onshore discoveries like Moonie in Queensland, Barrow Island in Western Australia and Mereenie in central Australia as well as gas finds at Gidgealpa and Moomba in South Australia, Roma in Queensland and, during the early 1970s, the North Rankin and Goodwyn fields off the country's northwest coast. Each of these finds have stories to tell — stories of overcoming horrendous logistics, of geological perseverance, and of engineering innovation. All are set against a changing political spectrum, ranging from a government bent on supplying the industry with tangible incentives, to one more interested in centralized national control, and to one preaching deregulation.

In the 1980s there were more significant finds both on and offshore, but Bass Strait still stands out as Australia's one real prize. It has never been matched and, more significantly, it is now in decline. Hence, in the 1990s, the country's explorers are standing at the crossroads wondering which route will best serve their needs in the twenty-first century. Of greatest concern, though, is the legacy left by the apparent ease with which Australia suddenly became self-sufficient in oil. At the time it seemed that all the explorers had to do was stab a hole in the Bass Strait seabed to make a discovery. Unfortunately, many outside the industry still believe this is the norm and that there really is no problem in finding future supplies. The truth is somewhat different. To fully appreciate the difficulties that lie ahead and the kind of spirit needed to combat them, the clock must be turned back almost two centuries to the time Europeans first nosed their way round the Australian continent...

PART ONE

1800–1950

HOPS IN THE DARK

Chapter 1
EARLY GLIMPSES

The first white settlement in Australia was barely 10 years old when adventurers and traders began to explore the coasts of the vast new land. Sealers and whalers were among the early comers and, encouraged by what they saw as an abundant sea harvest particularly along the southern shores, they became regular visitors from about 1800 onwards. It is probably to them that credit for the first discoveries of Australian oil should go, although it would be stretching the definition to call these men the country's first oil explorers. No precise records were kept, but these seamen saw and collected stranded seepage oil along the beaches of what is now South Australia and western Victoria. The origin of the black, bituminous material emanating from seeps on the ocean floor and washed ashore during storms caused long and heated debate in later geological circles. But among the early sailor-traders, practical men who had few comforts and faced a daily battle with the elements for survival, there was little speculation. The oil found a welcome use as caulking material for their boats and flooring for their huts.[1]

The British marine surveyor John Lort Stokes was the first white person to document the presence of oil during a voyage on the sloop HMS *Beagle* to chart the coastline of northern Australia in 1839. By that time he had long been associated with the famous vessel, initially joining it as a midshipman in 1825 and later sailing on naturalist Charles Darwin's voyage round the world. The ship briefly visited Australia in 1836 on the homeward leg of that journey. In 1837 the *Beagle* was back in Australian waters with Captain John Clements Wickham in command and Stokes a lieutenant.[2] Their orders were to accurately map the northern and northwestern coastlines — a task that took until 1843 to complete. At the time there was still speculation that the centre of the Australian continent contained a large inland sea, and part of Wickham's brief was to look out for 'the Great Northern River' thought to be the outlet to open ocean.

In October 1839, about 240 km southwest of Port Darwin, the *Beagle* steered towards a wide gap in the coastline and the crew immediately began to speculate on a strong tidal rip noticed in the vicinity. Stokes took a whale boat and confirmed the presence of a 'noble river' which he named the Victoria in honour of his sovereign. Back on board the *Beagle* he navigated the vessel some 80 km upstream. From this point the crew made several short land forays as well as preparing for a longer investigation in the ship's boats along the river's narrow upper reaches. The latter journey soon showed that the Victoria was not the hoped-for passage to the inland sea, however Stokes used his surveying skills to accurately chart the river's winding course through sheer rock gorges and tree-lined banks for 160 km from the *Beagle*'s anchorage. His powers of observation and attention to detail provide a fascinating description of the whole region, but it is a section of his diary entry for 20 November 1839 that is recognized as the earliest written record of oil discovery in Australia.

> I went ashore to collect a few geological specimens. The sandstone which prevailed everywhere was in a decomposed state, but there was a very decided dip in the strata to the south east of about 30 degrees. On the west side of Water Valley I found the same kind of slate, noticed before at Curiosity Peak, but what most interested me was a bituminous substance

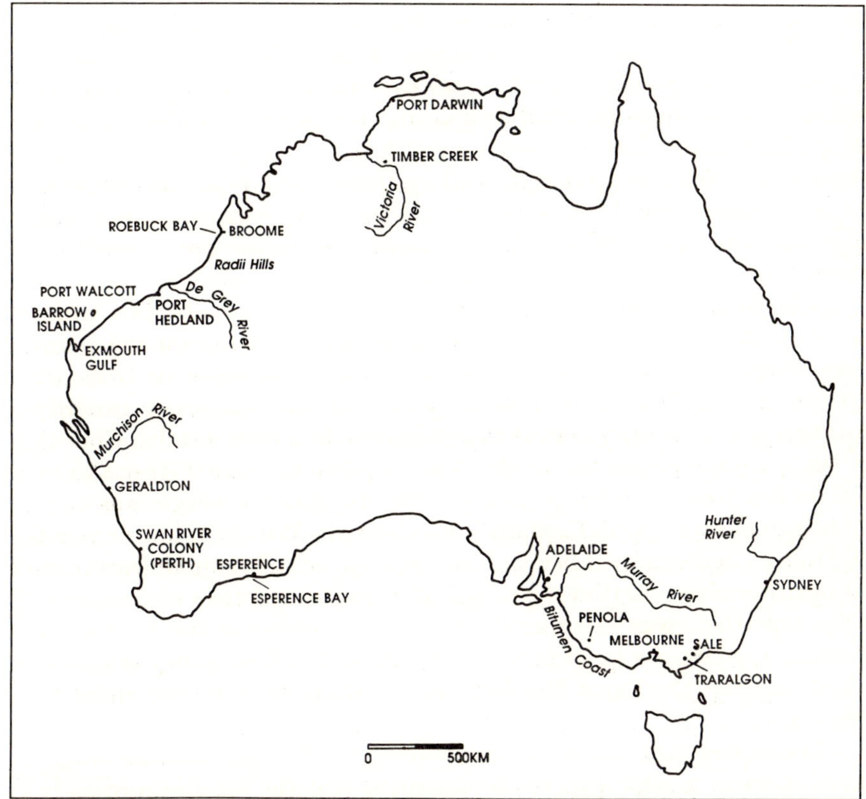

Location of key places mentioned in text.

> found near the bottom of the wells recently dug, and 23 feet [7 m] from the surface of the ground. It was apparently of a clayey nature when first brought up, but became hard and dark upon exposure to the air, and ignited quickly when put into the flame of a candle. The sides of Water Valley were very precipitous, and nearly 300 feet [91 m] high, a growth of palms marked the spot, and served to indicate our wells.[3]

Stokes does not mention the substance again. The site of the water wells later became the Victoria River Depot and is located about 15 km west of the present Timber Creek Police Station in the Northern Territory. The surveyor's diaries also note the presence of Aboriginal groups during his river journey and, only two weeks after the oil discovery, Stokes was severely wounded by a native spear. Fortunately the injury was not fatal and he was close to the *Beagle* at the time, however the incident terminated the party's exploration of the Victoria River region and it was over 60 years before any other reports confirmed the presence of oil in the area.

By the time the *Beagle* reached the central west coast of the continent, on its way down to the Swan River colony, Stokes had recovered sufficiently to visit Barrow Island. The island had been named by Lieutenant Phillip Parker King in 1818 after John Barrow, then one of the secretaries of the British Admiralty. The purpose of the *Beagle*'s visit in 1840 was merely to make observations of the fauna. There was no surface indication of the oil reservoir destined to become one of the country's most important finds 125 years later.

Throughout the remainder of the 1800s reports of Australian petroleum occurrences were scattered, sketchy and for the most part unreliable. Some were confused with, or related to, deposits of a rock known as 'kerosene shale' found during the early days of the colonies in New South Wales and Victoria. Others were related to a substance dubbed 'coorongite' and found on the ground in southeastern South Australia. Still others were the result of oil from drilling equipment finding its way into water bores either by accident or design. Experienced present-day geologists Professor Eric Rudd and Reginald Sprigg point out that drillers have always been noted for practical jokes, and that natural gas is a widely occurring substance[4] — a comment which suggests that dubious oil and gas results are not confined to nineteenth-century reports. An 1880 event, recorded 23 years afterwards by New South Wales Assistant Government Geologist Joseph Carne, illustrates the point.

> A report was circulated that oil had been struck in the Moore Park Bore (New South Wales) put down by Mr J. Coghlan to a depth of 1 860 ft [567 m] without reaching coal. The early discovery that the oil floating upon the water of the drill-hole was refined petroleum speedily arrested the progress of the sensational news.[5]

Nevertheless, as pioneering settlers followed the explorers and government surveyors, close attention was paid to the land and any new phenomenon was passed on by word of mouth even if it was not specifically written down. An early report in Victoria, for instance, came from whalers setting up a station near Portland in the west of the State. They commented that the water from nearby swamps was not potable and had an 'oily' film on the surface.[6] In a similar record from 1885, government analyst J. Cosmo Newbery told friends he had seen iridescent scum over water holes near Traralgon in the Gippsland district of eastern Victoria.[7] Further to the east an oil bore was reportedly put down in 1886 by John Buntine, a member of a prominent early Gippsland family, on his property in the Toongabbie area about 40 km west of Sale. Unfortunately there is no record of any borehole samples and a subsequent report from government geologist James Stirling gives the impression that the borehole was, in fact, a shaft sunk to evaluate suspected kerosene shale. Nevertheless Buntine, a member of the local Rosedale Council and a Justice of the Peace, had visited California and saw oil discoveries there that may have sparked his interest in the local scene.[8]

Another Victorian bore, this time attributed to John Gillam (apparently a publican), was put down prior to 1892 near his hotel, although the location of the building is not known. Mines Department reports indicate the work was terminated at 46 m after the discovery of a 1.5 m seam of good fire clay and indications of coal.[9] A postscript of sorts was added to the nineteenth-century Victorian chapter when, in 1916, Victorian Geological Survey official L.H. Ower pointed out that the presence of iridescent iron oxide films on the surface water was a common occurrence in marshy ground. He added that though it might look like an oil seep, it was really 'fool's oil'.[10]

In Western Australia too there were early references to the possible presence of oil in that State. The first, dated 17 August 1854, comes from the journal of Robert Austin, a government surveyor who reported finding traces of an oil-bearing substance in a cave at Mt Kenneth near Geraldton. He describes the material as a 'soft, black bituminous substance exuding from a cleft in the sandstone rock and presenting a discoloured line on the lower portion of the back of the cave'. A few years later another government surveyor, James Cowle, found oil indications when crossing the Radii Hills during a journey from Roebuck Bay near Broome to the De Grey River. Finding a native well in a valley between the hills, he cleared it out to quench his thirst, but found its contents undrinkable. Cowle reported that the liquid tasted like a mixture of kerosene and water.[11] A more definite indication of petroliferous deposits is contained in the catalogue of the Paris Universal Exhibition of 1866–1867. In the Western Australian section a series of 'bituminous resins, inflammable amber-like minerals, peat, and chalk' from Port Walcott (near the present town of Cossack) were exhibited. The entry continued:

> The inflammable substance is of a dark brown colour, resinous fracture, and soluble in alcohol. The bitumen or asphaltum is free from sulphur compounds, and is, therefore, of value if occurring in quantity. One of the resins closely resembled copaline [a waxy substance found as thin film or fragments in clay]. On distillation it yielded an oil like oil of amber; the specimen was marked amber. A second, also marked amber, seems to represent a transitional change from the former, which again seems connected with specimens marked 'peat', which consist of sand grains coated and cemented by a peaty substance soluble in potash.[12]

All were apparently found in rock formations near the coast. At about this time the leader writer for the *Perth Gazette* made, in the light of later events, an astounding prophecy:

> Our belief is that at one period the whole coastline from south of the De Grey River to Exmouth Gulf was similar to that lying north of the De Grey, until it was acted upon by the volcanic agency which has left so many proofs of its existence in the various ranges of hills. If this be so, Mr Cowle's kerosene and the existence of bitumen to the south [at Port Walcott] become very significant facts and we should not be surprised at one day hearing of rich stores of petroleum shale being discovered, or even of oil being struck.[13]

A little later again, during a brief period as Government Geologist of Western Australia between 1870 and 1872, H.Y.L. (Henry) Brown researched and produced 10 detailed reports and three geological maps covering various parts of the State. The last of these, complete with map, carried the cumbersome title *General Report on a Geological Exploration of the Portion of the Colony of Western Australia Lying Southward of the Murchison River and Westward of Esperence Bay*. Although gold was his chief concern, Brown opined that 'the colony is extraordinarily rich in lead, silver, copper, iron, plumbago (graphite or black lead) and many other minerals'. He added, somewhat cryptically, that 'indications of coal and petroleum are not wanting'.[14]

In South Australia, apart from the continuing reports of oil strandings along the southern beaches and the discovery of the bituminous substance called 'coorongite' (described in chapter 2), there were some similar, although isolated, 'oil' occurrences noted. An odd example is mentioned in a letter which appeared in the popular illustrated London monthly magazine *The Geologist* in April 1862. Written by chemist T.L. Phipson, it describes a sample found extruding from cavities in Tertiary limestone cliffs along the lower reaches of the River Murray about 80 km east of Adelaide (written about in the February edition of the same year) and given to him for analysis. Phipson writes that the caves were apparently inhabited by 'hosts of wallabies, kangaroo rats etc.' and the sample was of 'dark brown, sticky, odiferous matter in considerable quantities'. Chemical analysis showed it was about 40 percent bitumen and petroleum with debris of mosses, just over 30 percent lime carbonate, 22.5 percent sand and white mica, and the rest iron oxide and phosphates.

> It will be seen... that this peculiar substance is made up of natural hydrocarbons which have cemented together a certain amount of mineral matter. It has nothing to do with the animals which infest the warren, except perhaps that by boring into the rock they have given it a means of exit.[15]

One cannot help feeling that there is a certain amount of 'leg pulling' in this correspondence! But, putting aside any judgement about whether some 'discoveries' were bogus or not, the paucity of petroleum reports of any sort in the early days of Australia's colonization was due mostly to the relative scarcity of surface indications. The apparent lack of enthusiasm in geological circles in following up such indications during the 1800s — particularly in the first half of the century — was due to the absence of commercial need for the substance. The quest for the precious metals gold and silver, and the base metals copper, lead and iron was considered far more important to the viability and independence of the fledgling settlements perched precariously round the perimeter of a vast new continent. And it was not as if petroleum was unknown. The substance had been identified and used throughout the world for centuries — for caulking boats, fuelling lamps, preserving mummies, and even medicinally as a drink or applied to wounds. But these applications were haphazard and somewhat isolated instances.

Certainly, when Europeans first settled Australia there was not the ordered, commercial use of petroleum products that the twentieth century now takes for granted. Colonists went to bed soon after sundown and rose with the dawn. Streets were dark and dwellings were lit with the flickering light of tallow candles, perhaps augmented with oil lamps. But the oil was supplied by the whaling industry and, as the century grew older, by oil derived from kerosene shale. It was not until 1841 that the streets of Sydney were lit by gas, and that was derived from coal.[16]

Despite the preoccupation with commercial minerals in rocks of igneous origin like gold and copper, sedimentary geology was surprisingly well understood in the early days of the New South Wales colony, owing to the obvious seams of coal in cliff faces along the coast and the need to dig quarries for sandstone blocks for buildings and roads. The enthusiasm of men like the Reverend William Branwhite Clarke, who saw no conflict between his religion and the natural sciences of geology, botany and zoology, was shared by the early explorers, who honed their powers of observation and description of the land they traversed to a remarkable degree.

Clarke, who reached Sydney in 1839 at the age of 40, had already combined his clerical duties and geology with distinction in England. Referred to today as the 'Father of Australian Geology',[17] he quickly immersed himself in both pursuits in his new home. The coal seams of the Hunter Valley caught his early attention and he also traversed the Blue Mountains noting the evidence of kerosene shale deposits. Although primarily assigned by government to report on the gold potential of the colony, he found time to collect and describe fossils and rock specimens from a variety of locations. Equally importantly, he maintained a lively interest in the work of others and was a regular correspondent with prominent British geological figures of the time, such as Professor the Reverend Adam Sedgwick of Cambridge University, Sir Roderick Murchison of the Geological and Royal Geographical Societies, and Charles Darwin, author of *The Origin of Species*. He also valued the views and comradeship of visiting geologists like the American James Dwight Dana, the Englishman Joseph Beete Jukes and the Pole Count Paul Edmund Strzelecki.

Clarke's legacy, however, sprang from his contention, based on detailed field work, that the coal measures of the Hunter River and the Illawarra areas were a continuous, uninterrupted series of sediments. In this he was vigorously opposed by a number of the colonial geologists, notably by paleontologist Frederick McCoy, who said the fossil evidence indicated that a vast interval of time separated the individual rock units. The debate continued for three decades until Clarke was eventually proved correct. Transcending his local victory, the 'Australian Coal Controversy' of the mid-1800s undeniably tested, and finally proved, that the position of rock layers and their structural relationship to each other as seen in the field outweigh the evidence of fossils examined in isolation.[18] In other words, field work is a vital component in understanding the total geological picture.

Another noted geologist-naturalist in the Clarke mould was the Reverend Julian Edmund Tenison Woods, an Englishman who emigrated to Australia in 1855 after developing an interest in geology when working as a professor in the naval college at Toulon in France. Woods also had a parallel interest in religion and, after being ordained as a priest in 1857, took up his mission in the small town of Penola in the southeast corner of South Australia combining geology and religion on his travels through the district. His headquarters were a disused store, the shop front making do as a place to celebrate Mass. Like Clarke, he saw no conflict between scripture and the theories of Darwin. Described as a prodigious writer and a horseback traveller in all weathers (often in the company of the poet and local horse breaker Adam Lindsay Gordon) who never spent a week in any one place, Woods's treatise *Geological Observations in South Australia* was published in London in 1863. The book, which presented detailed data gathered during his missionary journeys over 57 000 km^2, won him praise and encouragement from the eminent British geologist Sir Charles Lyell and was still regarded as a valuable reference 60 years later.[19]

From that point his travels widened to take in much of eastern and southern Australia, and Asia, where he witnessed part of the eruption of Krakatoa. In 1865 he published the two-volume *History of the Discovery and Exploration of Australia.* Probably more an all-round naturalist and communicator than Clarke, his oratory prowess won Woods wide acclaim and he was well known for his popular science lectures — especially those for children. By the time of his death in 1889 he had penned over 155 scientific papers plus a number of other educational writings of a religious, grammatical and geographic nature. The scientific range covered geology, botany, paleontology (fossils), zoology and mining techniques. Woods often illustrated his work with his own drawings.[20]

Although not trained geologists, explorers like Thomas Mitchell, Charles Sturt, Ludwig Leichhardt, Augustus Charles Gregory and his brother Francis Thomas Gregory, Phillip Parker King and John Lort Stokes instinctively knew the value of careful and accurate field observations. All took careful notes of the rock outcrops they passed on their journeys and, where possible, accompanied them with samples, sketches, diagrams and maps. The botanist-explorer Allan Cunningham, for instance, is credited with recognizing that the changes in soil, and hence underlying rock type, are reflected in changes in vegetation. The important work of these scientists and explorers is made all the more amazing when one considers that they suffered a lack of adequate transport, a scarcity of roads, an unforgiving climate which sapped strength and undermined health, and the often hostile attention of Aboriginal tribes. Naturally enough, the net result of their efforts fell short of a comprehensive picture of Australia's geology complete with an understanding of the vast sedimentary basins destined to be the focus of the country's later oil explorers. That would have been an unrealistic goal in a land where so much evidence lay buried under desert sands and so few tools were available to aid investigation.

Instead, the legacy of the early workers was the raising of geology from an amateur pastime to a profession, and the formation of a solid base of detailed data upon which to build. Woods's work is an outstanding example and Clarke, although not as prolific, also published a number of papers including a valuable treatise on the sedimentary rock formations in New South Wales complete with fossil identifications and stratigraphic information from his ceaseless field work. It is interesting to note in today's environmentally conscious world that Clarke was also concerned about the unthinking destruction of forests and its effect on fauna, climate and human health. A number of his scientific papers address these subjects in strongly argued detail.

The second half of the nineteenth century saw the formation of government geological surveys in each State and the creation of geology schools at the universities. The emphasis was on 'hard rocks' because gold and base metals were still the driving commercial forces. But there was a continuing study of sedimentary 'soft rocks', with particular reference to fossil identification, classification and stratigraphic correlation. Overseas, 'Colonel' Edwin Drake's natural, or free-flowing, petroleum discovery (as opposed to oil extracted from kerosene shale) in Pennsylvania during 1859 introduced a new era both commercially and geologically. The commercial aspect had little immediate impact on Australia. Petroleum oil was first imported into the country from America during the 1860s, but still had to compete with oil produced locally from kerosene shale. It was not until the turn of the century that it contributed greatly to the development of the nation. On the other hand, news of the geological side of Drake's discovery soon reached Australia and, inevitably, reawakened interest in previously reported local 'oil' sightings. It also produced feverish endeavours to 'prove up' countryside judged to be similar to that surrounding the ongoing American discoveries. Initially, at least, mistaken interpretations of the phenomenon began a chain of events leading in the wrong direction.

Chapter 2

An Alternative Route and a False Trail

Oil shale and coorongite are diversions from the main course of oil exploration in Australia. Nevertheless, the pursuit of these substances did much to stimulate interest in the country's petroleum potential. Coorongite in particular prompted lively and colourful debate, and both substances provided an avenue for new commercial enterprise. Awareness of their existence in Australia began in the nineteenth century and spilled over into the twentieth. Oil shale enjoyed two periods of commercial production and there are renewed signs that the material will have commercial significance for a third time as an alternative to natural petroleum production in Australia during the twenty-first century. The coorongite quest, on the other hand, began with great vigour during the mid-1800s only to finally peter out as a false trail in the 1930s.

The presence of oil shale, or kerosene shale as it is called in the early records, was officially noted in Australia as far back as 1802. It is possible that the Aborigines already knew of its existence because later white settlers in the Blue Mountains region of New South Wales heard tales of local tribes burning the black-brown, soft rock material believing (correctly, in a way) it was some kind of fire stone.[1] The earliest published reference to Australian shale appeared in Paris during 1807, written by Baily and Depuch, two men who were part of a French Government sponsored round-the-world scientific expedition between 1800 and 1804. Exploring the foothills of the Blue Mountains and the Hawksbury River area during their Australian leg, the scientists found coal, sandstone and shale, the latter having a bituminous nature.

> Independently of coal (carbon de terre) which I suspect ought to exist beneath Parramatta, we have discovered my friend Depuch and I, at the foot of the mountains, considerable masses of bituminous schist which burns with a very lively flame, giving off thick smoke and an extremely pronounced odour of bitumen. The combustion does not cause the particles of schist to lose in any degree the character which they originally had. It renders them more friable, and deprives them of their

> colour. In a country deprived of coal, in which wood might be scarce, the schist would offer a valuable resource.[2]

The latter comment proved to be inappropriate for the new colony because rich coal seams were found in the region soon afterwards and wood was certainly in abundance.

Reviewing the history of kerosene shale 100 years later, the Assistant Government Geologist for New South Wales, Joseph Carne, noted that the effects of combustion recorded by the French group were characteristic

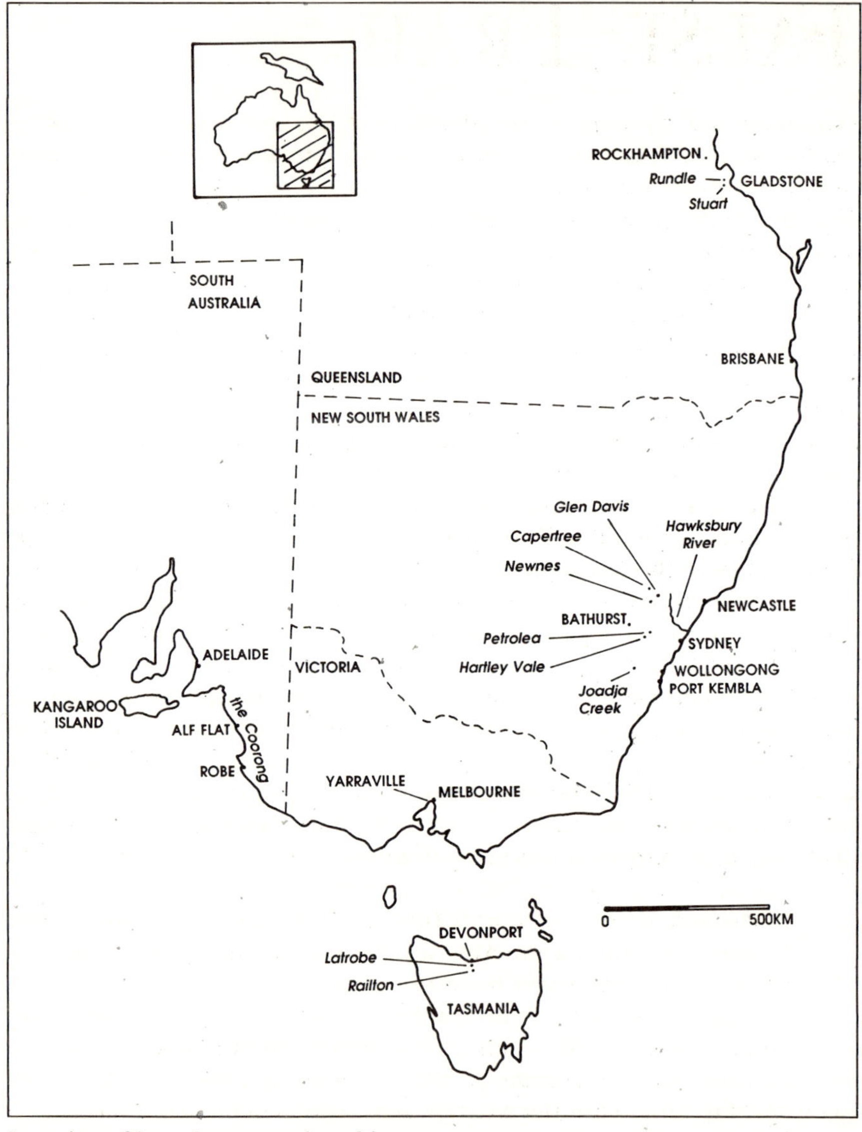

Location of key places mentioned in text.

of ordinary kerosene shale retort residues, whereas in 'cooked' or 'spent' shale, the white laminated, biscuit-like features were common. Carne also noted that the pieces found in 1802 were probably washed down from the mountains beyond. The actions of the first settlers pushing over the Blue Mountains in the 1820s seem to support this because they reportedly used as fuel loose blocks of kerosene shale scattered round outcrops further up the rivers.[3] At around this time the first road to Bathurst through the mountains descended the western slopes close to what became Hartley Vale — one of the main New South Wales mines during the second half of the century. Other early reports of the material came from visiting explorer-scientist Count Paul Strzelecki at Hartley Vale and Petrolea in 1840 and 1845, and from resident geologist the Reverend W.B. Clarke at America Creek near Wollongong in 1849 during their respective wanderings over the Blue Mountains region.

It is not surprising that the kerosene shale was quickly recognized in the colony, as its properties had been known in the United Kingdom and elsewhere in Europe for at least a century before the French scientific visit to Australia. The method of extracting oil from shale was recorded in England as early as 1694 via a patent taken out by a Martin Eele. However the industrial application of that process did not emerge until 1850 when a large deposit was found at Torbane Hill in Scotland, whereupon the name 'torbanite' was coined for the particular type of oil shale found at that location. The 'father' of the Scottish oil industry, James 'Paraffin' Young, rose to world prominence with his commercial development of this and other Scottish deposits.[4]

Meanwhile, in Australia, as settlers and explorers spread out during the mid-1800s, signs of oil shale were discovered in virtually every colony. However the richest known seams were confined to New South Wales and Tasmania, and these were the only deposits to be worked commercially. In the latter instance the earliest record of discovery appeared in a scientific paper read to the Royal Society of Van Diemen's Land in 1851 discussing the Latrobe district deposits in the Mersey Valley near Devonport.[5] Young's success in extracting oil from shale in Scotland was soon noted in Australia, but development of the colonial deposits was not viable until Young's British patent on the process lapsed in 1864. The terms of this patent covered the low temperature distillation of coal and similar bituminous substances, and a royalty was payable to the Scotsman on each gallon of crude produced via his technology. Any oil derived from the New South Wales deposits had to compete with kerosene imports from America, so the cessation of the royalty obligation gave an important boost to the local product.[6]

First off the mark to set up the Australian industry was J. Graham, at his Pioneer Kerosene Works at Port Kembla near Wollongong in 1865. Graham also owned the mine at nearby America Creek which fed the plant. In the same year the Hartley Kerosene Works were established, using shale mined from the Hartley Vale deposit first noticed by travellers on the Sydney–Bathurst road 40 years earlier. The initial mining of shale in New South Wales was noted in the *Quarterly Journal of the Geological Society* in 1866 which, in turn, paraphrased a report from the *Sydney Morning Herald* the previous year.

> The mineral is stated to be found near Hartley and also near Wollongong. The seam near Hartley is five and a half feet thick and worked through a tunnel. The mineral is of a dark brownish colour; it is very tough, so that if struck with a hammer the instrument will bound off as it would from a block of wood; it has a conchoidal fracture, and does not powder when broken. The Hartley mineral is stated to be superior to the boghead coal [the Scottish torbanite], in consequence of it yielding a larger quantity of gas, and therefore, of oil, and also of its freeness from sulphur.[7]

Enthusing further on the subject, the report goes on:

> The importation of paraffine oil into Australia is large (nearly one million gallons [4.5 million L] per year) and it is confidently believed that the oil can be produced at a lower price in the Colony than it can be imported from America.

On the same theme the Melbourne *Argus* opined in its 4 May 1865 edition that the 'discovery of an oil-yielding mineral in New South Wales is scarcely less important to that colony than was the discovery of gold'.[8]

Certainly the industry grew rapidly in the 1870s and a number of New South Wales mines were in operation by the 1880s. These included Petrolea on the opposite side of the valley to the Hartley mine, Genowlan near Capertee, and one of the richest of all, Joadja Creek near Mittagong. Unlike coal mines, none of the workings had problems with fire damp so naked lights, burning paraffin wax or heavy oil could be used. Ventilation was via air shafts to the surface and furnaces helped to stimulate the flow. However the shale mines presented their own hazards to the miners, as recorded by Assistant Government Geologist Carne.

> In some of the mines, notably Genowlan and New Hartley, explosives were unnecessary. On the contrary there was danger from cracking and shooting of the shale as the roof settled on the face.
>
> The miners were compelled to work behind breast boards, and cover their eyes with wire gauze to protect themselves against flying fragments of keen edged shale. The tendency of kerosene shale to conchoidal fracturing under stress at right angles to the stratification increases the development of sharp

> edges which cut like knives. Numerous accidents resulted — some with serious impairment of sight.
>
> The enormous weight of overburden is a major cause. At Joadja this overburden is less, and fewer troubles from 'shooting' occur. Here gelignite can be used for breaking down [the shale face].[9]

Joadja is a good example of the enormous physical difficulties involved in actually transporting the shale from the mine in the rugged and, in those times, isolated terrain. The seam was located high up the side of a steep valley which had an incline of 1 in 2, and initially all equipment had to be hauled up and down the slope using rope and pulleys. The way out to Mittagong was over the opposite side of the valley, so loaded trucks were lowered down the mine side by gravity and discharged into bins at the bottom. A system of twin cableways was used such that the descending full trucks would pull the empty trucks back up to the mine entrance. The bins would be dragged across the valley floor by bullocks, pit ponies or mules to the exit haulage incline and then be drawn up by horses yoked to a circular winding drum at the top. Again, empty bins would descend as the full containers ascended.[10] Soon after the Australian Kerosene Oil and Mineral Company took over the operation in 1878, a 40-horsepower engine was installed for the incline haulage and a steam locomotive running from the crest of the slope to Mittagong replaced bullock wagons. Later still retorts and a refinery were built in Joadja Valley itself.

On the social side, isolated mines like Joadja tended to spawn small, self-contained communities, the more so as workmen and their families often emigrated direct from Scotland, their experience in shale mining and processing being invaluable to the Australian companies. These emigrants tended to form close-knit versions of their homeland communities, retaining their broad accents and traditional dress. At first, houses were rudimentary shanties of bark and corrugated iron; more substantial brick homes were erected later. Water was scarce for most of the time at Joadja. A company store, post office, butcher and baker were supplemented by travelling hawkers and shopping visits to Mittagong when the rail line was completed.[11]

The cost of mining varied considerably depending on the conditions, ranging from three shillings and nine pence per ton for easily accessible material up to one pound per ton for the most adverse. This amount covered the cost of mining and placing in trucks only, not the subsequent freight and processing. Pay rates for the miners themselves were noted by Carne:

> At Genowlan rates were graduated according to seam thickness ranging from five shillings and three pence per ton for a 30 inch [762 mm] seam to thirteen shillings and nine pence a

> ton for a 15 inch [381 mm] seam. Payment was made on total output, the miners employing their own check-weigh man.
>
> Trimmers employed to remove splint-bands and inferior grades (for which small hatchets are used) receive seven shilling and six pence an eight hour day; wheelers six shillings to eight shillings; drivers (youths) five shillings; and firemen seven shillings and six pence per day.
>
> At New Hartley the cost of tribute, winning and placing export shale on the Great Western Railway at Torbane is estimated at one pound and six pence a ton. The freight cost from there to seaboard [210 km] was eight shillings and seven pence per ton.[12]

The exact origins and classifications of various coals, bitumens and kerosene shales exercised the minds of academics and researchers round the world for most of the nineteenth century, and spilled over into the twentieth century. Some early theories suggested an inorganic origin for oil shale, others favoured a natural distillation of hydrocarbons in coal seams by the heat of igneous rocks intruded into the coal measures. Still others suggested that oil-bearing shale might be produced by the absorption by clay of the petroleum rising through water from submarine or underground springs.[13] But, as the discoveries round the world became better known, the principal opinions (including those of noted Australian geologists the Reverend W.B. Clarke and, later, Professor Edgeworth David) were for an organic origin — plant spores, pollen grains and algae deposited with fine sediment in calm lakes and coastal inlets. In 1889 Professor David, then Assistant Geological Surveyor with the Government of New South Wales, presented a paper on the subject to the Linnean Society of New South Wales.

> Every one of the minute laminae of the kerosene shale may represent a spore shower, or a season of spore showers; so that it may have taken many hundreds of years to have admitted the formation of a seam of kerosene shale five feet [1.5 m] thick, as at Hartley.[14]

Heedless of the academic debate, shale mining continued from a number of New South Wales deposits and Carne reported that the approximate output to the end of 1901 was officially estimated at 1.1 million tonnes (1 073 468 tons). The discrepancy between this figure and the total calculated by addition of recorded tonnages from the individual mines reflects the loss or nonexistence of some early records.

Hartley Vale	382 560 tons [388 681 tonnes]
Joadja	374 957 tons [380 956 tonnes]
New Hartley	71 360 tons [72 502 tonnes]

Genowlan	61 621 tons	[62 607 tonnes]
Mort's Mine, Katoomba	59 606 tons	[60 560 tonnes]
Ruined Castle, Katoomba	20 341 tons	[20 666 tonnes]
Mt. Kembla, Wollongong	8 711 tons	[8 850 tonnes]
Murrundi	500 tons	[508 tonnes]
Greta	200 tons	[203 tonnes]
Glen Alice, Capertee Valley	50 tons	[51 tonnes]
Cottage Rock, Capertee Valley	50 tons	[51 tonnes]
Total:	979 956 tons	[995 635 tonnes]

Output recorded to 1901 — NSW Dept. Mines[15]

The grade, or yield, of the shale varied from place to place as well as within each deposit itself. But an average production from Joadja in its boom days was 350 L a tonne,[16] while Hartley Valley mines were closer to 680 L a tonne.[17] The oil extraction process began by heating shale in retorts in the absence of air to vaporize the kerogen (crude oil). The vapour was then cooled and the resulting liquid mixed with chemicals and finally put through finishing stills to yield the kerosene oil product. Often the process from mining to finished product would take about a month. It was soon noted also that gas could be extracted from shale, and it was found that mixing shale with coal in the town-gas retorts produced an excellent quality lighting fuel giving off a flame of high luminosity. Some of the Hartley shale was supplied untreated to Sydney and Melbourne gas companies who used it for coal-gas enrichment. Exports of shale were also sent to Germany, Spain, Italy, Holland, England and France for the same ultimate product.[18]

Towards the end of the nineteenth century, however, the known reserves of good quality shale were rapidly diminishing and the remaining commercial operations faced increasing competition from imports of petroleum-derived kerosene. In addition, the increasing use of the internal combustion engine called for a higher quality gasoline than that being distilled from shale; and the invention of the Welsbach gas mantle diminished the importance of the luminous flame in lighting applications.[19] A further blow came when the Federal Government decided, in 1903, to remove the fourpence per gallon import duty on kerosene. Carne opined at the time that:

> Free entry of kerosene oil into the Commonwealth... backed by keen competition between contending American, Russian, and Borneo oil interests for the Australasian market, render the immediate outlook of the local shale oil industry gloomy indeed.[20]

In fact, Hartley Vale and Joadja were the only mines to survive the change of centuries. Both staggered on for a few more years but, by 1906, they too were gone. Surprisingly, that year saw a new and, as it turned out, reckless venture, backed by the London-based company Commonwealth Oil Corporation, commence mining in the Wolgan Valley about 160 km northwest of Sydney. The township established round the mine and plant was called Newnes, named after the London publisher Sir George Newnes, who had subscribed a large proportion of the company's authorized capital of £800 000 sterling. About £1.3 million sterling was spent on the project, including a vast sum for a 51.5 km long railway into the valley and a special locomotive to negotiate the steep grades and sharp bends. Commonwealth Oil Corporation began retorting in June 1911, but lasted only four months before financial problems crippled the operation.

The following year John Fell and Company took over the reins, modified the retorts and recommenced operations in 1914. John Fell was a relative of James Fell who had been brought out from Scotland to manage the shale oil refinery at Waterloo, Sydney, in 1868. Newnes remained active till 1922, but struggled throughout its life against high production costs.[21] Even the Shale Oil Bounty Act, providing for payment of bounties for crude oil produced in Australia and passed by Federal Parliament in 1917, failed to halt the decline. Fell closed his operation early in 1923. Yet, as before, the closure of one shale oil venture seemed to encourage another to begin, and it was during the early 1920s that new efforts were made to set up an industry — this time in Tasmania. Two earlier attempts, in the Railton–Latrobe area near Devonport, had produced about 163 656 L of oil between 1910 and 1915.

The 1920s venture began with the ambitious idea of producing crude from the Latrobe deposits and shipping it across Bass Strait to a refinery set up at Yarraville in Victoria, on river land leased from the Melbourne Harbour Trust. The interstate connection was short-lived, although mining and oil production continued in Tasmania until 1935 under the wing of the Tasmanite Shale Oil Company Ltd. Products were sold locally, including long-running contracts for fuel oil to the Tasmanian Government Railways and the Mount Lyell Company totalling 25 000 L a month. Indeed, a Department of Mines survey of the industry published in 1932 was very hopeful that shale oil would hold its own against imported petrol.

> The petrol [refined from shale] is used generally in private cars, commercial vehicles, marine motor-engines, and motor cycles, and has given winning results in hill-climbing competitions and speed tests. The mileage obtained in tests allowing adjustment has shown an improvement over the imported petrol, the usual figures being from 10%–20%.[22]

This optimism quickly faded when it was realized that the high sulphur content of Tasmanian shale oil put the local product at a disadvantage when competing with fuels derived from petroleum. Like its fellows on the Australian mainland, Tasmanite Shale Oil Company was forced to shut down. Its demise in January 1935 spelt the end of the Tasmanian industry which had yielded some 1 623 445 L of crude shale oil from 42 237 tonnes of mined shale over a 25-year period. To all intents and purposes Australian shale oil mining had run its course but, like the dog that will not lie down, another attempt was made in New South Wales during 1937. Realizing the vulnerability of Australia's oil imports under the threat of a second world war, the Federal and New South Wales Governments subscribed debentures to aid the formation of a new operating company — National Oil Pty Ltd. The Chairman was Sir George Davis of Davis Gelatine Ltd, a well-known glue and gelatine manufacturer.

A town named Glen Davis was built in the Capertee Valley at the eastern end of the rich seam that had been mined in nearby Wolgan Valley at Newnes, and a number of retorts and other items of equipment were carted from Newnes to the new operation. Problems were soon encountered with water, power and housing shortages as well as unsatisfactory retorts for the rich shale — something that should have been detected before the venture began. The outbreak of World War II aggravated the problems but, at the same time, acted as a spur to production. The Federal Government declared the petrol output of 45.46 million L a year would be exempt from excise duty. However, National Oil could not muster the financial resources to keep going and, in 1942, the Federal Government took control of the operation, producing about three percent of the country's petrol requirements during the war years. Shale mining and oil production continued during the latter part of the 1940s, but it became increasingly obvious that the operation was not viable in the post-war economy. A long sit-in strike underground during 1951 by miners protesting about the government's resolve to shut down the mine and refinery failed to reverse the decision. The official closure date was 30 May 1952 and, to most, this represented the industry's final curtain.

More recent history, however, indicates the demise of Glen Davis may have been simply the close of an era in shale oil development. When crude oil prices rocketed skywards on the assumption of short or uncertain supply in the early 1980s following the Iranian Revolution, the world suddenly began reviewing alternative energy sources. Oil shale was one of the options and Australia was one of the main places with the potential for development. The focus of the new push was not New South Wales or Tasmania, but the sandfly- and mosquito-infested mangrove areas of the central east coast of Queensland at a deposit known as Rundle. Rundle and several other locations in the State had been known for almost as long as those in the Blue Mountains, although the occurrences in the early days were treated more as curiosities than potential mines. The favourable economics faded again with a fall in crude oil prices during the mid to late 1980s, but modern technology allied to commercial production of syncrude from tar sands in Canada has kept the possibility of a return to viable shale oil mining and production very much to the fore. Several major oil companies are still involved in research and evaluation of deposits in Queensland, while a rich deposit near Rundle known as the Stuart area is now targeted by the two Australian companies Southern Pacific Petroleum and Central Pacific Minerals as the most likely place for a demonstration plant. There is a body of opinion which suggests shale oil's third 'coming' may not be too far into the twenty-first century.

The coorongite question, on the other hand, has long been put to rest. Some experimentation with the substance continues, but today it is seen mostly as just an interesting local phenomenon in the southeast corner of South Australia. In hindsight, it is hard to imagine the passions aroused by its initial discovery. Yet aroused they were, so much so that the debate over the existence of a 'mother lode' sparked the first deliberate attempts at oil exploration in Australia.

Coorongite takes its name from the Coorong — a low, thin spit of sandy land located along the South Australian coast between the mouth of the Murray River at Lake Alexandrina and the town of Kingston nearer the Victorian border. Passing through the region during 1852 in charge of a gold escort travelling to Adelaide from the Victorian diggings, Police Inspector C.W. Stuart picked up a piece of odd, dark-coloured rubbery substance which appeared to be deposited on the ground in extensive sheets. Although puzzled by the occurrence, he did no more than make a note of the find and the place, near a stream called Salt Creek which drained into the Coorong. During the 1850s settlers were just beginning to try out the grazing potential of the district and they too came across the substance, particularly in low-lying areas after winter flooding had subsided. Shepherds soon discovered it was flammable and cut off strips to use as candles and lamps, referring to it as 'oil cake' and 'elastic bitumen'.[23] The Aborigines encountered in the region during those years were apparently well aware of its lighting properties.[24]

The phenomenon remained a local curiosity until the early 1860s. However, by that time news of Edwin Drake's discovery of oil in Pensylvania during 1859 had reached the Australian colonies and, when the first refined 'kerosene' from this free-flowing crude arrived, it generated a great enthusiasm to find indigenous deposits. Suddenly the sketchy reports from the Coorong took on a new perspective. Perhaps this strange india rubber-like substance was the surface expression of a vast underground reservoir not far below. People had heard that Drake's discovery was made at a depth of only 21.2 m. Adding weight to the 'wisdom' of the times were the continuing, although intermittent, reports of bitumen stranded along the beaches of Kangaroo Island and along the Coorong right round to the settlement at Portland in Victoria.

During 1865 a particularly large deposition of oil cake occurred in the Coorong region following exceptionally heavy rainfall and extensive flooding. This was brought to the notice of the Adelaide press and the resulting articles sparked a number of letters from correspondents expressing confidence that the material was a true indicator of petroleum. The newspapers also published opinions from those who took the contrary view that oil cake was definitely of vegetable origin completely unconnected with the sort of oil that Drake had found.[25] This exchange of ideas, which began during 1865, swiftly grew into a long-running international debate waged well beyond the pages of the Adelaide papers, but, at the same time, the news prompted Australia's first oil exploration programme. On 23 June 1866 the *Adelaide Observer* was able to report:

> It is rumoured that a valuable discovery of petroleum has been made somewhere in the vicinity of the Coorong; if such is the case the lucky individuals connected may rejoice. However, it is a well known fact that some of our influential citizens have equipped a party with boring apparatus and everything necessary for a six month's search, and should they succeed they will be well paid indeed for their trouble.

The 'well known businessmen' were not named, but they did form a company called Coorong Kerosene Company and began drilling what is recognized as Australia's first oil exploration well soon after the article appeared. The well's location on Alfred (now Alf) Flat near Salt Creek must have been close to the original sighting of oil cake recorded by Police Inspector Stuart. Little is known of this attempt except that two years of presumably intermittent drilling took place before the well was abandoned in 1868.[26] Thomas Scrutton, an early entrepreneur, man of some influence[27] and quite probably a backer for Coorong Kerosene Company, expressed the disillusionment which followed this and other attempts.

> Much labour has been uselessly expended and useful tools misapplied, with 25 feet [7.6 m] being the greatest depth attained for all the time and money expended in the search.[28]

Nevertheless, the failure did not kill any optimism for, in the following year (1869), a group of entrepreneurs gathered a ton [1.02 tonne] of the oil cake and shipped it to James Young in Scotland for analysis in his world-famous oil works. The result was astonishing, and certainly boosted confidence that continued exploration was not only justified, but also bound to be successful. The ton of material had yielded 545.5 L of crude oil from which further refining produced 318 L of kerosene, 59 L of liquid paraffin and 32 L of black varnish. Soon after this report arrived in Adelaide the search resumed with vigour, a number of companies and private operators renting tracts of Crown Land in the Coorong area from the South Australian Government.

The proponents of a vegetable origin for the oil cake were not dissuaded by Young's analysis, however, and they continued to voice their opinions based on scientific observations similar to those made by researcher and experimenter G. Francis as early as 1866. Writing in the *South Australian Register* of 8 May that year, Francis pointed out that the material had a cellular structure which he considered to be neither caoutchouc (unvulcanized rubber), elastic bitumen, asphalt or petroleum, but a peculiar fungoid growth, and that it had no connection with coal or any other combustible mineral. However he did suggest the substance might have commercial value if sufficient quantity were available — a remarkably perceptive comment for those early years.[29] But the vegetable theory made no impression on men like Scrutton. In 1874 he addressed a meeting of the South Australian Chamber of Manufactures and, playing for maximum dramatic impact, he read his text by the light of a lamp fuelled by kerosene distilled from oil cake — the substance which by that time had been dubbed 'coorongite'. His verbose speech covered all the world occurrences of petroleum known at the time and the theories of their origin, including reference to the classical scholars of ancient Greece and Rome. Directly concerning coorongite he said:

> This cake deposit has evidently risen from below, as oil, forced up by the superincumbent pressure of a vast body of water — the result of very wet seasons.[30]

Scrutton reasoned that because of the high yield of oil from the substance on heating, it was likely that millions of tons of oil had been projected from subterranean sources in the past, and that it only required money to find it.[31] Quoting the eminent Dr Hooker of London's Kew Gardens who believed that coorongite was not of vegetable origin, Scrutton went on to compare the substance with asphaltic 'gum beds' formed by natural springs rising to the surface in Wyoming in America.[32] One intriguing aspect of his talk was the scathing comments he made about some elements of the early exploration industry. He pointed out that, since 1866, £480 had been paid in lease rental by 14 companies and private individuals, but that some were merely shepherding the land and hoping to profit from the labour of others. More than £1 300 contributed to the search had been mostly frittered away because of the want of skill, or integrity, or both, practised by field crews.[33] These sentiments have been repeated with varying intensity throughout oil exploration history, particularly during boom periods.

Whether Scrutton's criticisms of industry practice discouraged would-be explorers or whether the proponents of vegetable origin for coorongite held greater sway is not clear, but no further attempts to drill in the Coorong were recorded until 1881. That year saw the registration in Adelaide of the Salt Creek Petroleum Oil Company and, during the next three years, this company drilled three wells on Alfred

Flat, engaging an American tubewell sinker, a six-horsepower engine and four other employees for the work.[34]

Initial reports of good progress and apparently favourable indications of oil and coal gave way to mechanical problems with the equipment and caving of the formations being drilled. With finances rapidly diminishing after two failures, the third well was spudded (begun) a mere 457 mm from the second so that there was no need to move the machinery. However this last attempt was also doomed and the company called in the liquidators in December 1883.[35] Several other wells during the 1890s probed the Coorong and surrounding area to varying depths down to 274.5 m without success. By the turn of the century most opinions favoured a vegetable origin for coorongite despite the acknowledgment that liquid fuels could be distilled from the substance.

A notable exception was that of Dr Herbert Basedow, a scientist whose qualifications and vocations included geology, anthropology, medicine and South Australian politics. Consulting for the South Australian Oil Wells Company in 1914, he selected the site for a well known as Richmond Park Bore near Robe on the southeast coast. This well achieved a depth of 137 m, encountering seams of black bituminous coal and shows of natural gas before being abandoned.[36] Preoccupation with World War I and a complete lack of exploration success combined to dampen the coorongite debate during the second decade of the 1900s. An apparent epitaph to the controversy was given by the South Australian Director of Mines, L. Keith Ward, when he charged that pieces of the substance found on the crests of calcareous (limestone-rich) sand dunes were evidence that the material was deliberately placed there with fraudulent intent.[37] Ward was also of the opinion that there was as much chance of finding oil in the southeast of South Australia as there was of finding the letter 'W' in the French language![38]

But the weight of Ward's sarcasm was shrugged aside by a renewed rash of oil promoters during the 1920s following another thick deposition of coorongite on Alfred Flat after heavy rain and flooding at the beginning of the decade. Prominent among those urging resumption of drilling in the region was a character named Eugene de Hautpick MRES, MIRE, a Russian who arrived in Australia in about 1921 with the self-styled title of 'Captain' and claiming to be a geologist. The newspaper *Smith's Weekly* recorded the event in scathing terms six years later.

> He was immediately established as an authority on oil by reason of letters after his name, the meaning of which no-one understood, and by the fact that he was enthusiastically backed by the 'Mining Standard'. That journal published a series of learned articles from his able pen which, like his equally abstruse reports, were cribbed verbatim et literatim from standard works on oil geology.[39]

It is possible that *Smith's Weekly* confused the title of the publication running de Hautpick's reports with the *Mining Journal* of London. The latter described him as 'our well known correspondent' in the introduction to an article by him on coorongite in 1923. In it de Hautpick noted the work of well-known geologists and chemists of the time like Professor David Day (USA), Professor L.T. Dalton (London), Professor Hoefer (Vienna), Dr Reinhardt Thiessen (USA) and M.J. Berkley (Kew Gardens, London), all of whom had studied the substance and concluded it had an organic origin unconnected with petroleum. De Hautpick then dismissed these opinions by saying that none of the researchers had ever set foot in the Coorong to see for themselves as he had done.

> The idea that it 'grows' there is made ridiculous by a five minute's examination of the locality. It represents undoubtedly the solidification of some substance which previously floated on the surface of the water, and then was deposited on the sand, and gradually dried up, absorbing and uniting with, during the drying process, all matter that happened to come in contact with it.[40]

De Hautpick went on to give an interesting, if somewhat colourful, description of the deposit.

> It is of various thickness from an inch to one foot... When moist the material breaks, much after the fashion of 'green cheese'. Although rubbery under compression, it does not resemble rubber in tenacity or cohesion... It is soft, flexible, and easily cut, clammy to the touch, yet does not soil the skin. It has a characteristic 'swampy' or 'mucky' odour when wet and a characteristic odour of humic soils when moist. In thin strips it burns like a taper, melting before the flame, which is smoky.

Citing his discovery of 'oil seepage' not far from the mouth of Salt Creek as convincing proof, de Hautpick then delivered his verdict on coorongite.

> In times of exceptional floods the hydrostatic pressure of the artesian basin acquired power to force petroleum in large quantities to the surface through the fissures of the faults. This crude oil floating on the surface of the flooded depression produced complete devastation of the vegetable growth within the boundaries of the basins.
>
> The vegetable origin [theory] must assume that the appearance of this new 'plant' killed all the other plant on an area of thousands of acres. To discuss seriously such an idea would be to confess the need of the services of an alienist.

> The vegetation which has been killed by the intrusion of liquid mineral oil covered a great surface, therefore this is visible evidence of extremely rich oil deposits underground in the Coorong district.

The *Mining Journal*'s well-known correspondent closed his article with a lofty declaration.

> All these oils [produced from coorongite] are of paraffin base, the generation of which is the exclusive royal prerogative of Her Majesty, Nature, operating in the majestic solitude of her deep mineral kingdom.

It is unlikely to be a coincidence that at about the same time as the *Mining Journal* article appeared, de Hautpick was busily setting up an oil venture to drill in the southeast of South Australia. A backer was found in the person of an Adelaide businessman who, with others, invested heavily in the venture, according to *Smith's Weekly*. Three bores were sunk at Robe to a depth of 915 m. The paper went on to describe the captain's exit from the Australian scene.

> The Coorong yielded no petroleum, and as the subject of oil was becoming distasteful to the public, de Hautpick returned to Europe, leaving the mining authorities to reflect... on a cable that was sent them from London early in the piece advising caution in dealing with him.[41]

There is a certain irony in the fact that just when de Hautpick was chastising the leading scientists of the day for not visiting the Coorong to see coorongite forming, there was a qualified geologist quietly studying the 1920 phenomenon at first hand. Arthur C. Broughton was consultant to the South-East Oil Prospecting Syndicate at the time,[42] and the publication of his observations was the first major step in proving beyond doubt an organic origin for the substance.

> ... on these new lagoons [floodwaters] a thick scum, like green paint, is forming. This scum is drying on the water in places to a semi-elastic substance... The drier portion can be scraped in with sticks... This skin in places is yards in area and can be dragged along in sheets... Scooped with the hand from the surface of the lake this substance within a few minutes changes before the eyes from a green liquid which drips from the fingers, to a brown, plastic solid.[43]

Broughton concluded:

> Conditions are now very satisfactory for field work from a geological aspect, and for microscopic work from the biological side to determine the facts in connection with this occurrence...

Although the die-hards and company promoters like de Hautpick refused to give ground and continued to drill wells right up to the early 1930s — Coorong Oil Company and Enterprise Oil Company in particular[44] — the microscopic evidence for a vegetable origin began to stack against them. Perhaps Australia's greatest geologist, Professor Sir Edgeworth David, then in the chair of geology at the University of Sydney, added his opinion to the 'vegetable origin' camp, and even Dr Herbert Basedow became a convert after reading the report of US chemist Reinhardt Thiessen on the subject. Basedow went even further and grew the material in his own laboratory.

> If we can determine the conditions congenial to South Australia's oil plant, we should be able to produce unlimited supplies of coorongite which, containing up to 97.4% of volatile material, could be as valuable to the state as a gusher of oil.[45]

By the mid-1920s the substance had also been reported from Murray's Lagoon on Kangaroo Island and from the swamps of Martagallup in Western Australia. The latter occurrence prompted the Western Australian Government Geologist of the time, E.S. Simpson, to make a late entry into the debate and to sarcastically remark that oily liquids could be obtained by the distillation of almost any substance from coal to cucumbers.[46] More recently, the same material has been noted in Siberia, East Africa and the Northern Territory of Australia. In the 1980s, modern researcher R.F. Cane reports, it is now established that coorongite is essentially a hydrocarbon polymer arising directly by metabolism of the alga called *Botryococcus braunii.*

> The alga can exist in three physiological states, two of which produce large quantities of polyene hydrocarbons. On the death of the colony, the hydrocarbon metabolites oxidize and polymerize into a dark coloured rubbery mass from which a hydrocarbon oil can be obtained by pyrolysis.[47]

It is now clear that the establishment of early coorongite-prompted wells was based on a misconception, perhaps deliberately prolonged by some. But two things have come from it: the first is the plausible, but still futuristic, possibility of developing a source of renewable energy by growing the alga; the second, and more important thing from a historical stance, is that it gave early birth to the practice of drilling for oil in Australia. The fact that it also created long-running scepticism about the country's oil potential is another part of the story.

Chapter 3

PROMOTERS AND EXPLORERS TAKE TO THE FIELD

Apart from the early attempts to find oil in the Coorong District of South Australia (and perhaps even because of the drilling failures there) Australia was slow to begin a serious search for its own petroleum reserves as it entered the 1900s. Certainly there was no coordinated effort. Nevertheless, spurred on first by reports of strikes in America, like that of the great Spindletop gusher on a hill outside Beaumont in Texas during January 1901, and later by the rise of the motor car and the demands of war, promoters and explorers did take to the field — albeit on a hit-and-miss rather than a scientific basis. A glance at a map of the period shows the settled areas were still little more than isolated colonies despite the new umbrella of Nationhood. Thus, for the first half of the new century — longer, in some areas — the oilmen, like the gold seekers before them, were often pioneer settlers, surveyors and route makers as well as fortune hunters.

WESTERN AUSTRALIA

Petroleum and asphalt strandings near the Warren River between Cape Leeuwin and Point D'Entrecastreaux in the southwest of the State sparked the first concerted effort to prove up oil reserves in Western Australia. Bituminous deposits had been seen in the area over a number of years stretching back into the 1800s, but it was not until overseas reports filtered in concerning liquid fuel developments that any real interest was taken. Two prospectors, Messrs Boyd and Sinclair, were dispatched to examine the country. The *Sydney Morning Herald* of 5 February 1902 recorded the story:

> At one point indicated by residents oil could be seen floating on the sea a few hundred yards from shore. Samples of sea water were obtained by the prospectors and subsequent analysis showed that the oil was of mineral and not vegetable origin as at first had been feared... Traces of mineral oil were found further inland and, three miles [4.8 km] from shore, sandstone

> saturated with petroleum oil was found in the water course of the Warren River.

The article reflected the American influence prevalent in the early part of the century as it continued:

> A curious feature... is that the geological conditions appear to be identical with those of the oil regions of the Californian coast where the same phenomenon of oil on the sea near the shore and the presence of bituminous asphalt thrown up by the tide are observed.

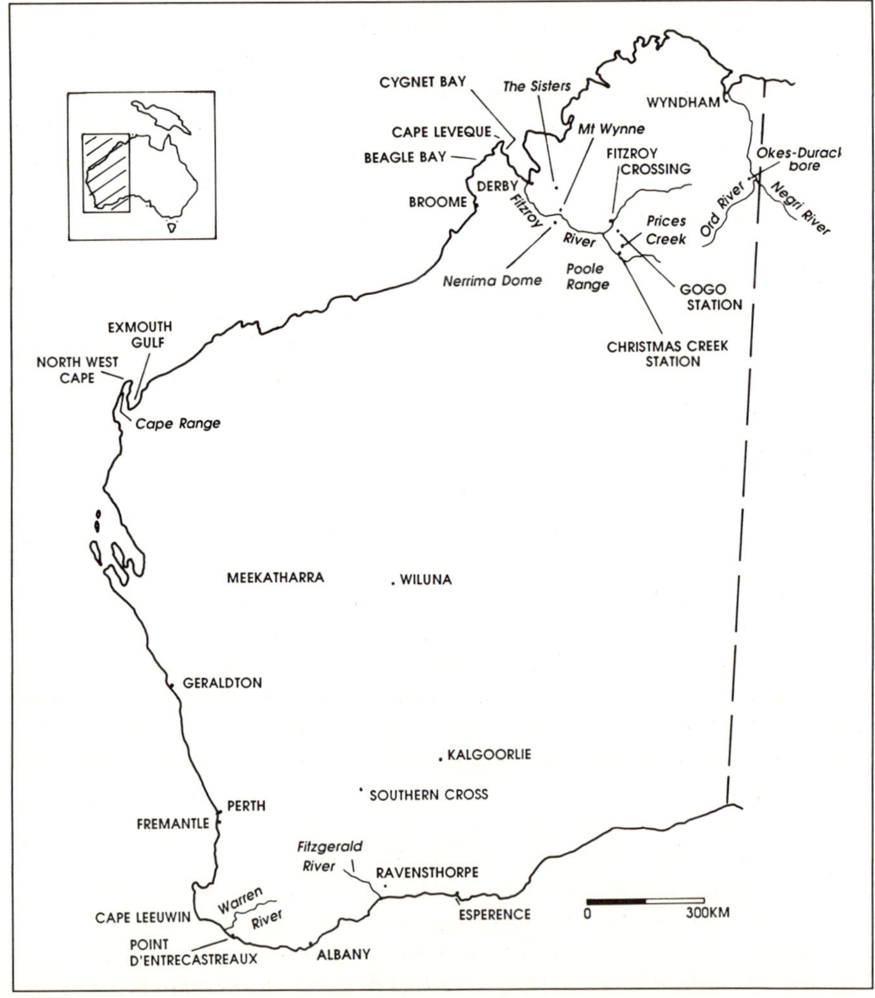

Western Australia — location of key places mentioned in text.

The report even attributed the geological name 'Warren-Blackwood Basin' to the area.[1] Intrigued by these findings and spurred by the thought of a new industry, a group called the Westralian Mining and Oil Corporation Ltd signed a contract with Diamond Drilling Company of Kalgoorlie to drill a 500 m deep trial bore in the region. Not surprisingly it found nothing, yet the company persisted. In the end, three wells were drilled during the period 1902–1904, the deepest reaching 524 m. None had any success.[2] After this shaky start the field was left to the artesian water borers for the next 15 years or so. Ironically, these men, who spread out over the State with their horse, camel and donkey teams, unknowingly played a part in later oil exploration by virtue of the painstaking geological well records which they, the station people and the Mines Department kept intact. The data was invaluable to companies who followed, even as much as 50 years later.

The southern part of Western Australia attracted two more well-publicized abortive forays in about 1920, before the focus of attention swept up to the State's Kimberley region where it stayed for the next three decades. At Jonacoonup Hill on the Fitzgerald River about 80 km from Ravensthorpe, a prospector named Perkins and an oil company chemist named Martin excitedly noted oil floating on a tank holding water coming from a bore they were drilling. Not only that, an oily scum was seen on water holes in the river and a bituminous substance was reported from the nearby mountainside. Messrs Perkins and Martin were brought back to earth with a thump, however, when a government geologist declared the oil in the settling tank to be lubricating oil used on the drill, the scum on the waterholes to be iron oxide, and the mountainside 'bitumen' to be animal excreta. The second 'rush' was begun by a farmer who noticed oil scum and gas bubbles rising to the surface of a creek on his property, but the stampede was stopped in its tracks when an investigating expert found that both oil and gas were 'emanating from a large and very defunct pig wedged in the bed of the creek'.[3]

This sort of farce did little to enhance the credibility of the fledgling oil industry but, fortuitously, at around the same time two almost simultaneous reports from the far north did stand up to expert geological inspection. In 1919 a stockman named Walter Okes employed on the Durack family's Argyle Downs cattle station reported finding a type of pitch in basalt near the junction of the Negri and Ord Rivers about 180 km southeast of Wyndham. The same year a water-well sinker named Harry Price found what he considered to be oil traces at shallow depth in a well on Gogo Station not far from the Fitzroy River. In fact, Okes's find was probably earlier, as notes from the Durack family archives in Perth indicate the stockman mentioned the existence of a 'coal' prospect on the property before he enlisted in the army during World War I.[4] At any rate, Okes took Michael Patrick (M.P.) Durack back to the spot on his return and the subsequent report, sent to Perth, brought Assistant

Government Geologist Torrington Blatchford onto the scene to investigate. He confirmed the discovery as bitumen and Durack lost no time in forming the Okes-Durack Kimberley Oil Company NL on the strength of it.

Durack, at that time member of parliament for Kimberley, was instrumental in securing a drilling plant and sufficient drill casing for 1 068 m of hole. Hopes were high as the rig left Fremantle by ship for Wyndham. At one point, influential Sydney-based financier Sir Owen Cox offered to buy Okes out at the high price of £10 000, but Durack's conscience would not allow him to assist such a deal as the sum would have been small if expectations were realized. The cattle baron was sure that the port of Wyndham would become a gateway to the development of trade with Asia, the more so if oil was found in the region.[5] The rig was expected to drill at a rate of 45 m a week and reach the likely oil horizon somewhere between 430 m and 760 m. However the Okes-Durack bore was a grave disappointment. It drilled to about 366 m, nearly half of which was in basalt, and the effort was rewarded only with 'petroliferous and sulphurous odours'.[6]

While in the Kimberley area during 1921, geologist Blatchford also confirmed the presence of oil in the Price well by putting down a 27 m hand auger bore beside it. On hearing this, Richard Freney, a Broome pearler and a director of the Okes-Durack Kimberley Oil Company, set up a new exploration enterprise called Freney Kimberley Oil Company (in which M.P. Durack was also a director). Four wells were drilled in the immediate vicinity during 1922 and 1923, all to depths of less than 300 m. Unfortunately, although finding traces and odours of petroleum in the drill cuttings, there was nothing to indicate a commercial discovery.[7] Perhaps surmising this might happen, Blatchford had already noted several other structures in the region which he considered favourable for oil accumulation. Acting on this advice Freney Kimberley Oil moved one of its two Calyx rotary rigs (rented from the Western Australian Public Works and Water Supply Department) to a place called Mt Wynne and there drilled two wells. The hard limestone made progress slow and it was decided to change to percussion drilling. After a lot of downhole mechanical difficulty, samples of bitumen and even globules of oil were recorded, but again there was no commercial discovery. In the mid-1920s English consultant geologist Dr Arthur Wade visited the area and recommended another site in the nearby Poole Range. Once more following expert advice, Freney Kimberley drilled two further wells with the same teasing, but frustrating result. One was eventually abandoned because the drilling cable broke and the tools were lost down the hole.

Although Freney Kimberley was the most active company in the area during the 1920s, it was not the only one. The Locke Oil Syndicate, headed by Alfred Locke, who gave his status and address as 'investor, the West Australian Club, Perth', was formed with a capital of £12 000 to prospect for oil over 124 320 km² of what was then known as the Desert Basin (now the Canning Basin).[8] Geological investigations by a six-man party were led by Leo Jones, a geologist on 12 months leave of absence from the Geological Survey of New South Wales. He and his men crossed the basin by camel train from Wiluna in the south to Fitzroy Crossing and then Broome, tragically losing one member speared to death by Aborigines on the Canning Stock Route.[9] Another company, Australian Petroleum Development Company NL, took up an area along the coast between Broome and Cape Leveque, prompted by alleged bitumen finds by Aborigines on an island in King Sound and in springs round Beagle and Cygnet Bays. One Thomas Esdaile undertook a geological expedition through the area for the company, equipping himself for the journey with 22 mules, two pack donkeys, a four-wheel cart and five saddles. His companions were three white men, two half-castes and three Aboriginal full-bloods, one of whom knew the country intimately.[10]

Despite all these bona fide efforts to find oil, the 1920s was not without hoaxers. One prominent example was reported in the *Mining Journal* of 1 December 1923 under the headline 'Oil Fake in Western Australia'.

> Upon the government analyst reporting that a sample of oil from a bore on the edge of Lake Eva, near Southern Cross [in the goldfields 400 km east of Perth] consisted of vaseline, soap, and lubricating oil, an inspector of mines and detectives visited the bore. They reported finding under the bunk of the supposed discoverer a large hole in which was a 100 gallon [455 L] tank containing 80 gallons [364 L] of oil, from which pipes led 46 feet [14 m] into the bore. The man fled after the first analysis and warrants have been issued for his arrest and that of his companions.

By the 1930s all the exploration activity was centred in the northern half of the State. In what became known as the Carnarvon Basin, the region round Exmouth Gulf was taken up by a newly formed company called Oil Search Ltd as part of its ambitious and widespread plan to discover oil in Australasia. Thanks to its manager T.W.H. Dee, a former director of National Mapping and one of the surveyors who had laid out the city of Canberra, the company was the first to attempt a coordinated approach to exploration. An example of Dee's efficiency was the issue to all his field officers of a manual setting out a complete list of procedures to be followed when reporting to, or requesting items from, head office in Sydney. This included the usual forms for expenses and material orders, and also the correct filing of field data, a numbering system for correspondence, the use of index maps and standardized coloured inks to distinguish various details, like geology from topography. There was even a section covering the adjustment of levels and theodolites, and another describing collection procedures for field samples.

As it turned out, Oil Search's time in Western Australia was cut short by a management decision in the late 1930s to concentrate all its efforts in New Guinea. However this was not before the company had run several reconnaissance surveys in the Carnarvon Basin, one of which included the work of Harold Raggatt — then a young geologist seconded to Oil Search from the Geological Survey of New South Wales. Raggatt's favourable impression of the oil potential round Cape Range during that period was to lead to his later recommendation of the area to Ampol and the subsequent discovery at Rough Range in the 1950s. Another recruit in the early days was Eric Rudd, then a young graduate from Adelaide University, who is now Emeritus Professor of Geology there, and a well-known figure in mining and oil circles.

> My first job with Oil Search paid £250 a year which was good money in those days. I was often teamed up with 'Snow' Atkinson, a surveyor, and one of the first trips was to the North West Cape - Cape Range area. We were out in the field for eight or nine months without a break. Mostly it was in thorn bush country and our clothes became so ragged that we weren't respectable enough to walk into town. We used to come in at night to do our shopping.
>
> All the work was on foot then, mapping as we went. A tent fly was all we had for cover each night. We were living mostly on kangaroo and bustards (bush turkeys) and I can remember one time the cook struggling with the slaughter of an emu. Eventually he killed it and cooked the legs. They were like bundles of barbed wire round plastic.[11]

Oil Search departed without drilling a well and it was to be another 15 years before the Carnarvon Basin returned to the limelight. Not so the Kimberleys although, despite the continuing enticement of oil shows during drilling, by the late 1920s it had become clear that commercial discoveries in the Desert Basin would not be as straightforward as first thought. The original Okes-Durack company was amalgamated with Freney Kimberley Oil (M.P. Durack became a director of the new entity, but Richard Freney had apparently quit the scene) and arrangements were made to return to Poole Range for a third try. At this point one of the larger-than-life characters who seem to frequent the oil industry arrived in Australia to take charge of the drilling operation. Introducing him as F.M. Fox, an American driller, the *West Australian* of 9 January 1929 said he was an enormous, muscular man in his middle years who would probably tip the scales at 15 stone (95 kg) without carrying an ounce of extra flesh. Referring to the prospects at Poole Range, Fox exuded confidence, even though he had not been to the site.

> You can bet your last dollar it [oil] is there. Tell you another thing. I've just finished a job in the bush on a little island — Iram Island — which is way off in the ocean westward of British New Guinea. It's dead above the Kimberleys, only about 600 miles [965 km] to the northward. Now if you had oil 600 miles overland from Perth you'd think it was pretty close wouldn't you? It's only because there's a patch of ocean in between that Iram Island seems such a whale of a long way from the Kimberleys.
>
> There's another thing too. The Kimberleys are in the right belt — look at this map. Peru's awash with oil round about the same latitude. The Gold Coast of Africa's got oil a bit more northerly. New Guinea's pretty close to the same latitude too and of course there's a belt about the same distance from the equator on the other side which almost circles the earth. I've found oil in California, Texas, Peru, Angola (that's Africa), Borneo, New Guinea, Java and quite a few other places.

Not everyone was convinced by this heady talk. M.P Durack's sister-in-law Eva was one such doubter.

> Freney's [new well] is going to be just another bungle, a waste of time and money. They'll find oil in this country, but they won't find it by getting Fox at 50 pounds a week to help them. They will find oil here when, and only when, the Americans want to find oil and it will probably not be in my life time.[12]

Eva Durack went on to describe the day when the crucial test was to be made. All fires, even cigarettes, had to be extinguished and the party was told where to take refuge if a gusher inadvertently occurred. Then, just before the 'Great Thing' was to happen, a tool fell down the bore and all work was indefinitely postponed. Soon afterwards Fox returned home. Another well was started, but from that point there were constant delays — not enough wood for the boilers, breakdowns in the vehicles and wrong replacement parts sent up from Perth, problems with cartage contractors, and personality problems within the camp. To top it off, there were accusations that the well was not drilled in the proper place because of the money that would be needed to build a road to the correct site. All of these things were magnified by the long line of communication to the rig from head office in Perth, a difficulty which was to be experienced in most of the oil drilling in central and northern Australia during the years to come.

In October 1932 Freney Kimberley Oil was wound up after it had spent £150 000 over a 10-year period and drilled 2 440 m of hole without commercial success. Nevertheless, oil indications in nearly every well enticed the same management to continue and a new entity, Freney Kimberley Oil Co (1932) NL, was formed with an authorized capital of £25 000. In a bid to find promising new structures to drill in the Kimberley region, the company secretary in Perth, an accountant named Carcary with the firm Carcary and Halvorsen, was authorized to hire noted geologist Dr Arthur Wade. Wade had been in the region briefly 10 years earlier for the Federal Government, but for this second visit the Freney directors had in mind an extended stay so that he could geologically map the northern part of the basin within the company's leases. One of Wade's assistants in this task was Eugene O'Driscoll, who had just completed the third year of a four-year Engineering-Geology degree at the University of Western Australia. It is he who provides a first-hand account of the survey and the conditions in the field at that time.

> We went on the coastal freighter *Kangaroo* up to Derby where Freney had a jump-off base. I paired with a bloke named Ross Weatherburn to do the broad regional mapping, and Wade took Leo Waterford, an excellent surveyor from South Australia, to do the more detailed work.
>
> We went out to Poole Range first, not far from Christmas Creek Station which was a pretty rough and ready place then. The beds were made of bush timber with raw hide thongs stretched across as webbing. Water was scarce too and many times when there I saw long lines of Aboriginal women coming from Christmas Creek itself bearing water in tins on their heads to supply the station needs. We were lucky at our camp because there was permanent water in a rock hole. It teemed with desert finches — all kinds and colours, and so tame they would perch on your boots as you sat by the pool.
>
> The preceding drillers had left a boiler and tank on a stand plus a bit of a workshop — what the white ants left of it that is — and we used the place as a regional base. Those white ants were bad. One time Weatherburn left a snazzy pair of black and white pyjamas in the camp while he went off for a week further afield. He came back to find there was no piece of cloth left bigger than a two-bob bit.[13]

O'Driscoll's job was to use the plane table method to map the region for topography and rock outcrop at one inch to the mile scale. This involved setting up a horizontal drawing board on a tripod, taking readings of the main features of the landscape through a telescopic sight, and plotting them directly onto the board. He and Weatherburn mixed a flour and water paste to stick the drawing paper sheets to linen backing. Control stations 50–60 km apart were used as reference points in the work. They had been set up in Lord Forrest's day about 30 years previously and, while the altitudes were inaccurate, the latitudes were very reliable. Some areas had been mapped before, although O'Driscoll found there were a number of inaccuracies.

> Old H.W.B. Talbot from the Western Australian Geological Survey had been in that area, but on one bit of range he had seven contour lines on one side and nine on the other. He'd crossed them over somehow, so I had to remap it.[14]

O'Driscoll would report progress and hand over the maps to Wade once a month. Often the party leader would operate out of the regional base, visiting specific areas that O'Driscoll and Weatherburn had seen. In one area that Talbot had previously marked as 'undulating country — no outcrop', the plane table team found and mapped Nerrima Dome, a low-relief anticline that became the focus of Wade's drilling recommendations in his final report to Freney.

> Most of the work was done on foot — in that limestone country I wore a new set of boots off my feet in a week — but the parties did have vehicles for access to the work places.
>
> A truck carried all our supplies including a 44-gallon [200-litre] drum of water, although most of that was needed for the radiator, as holes were continually being punched in it by boughs or spinifex. We plugged them with cement so eventually the radiator was like a hollow concrete block. We used to take out the engine every week to decoke it and do a valve grind. That was a Sunday job. We'd have to do all our own repairs, improvising with bush timber or whatever we could find a lot of the time.[15]

In total, the Wade parties spent three winter seasons in the Kimberleys for Freney, surveying just under 260 000 km^2 of country. To do this they set up regional base camps and then worked out from them in different directions for a week at a time using single-night camps located where mapping for the day finished up.

> Most of the time our one-night camps were just a bed roll on the ground and a bough wind break. Water was always a concern and we relied on directions from the station people a lot. They'd let us have some fresh meat every now and then too. But you couldn't keep it long with no refrigeration. Most of the time we had tinned stuff or 'salt horse' — a black salted beef that was soaked for 24 hours before cooking. Pumpkin, biscuits and stale bread usually made up the rest of the menu.
>
> A normal day would start at 6.00 a.m. and finish at 8.00 p.m. with lunch and a sleep under the truck for 20 minutes

> or so in the middle of the day. But not if there were any sugar bees about. They were a bugger, crawling over every inch of skin to drink your sweat.
>
> Mosquitoes could be bad too and I got fever once from them. There was a bit of leprosy about among the Aborigines then, but few whites ever got it. I did have sunstroke one time though — laid me up for a week.
>
> We'd have to watch out for spinifex fires too. They could be bad. The stuff would generate a resin build-up and it burnt with a ferocious heat, sometimes going through miles of bush before it petered out.

O'Driscoll says the work was hard, but never dull or even lonely. There were any number of characters on the surrounding cattle stations.

> Weatherburn and I came across old Gus Peglow the hatter one day under a boab tree. He'd slung a tent fly up to get a bit of shade and invited us in. Just made some damper, he told us, using Eno fruit salts because he didn't have any baking powder. And he had a huge cheese which he reckoned was pretty strong. He'd poisoned seven dingoes with it during the previous week — the hides were nailed to the tree to prove it. Needless to say we declined the offer of a meal.

At the end of the third field season Wade's report to Freney in Perth recommended two wells — one to 1 525 m on Nerrima Dome 210 km east of Derby, and one to 1 220 m, 50 km further north on a feature called The Sisters. Several other areas, he said, were promising targets, but they needed further survey before drilling. In 1937 Freney acted on this advice and marshalled its capital to begin what turned out to be one of the longest sagas in the whole of Australia's oil search history — the drilling of Nerrima Dome.

Leo Waterford, who had played an important part in Wade's survey, stayed with the company as field superintendent at a salary of £120 a month. The rest of the drilling crew were carefully costed. A head driller would receive £80 a month, a second driller £50 and a third £40. Cathead man and derrick man would each be paid £23, as would a lead tong man and a mechanic, while a fireman to tend the rig boilers would receive £24. An arc welder could expect £32, a mechanic's assistant/wireless man £20, a cook £26 and cook's offsider £20 a month. Thus the company's estimated field wages bill would come to around £747 a month — quite an outlay when added to the cost of supplies and equipment. Fuel and lubricants were estimated to be £63 a month, firewood a hefty £400 a month and freight charges £60 a month. The all-up costs, including rig hire, came to an estimated £23 740 a month.[16]

Drilling began in 1939 and progressed slowly in the mounting supply difficulties of wartime Australia. Then, in 1942, the Japanese attacks around Broome on the west coast caused the Federal Government to step in and stop the work, declaring that no risks could be taken. A military guard was placed at the bore and all precautions taken to protect Freney's property. From the company's point of view the cessation of operations could not have been more frustrating. The well was at 1 303 m and the objective was reckoned to be between 300 m and 600 m lower. In September 1943 company secretary Carcary wrote to the Minister of Supply and Shipping in Canberra saying that Freney was anxious to resume drilling. He pointed out, however, that some necessary equipment had to come from the USA, and it would take up to six months to prepare for a new start.

The government, still concerned about invasion, was unmoved. In June 1944 Carcary wrote to Prime Minister Curtin with the same plea, adding that it was in the national interest for the company to try to find its own oil. But nothing was done until August the following year. The war over, Ben Chifley (who became Prime Minister after Curtin's death in July 1945) wrote to the West Australian Premier (Frank Wise) and this resulted in a joint federal-state inspection team of petroleum geologists and engineers going to Nerrima. Their report, delivered soon afterwards, was unexpectedly severe. It said that poor care and maintenance during the war years had left the machinery badly rusted and threw a question mark over the state of the rig's internal mechanism. They recommended that the whole rig and its engines be entirely dismantled and examined.

A protracted set of negotiations then followed between Freney and the two governments over funding of the Nerrima revival before a rig was purchased from Papua New Guinea and drilling finally recommenced at the site after a hiatus of seven years. Experienced driller L.L. McKillop had been hired to head up the new effort, but his task had not been easy. He was beseiged by problems ranging from poor road access to wood and equipment shortages. Then, just as work finally began in earnest, a downhole obstruction caused an abrupt halt to proceedings. In September 1949, after repeated attempts to remove the obstruction, Freney directors decided to close the bore once again hoping to resume operations the following year. It was not an auspicious end to the first half of the 1900s oil effort in Western Australia.

A postscript to the Nerrima episode was written in the mid-1950s when the Queensland-based Associated Group took over Freney Kimberley Oil and embarked on a new Canning Basin drilling programme. The first well was a return to Nerrima Dome, where the target depth had never been reached. A rig was brought in from the USA and, setting up camp in an old shearing shed, the company set to work. Drilling in the hard formations was a tedious affair, the drill bit sometimes advancing only 1.5 m in an hour. So it was both a relief and an anticlimax for everyone involved when the well finally reached its total depth and was declared a dry hole nearly 20 years after Dr Wade had made his original recommendation to Freney in 1935.

Two other companies came into the northern half of Western Australia in the 1940s, but they were gone again before the end of the decade. The first was Caltex Oil Development Pty Ltd, which acquired a prospecting licence in 1939 over the whole of the Canning Basin except the small area held by Freney around the Nerrima and The Sisters prospects. Hampered by wartime difficulties Caltex managed some reconnaissance geological and aerial surveys, but the company decided to relinquish its permit in 1946 without drilling a well.[17] The second was the Zinc Corporation which, at the urging of its Chairman, Maurice Mawby, acquired Canning permits in 1947. Dr Frank Reeves, a tough, rough diamond from the United States Department of the Interior, was commissioned to study the area. Although he presented his detailed report in 1949, at the end of what can be termed the first phase of the Western Australian oil search, the work really belongs to the beginning of the next phase. Zinc Corporation did not maintain its presence, but the lure of the region was undiminished and it became the first focus of Federal Government initiatives which ushered in a new era of Australia-wide exploration in the 1950s.

QUEENSLAND

An unconfirmed report that brothers named Bonney drilled two exploratory wells at Toorbul near Brisbane in 1887–88 would, if true, make this the first deliberate attempt to find oil in Queensland.[18] There were also reports of oil and gas in a water bore near Longreach in the west of the State during the 1890s. However most sources cite the accidental encounter with natural gas in a water well being drilled on Hospital Hill, outside the town of Roma in the southeast of the State during 1900, as the beginning of the Queensland oil search. Certainly it was the earliest confirmed hydrocarbon discovery. Yet, ironically, at the time it was considered a confounded nuisance.

A similar bore drilled three years previously to supply reticulated water to Roma had been found to be inadequate and the town council petitioned the State Government for a second attempt. The new bore, designated Queensland Government No. 2 (QG 2), did strike an aquifer at 1 046 m, but the council decided the water flow was still not sufficient for Roma's needs and convinced the Premier (and Treasurer) of the day, Robert Philp, to deepen QG 2. Drilling was continued, and at 1 123 m on 16 October 1900, workmen noticed there was more water than usual bubbling from the well. Then, with a whoosh, a column of mud and water shot 18 m into the air, drenching everyone present as it rained back to earth. Apparently familiar with the phenomenon from other parts of the world, one driller spat out a mouthful of mud, licked his lips and announced: 'S'truth, it's gas!'.[19] For several days afterwards workmen tried to contain the flow, but were unable to do so. The displeasure of the Roma Council at having their new water source soured by sulphurous gas was easily matched by that of the State Government which had outlayed £6 000 in now totally wasted drilling costs.

The first man to analyse the petroliferous gas, realize its possibilities and recommend it for commercial use was the State's hydraulic engineer, J.B. Henderson, who had been associated with the QG 2 project from the beginning. He sent an assistant from Brisbane to collect samples of the soured water and then, in February 1901, he went to Roma himself. A few months later he reported back to the town council.

> Its [the gas] value as an illuminant [24-candle power] is quite equal, if not superior, to that of any of the gas supplied by any of the Queensland [coal] gas works. It is superior to the London Gas Standard, hence if the volume issuing from the

Queensland — location of key places mentioned in text.

> bore proved sufficient, the result increases the value of the bore as an asset. The gas could be stored in a holder, but its permanency is a question that can only be solved with the lapse of time.[20]

A subsequent test on QG 2, ordered by Henderson, recorded a gas flow of 1 250 m³ per day. Then, when back in the Roma district during December 1901 to inspect a water bore a short distance away at Pickanjinnie, the government engineer saw gas bubbles in the water. Scooping up a handful he sniffed it and made a remarkably accurate prediction.

> It is probable that either oil, natural gas, or both might be tapped were the boring here and also at Roma carried deep enough. I take a deep interest in the possibility of oil and gas in Queensland and consider it to be the most important question for the state and worthy of full investigation.[21]

Henderson was soon proved correct at Roma, while the Pickanjinnie structure did flow gas at 182 000 m³ per day when finally drilled by AAR Ltd in 1961. At Hospital Hill the QG 2 bore was tested again in 1904 and found to have increased its flow to 1 960 m³ per day. The Roma citizens, now awake to the commercial potential, erected a gasometer, laid pipes and set up lamps in the main street. For ten days in 1906 the thoroughfare blazed with light supplied by natural gas, but plans to supply the whole town were dashed when the flow suddenly ceased due to a blockage in the well. By that time, however, moves were already afoot to increase the development. Roma Mineral Oil Company was formed in 1905 and, after some initial delays, the company began drilling QG 3 (otherwise known as ROM 1) in 1907 about 80 m from the QG 2 bore, little knowing the sensation it was about to cause. A big flow of gas was encountered during October the following year, but before the flow could be controlled, gas was suddenly ignited by the fire from the rig boilers.

The subsequent 'holocaust' melted the derrick and equipment within five minutes and continued to blaze for six weeks, turning Hospital Hill into a deafening inferno that attracted headlines and sightseers from round Australia and even from New Zealand. The stationmaster at Charleville, some 250 km west of Roma, began issuing train excursion tickets at £1 each — a move which annoyed some who thought 12s 6d a fair price.[22] During the first few weeks there was little anyone could do and the workmen downed tools to return to town. When asked how he intended to put out the fire, the rig manager, a man named Taylor, shrugged saying, 'There are only two of us here. What am I supposed to do — spit it out!'.[23] Roma township itself was only saved from destruction when the mayor, Hector Care, had the presence of mind to drain the contents of the gasometer into the old reticulation system soon after the fire began. Eventually, a huge valved stove-pipe hood, designed and engineered by the manager of the Intercolonial Boring Company of Brisbane, B.L. Schooley, was lowered over the wild well and the flame gradually extinguished.

During its fiercest there had been a strong smell of oil in fumes from the fire, sparking a lively trade in Roma Mineral Oil shares. Once the gas flow had been stabilized, the thought of reaching an oil reservoir encouraged the company to drill on. However a series of mechanical downhole problems finally forced abandonment of ROM 1 a mere three metres below the point where gas had first been encountered. The State Government then took up the challenge at Hospital Hill after consultation with a number of world experts, but its wells also ended in mechanical failure during the years leading up to 1922. A short-lived company called Lander Oilfields of Australia tried its luck with three wells near the township of Orallo, about 40 km north of Roma between 1923 and 1926. Although Lander recovered a few barrels of waxy oil from one of them, there was no hope of commercial application for the discovery and activity waned. In 1926 the State Government withdrew its reservation round the Roma district and a new company called Roma Oil Corporation moved in, taking over the plant and operations of Lander in the process. The location for its first well, ROC 1, was chosen by Queensland Geological Survey geologist L.C. Ball back on the old stamping ground of Hospital Hill and close to the scene of the spectacular ROM 1 fire 18 years earlier.

On 15 September 1927 ROC 1 created its own sensation when it hit the gas reservoir at 1 129 m, shooting a stream of mud and gas into the air. But, mindful of the previous disaster, the well was quickly brought under control. To the company's joy, these efforts were rewarded when small amounts of light oil condensed (hence the term *condensate*) from the gas as pressure lessened at the mouth of the blow-off pipe. The site manager measured the liquid flow at 2.6 pints per 1 000 cubic feet of gas. In reality it was not much (about 0.32 barrels or 51 L per 1 000 m^3), but that did not stop a stampede of share speculation that rocked the small Roma community and blossomed into a fully fledged boom which quickly spread to Brisbane, Sydney and Melbourne. Roma Oil Corporation's 10-shilling shares shot up to £10 within a few days while, in the cities, there was a sudden rash of new company flotations — all claiming some association with Roma, however tenuous. During the next two years the Queensland Government approved prospectuses for 45 of these new ventures and 120 prospecting licenses were granted out of the 225 applications made during that time.

Believing they had the benefit of local knowledge and being closest to the action, Roma people plunged into the buying spree — many selling their homes and mortgaging their sheep and cattle properties to raise the cash to spend on shares. Small investors were being coaxed with prices of sixpence or even as low as a penny a share on application, but then asked to pay regular monthly calls of similar denomination. In many instances big names headed up the newly floated companies. Nowhere was this more apparent than in the board of Australian Oil Industries, registered in Sydney during May 1928 with a capital of £2.6 million — the largest exploration company formed to that time. Among the principals were Sir C. Graham Waddell, president of the Grazier's Association, Sir Samuel Hordern, described as a 'millionaire sportsman', Sir Keith Smith and Sir William Vickers of Vickers Woollen Mills.[24] In no time the whole of Australia was caught up in the country's first oil boom. The Queensland Government endeavoured to cut down the racketeering of so-called 'penny dreadfuls' by insisting each prospectus have government approval and that companies drill within six months or show cause why they had not done so.

One of the more respectable broking houses of the time which gained a reputation for oil and mining floats was Wallace H. Smith and Co. Based in Melbourne, this firm began during the 1920s, setting up companies to explore the Lakes Entrance area of southeastern Victoria, expanding into Queensland with the Roma boom. The founder was Wallace Hugh Smith, a short but solidly built man from Bendigo who began his career there as a broker's clerk on the gold fields. Clarence Byrne (later Sir Clarence), a mining and oil entrepreneur who knew him in those early days, says he was a great cyclist who used to pedal round the town taking orders for scrip and shares.

> He was a modest man, but he didn't like technical people and was always suspicious of experts. He would hire someone else to look after well test and assay results. I can remember he bought 360 Collins Street, Melbourne just after the Second World War. I visited him there when the pink granite facade had been steam cleaned one time and remarked on the rich red colour. Wallace Smith replied that 'it was stained with the blood of Bailieu's clients' referring to a long-time rivalry with that other big broking house.[25]

Co-associates in the original Wallace H. Smith firm were Martin Cordier of the Wilhelmsen Shipping Line; Ralph Randell, a Melbourne clothing manufacturer of 'gent's attire'; Herbert Tucker, a 'retired gentleman'; and one other described as a Norwegian 'urger' named Christiansen. Clive Wallace Smith, son of the founder, who later joined the firm himself, remembers some of the excitement of those early days.

> My father went to California in the 1920s and saw all the mining and oil activity there. He came back saying there was no reason why the same thing could not happen on this side of the Pacific. After that visit he always had his nose in a book entitled *Oil Fields in the United States* [by Walter A.

> Ver Wiebe — Professor of Geology at the University of Wichita]. He was keen to learn everything he could about the industry.
>
> Most of the companies he floated were for small amounts — £50 000 maybe — but in those days there was no institutional backing as there is now. There was a sort of code of underwriting where brokers would pass it around among themselves and other firms to send on to their private clients. Generally the subscription was around a shilling a share, but there were frequent calls whenever any expenditure was needed.[26]

It was usual for one or more of the brokers to be on the boards of the floated companies and the other members were usually prominent businessmen of the day.

> My father devised a way of ensuring good attendance at board meetings. Directors were only paid if they turned up. A pile of sovereigns was placed on the table and split up among those directors actually at the meeting. When I joined after the war £500 a year was a big salary for a director. Most were paid much less.[27]

One of the most enduring floats by Smith was Roma Blocks (later to be incorporated with three others into the original Associated Oilfields Group) which brought out one of the early rotary rigs to Australia along with two American ex-Shell drillers to run it.

> It was a steam driven thing and the first well gave quite a smell of oil which caused terrific excitement. My father, who was chairman, and other directors wanted to go up and see it. But it would take 10 days from Melbourne by road, so they hired an old de Havilland aeroplane for the journey which would then only take two or three days. He wrote back ecstatically saying that flying was the only way to travel.
>
> On the way back the first day it was getting dark when they reached Coff's Harbour on the New South Wales coast, so they decided to land for the night. The drome keeper had just cut the grass and left it in heaps by the runway, and unfortunately the pilot couldn't see them in the dusk. He clipped one of these with a wing and badly damaged the fuselage. They all climbed out unhurt and finished the trip home on the train. My father had formed a different opinion of flying by then.[28]

The first days of the boom in Roma bordered on the frivolous. When the State's Minister of Mines, A.J. Evans, visited the ROC 1 bore, the Roma Oil Company manager, an American named Clarrie Evans (no relation), filled the minister's car directly from the wellhead. Minister Evans then drove off down the main street of town. Two months later, State hydraulic engineer Henderson, still on the scene 30 years after he had first pointed out the commercial potential of Hospital Hill, bought 728 L (160 gal) of condensate — believed to be the first ever commercial sale of Australian oil. Then, early in 1928, the company gave Roma town council 510 L of oil for spraying mosquitoes in the street drains. About this time too there was much public gossip and suspicion about sabotage to some of the wells. The oft heard allegation was that overseas 'interests' were bribing people to drop objects down the holes. Nothing was ever proved, and it should be remembered that the art of drilling was very new at that stage with experienced rig hands few and far between. Nevertheless it was semi-officially admitted that there had been some 'unfortunate occurrences'. But there was no ambiguity in the feelings expressed on the subject by Roma Oil Company's manager, Clarrie Evans.

> The ROC 1 hole is clear, and there is not enough money in America or elsewhere to pay me to 'junk it', and anybody making such a suggestion is going to buy a lot of trouble.[29]

The Queensland Government stepped in to halt the gimmick sales by promising a grant of £7 000 to Roma Oil Company provided the ROC 1 well was sealed until an absorption plant could be set up, and provided also that a new well be drilled to confirm and delineate the discovery. The company agreed, little knowing that ROC 2 would spell the beginning of the end of the boom. Fraught with mechanical delays, the absorption plant was finally in operation during the second half of 1928, and by the end of the year it had produced 4 550 L of oil. However this was used in the company's cars and not released commercially. In the meantime, ROC 2, located 180 m from ROC 1, drilled slowly ahead. But on 14 February 1929 came the body blow from which the company never recovered. The well struck granite and was abandoned. Prices plummeted £7 overnight as shareholders panicked. At the time about 20 wells were drilling in the Roma district and the news of failure caused consternation everywhere. Bringing the absorption plant officially on stream in August 1929 did rally the Roma Oil Co's shares, but only briefly. Gimmicks were reintroduced to try to lift the company's profile, such as when the company's managers drove a car from Roma to Sydney running on oil from ROC 1 but, while the plant continued operating for a while, the company's drilling days were over. What is more, as if on cue, most of the other companies in the area also quit the scene. By the end of 1929 only four wells were still being drilled.

One of these was at Mt Bassett, 11 km northeast of Roma. It had taken the Roma Blocks Co nearly six months to transport the imported steam rotary rig into the roadless, virtually inaccessible site, assemble it and begin work — all to no avail. The well found oil, but not in commercial quantities.[30] Several other attempts in the same area by the company were equally frustrating. On Hospital Hill the ROC 1 well struggled on, silted up, then revived in the early 1930s when its flow was combined with a new well, ROC 3. This was drilled by another set of entrepreneurs including Clarence Byrne, who later became a director of the Associated Oilfields Group and Conzinc Rio Tinto Australia.

> A few of us took over the ROC 1 well and sold petrol in Roma. But we lost about sixpence a gallon on it. Nevertheless we were all convinced oil would be discovered in the region before too long.[31]

The three-year boom between 1927 and 1930 was contained within a 50 km radius of Roma. Virtually the whole of this region was leased by companies, syndicates and individuals. Roma town itself was a hive of activity with surveyors, geologists, drillers, financiers, lawyers, speculative adventurers, journalists and quite a number of the merely inquisitive appearing on the scene. But the arrival of the new Sydney-based company, Oil Search Ltd, in 1932 signalled the end of the localized Roma stage of Queensland's exploration. It is true there had been some work round Longreach in the western part of the State and some in the Moreton Basin on the east coast, but it was spasmodic. Oil Search had noted that the local Longreach water bore No. 2 had been bringing up blobs of paraffin wax for some years. The company also knew that an attempt to find an associated oil reservoir in 1928 (by Longreach Oil Wells Ltd) had failed. Nevertheless the newcomer went ahead with a series of scout holes in the region between 1932 and 1934 in the hope of a more successful outcome. But the programme was fruitless and Oil Search returned to the Roma area, where a subsidiary called Drillers Ltd drilled Warooby 1, finding noncommercial gas. At that point the company management brought out to Australia two noted American petroleum geologists — Dale Condit and Dr Frank Reeves — both of whom had been with the US Geological Survey and had extensive oil field experience. Their task was to look at Roma area with fresh, but critical eyes. Professor Eric Rudd, who was working for Oil Search at the time, remembers both men (particularly Reeves) as crusty characters.

> Reeves had written a paper on the carbon ratio theory long before maturation studies in oil genesis was a topic of interest, but he wasn't a great communicator. I asked him to explain the theory to me one day. He just said: 'Son, if I can find the time when I want to tell you and you want to listen, OK, we'll talk'.
>
> The interesting thing was that Reeves and Condit didn't get on personally. When forced into the same vehicle, they'd both want control. What's more, they were both lousy drivers![32]

Despite this animosity, the two consultants quickly realized that an answer to the frustrating drilling results could not be found with local studies. Their work took them much further afield in the belief that oil and gas seen at Roma had migrated from source rocks deeply buried to the south, west and even the north. Under their guidance Oil Search's sights were focused north of Roma. The next five years saw Drillers Ltd occupied first with Hutton Creek prospect near Injune, and then with Arcadia, 40 km further north. A well on the latter struck gas but, like Warooby, it was still not considered commercial and Oil Search bowed out of Queensland in 1939 to concentrate on its permits in New Guinea.[33] Despite the company's departure, the Arcadia well was considered an important test because the result showed gas and oil were present in much deeper and older sediments than those at Roma, thus vindicating the opinions of Reeves and Condit and opening up new exploration possibilities in the State.

Into the vacuum left by Oil Search stepped the giant Shell Company to take up a permit area of some 350 000 km^2 in central and southern Queensland. Encouraged by amended State Government legislation which made it possible for companies to obtain prospecting rights over much larger areas than before, and hence make exploration possible on a large scale, Shell made its entry in November 1939. For the first time, an explorer with wide experience, the latest technical resources and adequate funding had embarked on the hunt for oil. Forming a Queensland subsidiary, an English geologist named Dr C.W. Creek arrived to organize and manage the programme. With him came a staff of thirty-five, in the main comprising geologists, geophysicists and surveyors, all of whom had experience in oil fields elsewhere in the world. Although some geophysical (magnetometer) and air photo surveys had been done previously, they were nothing compared with the wide-ranging and coordinated effort that the Shell team now employed. The Sydney-based Adastra Airways Pty Ltd was contracted to take aerial photos of much of the region. Specialists were flown in to interpret the data gathered and to prepare detailed topographic, geological and structural maps. In addition, all records of wells already drilled for artesian water or oil in the region over the previous 70 years were carefully examined along with any cores or cuttings that were still available. Arthur Wade, himself a consultant to Shell during this period following his work for Freney in Western Australia, described the magnitude of the work done.

> By the end of 1942, 250 000 square miles [650 000 km^2] of country had been geologically investigated (including adjacent areas outside the permit); 12 600 square miles [32 630 km^2] examined closely; 200 000 square miles [518 000 km^2] covered by geophysical surveys; 10 000 square miles [26 000 km^2] photographed from the air and mapped; 3 280 water and other wells had been studied. The amount expended on the work so far

> was 140 000 pounds. About one-quarter of the original [permit] area had been surrendered in 1942, but by 1944 two-thirds had reverted to the government.[34]

The programme was long and thorough and, as contemporary Shell geologist Bruce Thomas points out in a paper commemorating the company's 50th anniversary of exploration in Australia, some of the methods used were innovative. He quotes from a letter in which 1940s English field geologist John Woolley describes a technique employed for air photo interpretation. The photos were worked out in the Brisbane office, but not entrusted to the field parties in case they were damaged.

> [instead they] sent me a stack of huge [photo] mosaics, gummed onto great sheets of 3-ply. These were useful, but very cumbersome in the bush, and they frightened the horses. I used to lay the mosaic at the foot of a tree, climb up a short way, and view it through field glasses. You get a surprisingly clear stereo effect...[35]

Then came the task of selecting a drilling site, made all the more difficult knowing that depths of 3 000 m might have to be reached. Such a hole would be well in excess of any attempted in Australia to that time and would need special rigs and equipment from the US or Britain. In the event this quest proved hopeless. The war effort, especially after Pearl Harbour, mobilized the Americans, taking up all available steel, and so the Queensland Government granted Shell a moratorium on operations. The company returned in 1946 full of enthusiasm. General manager in Australia Eric Avery announced that £1 million would be spent on the company's exploration in Queensland. A preliminary programme of 61 shallow test holes was then carried out before Shell decided its main chance lay in the Morella anticline, 225 km north of Roma and 50 km northwest of Oil Search's noncommercial Arcadia find. Again Arthur Wade takes up the story.

> By early 1950 more than 1 500 tons [1 524 tonnes] of heavy gear, including a 136 ft [41.5 m] steel derrick, engines, transmission pumps, and all the rest of the machinery that is necessary for the drilling of a deep well had been transported to the drilling site. A small village was almost completed near the well with its own water supply, electric lighting, stores, mess room, recreation hall, workshops, laboratories and offices.[36]

After such a build-up for one exploration well great things were naturally expected. But nine months later, in February 1951, the Morella well was abandoned as a dry hole at 1 413 m — nothing like the

3 000 m once contemplated. Further drilling was suggested for the nearby Comet anticline, but that project was abandoned too after a survey indicated basement rocks at shallow depth. Shell, disillusioned after contributing about half of the State's total 1900–1950 oil exploration expenditure of £2 million during only six years of full-time work, decided to cut its losses. It relinquished its acreage and quit the scene in April 1951. Despite the company's failure and abrupt departure, Shell's painstaking approach to exploration set a standard for those explorers who followed. It was in marked contrast to the somewhat casual programmes of local companies immediately pre- and post-war, two of which in particular produced amazingly inept results. No doubt the humour inherent in these activities was only appreciated by shareholders in hindsight.

The first was the Murilla Oil Company's well spudded with a percussion rig near the town of Miles, 200 km east of Roma, in 1935. After drilling for nearly six years the site was visited by government inspectors who found that the drill bit was going up and down in the hole without ever touching the bottom! The cost to shareholders in wages for five men to 'dance the drill rods in mid air' for that length of time was estimated to be £11 000.[37] The second ineptitude began in 1948 and, amazingly, was not halted till the late 1960s. John White, who played a part as downhole log analyst with Schlumberger towards the end of the saga, tells the story.

> Condamine Oil Ltd, a local Roma company, began drilling a well called Speculation No.1 about 40 km northeast of Chinchilla during August 1948. The site was apparently selected by a diviner and confirmed by a clairvoyant.
>
> Drilling began with gusto using a steam driven percussion rig, but after a while progress slowed and then became intermittent. Drilling only continued when the local wood cutter brought fuel to burn in the boiler furnace. This went on for nearly 20 years until the advance of exploration elsewhere in the Surat Basin finally caught up with the operation during the 1960s. The small permit was found to be right in the middle of the major exploration area known as Authority to Prospect (ATP) 145.
>
> Burmah Oil eventually bought out Condamine Oil and decided to find out what had been going on for all that time. The big company whistled up Schlumberger for a log programme to be run and I was surprised when they told me there was no hurry. Normally logs are wanted yesterday.
>
> Burmah explained that after 20 years a few more days wouldn't matter. Anyway I went up there, ran the logs and found that the well was down to about 4 000 ft [1 220 m] having been drilling in basement conglomerate for most of its life![38]

Condamine's Speculation No.1 can be seen as an isolated throwback to times past in Queensland because the new era of post-war exploration in Australia swept the rest of the board clean at the beginning of the 1950s.

PAPUA and NEW GUINEA

Oil exploration in Papua and New Guinea has always been an integral part of the Australian story because of the Australian Government's jurisdiction in Papua and, after the First World War, the administration of the Mandated Territory of New Guinea. However, as the companies who took part in the early search soon found out, work in the region demanded a totally different approach to that followed back in Australia. Inaccessibility was, and still is, a major problem. Maps of any sort were primitive, while infrastructure and lines of communication outside Port Moresby were virtually nonexistent. In some cases the geologists and surveyors were the first whites to travel through an area. Some movement of men, supplies and equipment was by canoe and launch, but the greater part of the work was on foot. In a country where the average rainfall is over 6 350 mm a year, exploration was very slow, costly and arduous.

Apart from the physical difficulties of penetrating the rugged terrain, the thick jungle growth made examination of the surface geology extremely difficult. Bare rock outcrops were few and far between and, for a lot of the time, traverses had to follow the rivers and streams, or at least be frequently correlated with them. In the face of so many daunting problems, it is reasonable to ask what brought the explorers to New Guinea in the first place and, more to the point, what made them stay? The answer is given by the New South Wales Assistant Government Geologist Joseph Carne, who first went to the region himself in 1912 to look at some reported coal deposits.

> The probability of the existence of oil suggests itself whenever the latitude of British New Guinea or Papua is considered in relation to the existing oil fields of Java, Sumatra, and Borneo and the oil and gas occurrences in Celebes, Ceram, Timor and other islands adjacent to the Malay Archipelago embraced within 10 degrees north latitude and 10 degrees south latitude.[39]

The idea of attempting proof apparently never assumed concrete form until shortly after the arrival of planters Garnet Thomas and Lewis Lett on the Vailala River (about 250 km northwest of Port Moresby) during 1910. At the time these men were employees of the British New Guinea Development Company and their job was to reconnoitre the area for potential sawmilling operations. However, after a short period they struck out on their own to try to establish a coconut plantation in the region. While seeking additional land for this purpose Thomas accidentally stumbled on a natural gas seep near Opa village on the east bank of the Vailala River in August 1911. He and Lett followed up the clue with the aid of local natives who pointed out similar occurrences in the locality.

Like many discovery stories this 'beginning' was disputed. A prior claim was made by a gold prospector named E. McGowan and his partner H. Swanson, who said they noticed a 'smell of kerosene' in the area during June 1911 and allegedly mentioned it to Thomas and Lett. In any event it was Thomas who reported the finds, in writing, to the mining warden at Kerema in September 1911 while Lett journeyed to Port Moresby with a sample of frothy scum from one of the seeps which he showed to Lieutenant-Governor Hubert Murray. Lett went on to Melbourne (then home of the national Australian Parliament) with it and informed the Secretary of the Department of External Affairs of the discovery during November 1911. The sample was then passed on to Carne in Sydney for analysis, which revealed that the material contained a high percentage of iron oxide plus a small amount of petroleum substance, although the sample was not large enough to make a definitive judgment. In the meantime, the Acting Resident Magistrate and Mining Warden at Kerema, L.F. Henderson, was ordered to investigate the Vailala region

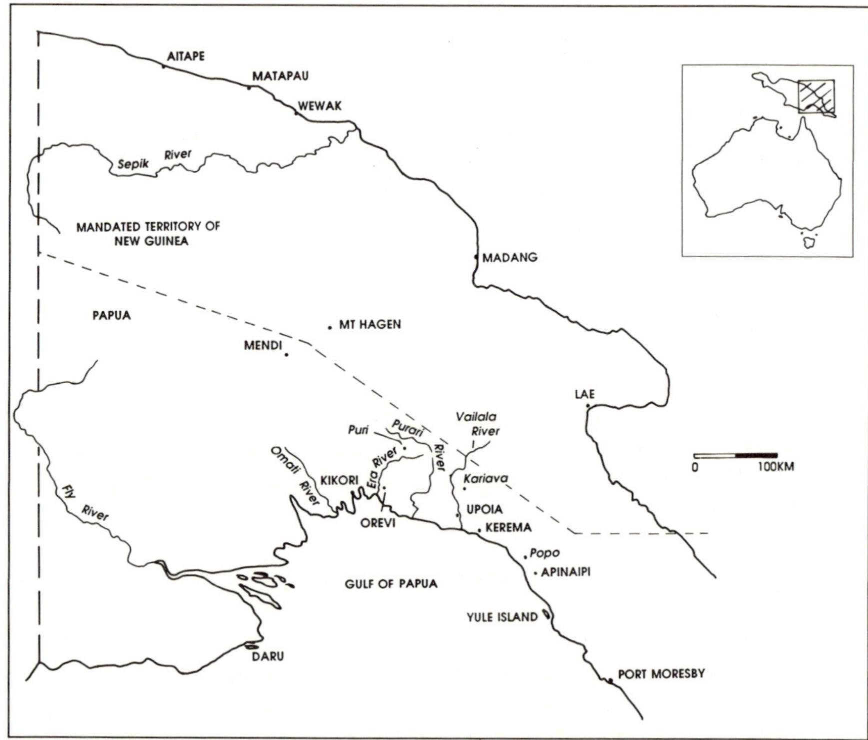

Papua and New Guinea — location of key places mentioned in text.

without delay. Accompanied by Thomas as guide, he visited a number of the seeps during the last weeks of 1911 and furnished a descriptive report of his findings.

> These [the indications which had first attracted Thomas's attention] consist of a series of minute craters on a level, and located in wet clayey ground, at intervals of a few yards, for about a quarter of a mile in an easterly direction. The mud and water in these craters was churning, as if in the process of boiling (but was quite cold); and the ground in places was working as if some great pressure from below was causing it to do so. On applying a match to these craters, I found that the gas which was being exuded (and causing the bubbling action in its upward course) ignited readily as it escaped from the water, and gave a red and bluish flame. There was a slight odour arising, faintly resembling kerosene, while the smell of the fumes resembled more than anything those of methylated spirits.[40]

At another place about 50 km up river Henderson saw seeps in drier ground.

> The noise of the gas forcing its way upward could be heard for some distance. Once lighted, these jets of gas would burn brightly for an indefinite time. One remained alight for well over two hours, and was still going strong, after having boiled a billy-can full of water in quick time.[41]

The lighting of seeps also had its humorous side.

> Another larger fissure which I opened up, and sunk a hole on about eight feet [2.5 m] deep, gave off a powerful jet of gas, which when ignited I found difficult to extinguish by beating it with a branch of a tree; as a matter of fact we had to smother it with earth, and pour some water on it to put it out. The native police and other natives with us were very frightened, and it took some sternness to prevent them taking to the bush.[42]

Henderson also made an interesting observation about the local knowledge of the seeps.

> The local natives informed me that this phenomenon has been known to them for many years, and, as far as they know, has always existed. They call it 'havila' which appears to be their name for the peculiar boiling action, but they had no

> idea of the existence of the gas until Mr Thomas came along and demonstrated its presence by burning it. The fact that they were very much afraid on seeing this would indicate their ignorance of its inflammable nature.[43]

Despite his readable report Henderson was not qualified to give a professional opinion on the discoveries so, when Carne arrived in Papua during January 1912 to join the official coal expedition to the Purari River, the New South Wales geologist was immediately sent to examine the Vailala gas finds. During this side trip he confirmed the presence of gas and went on to find and collect samples of oil at Orevi and Upoia in the same region. In his report of the whole expedition Carne was enthusiastic enough about future commercial oil discovery in Papua to make some concluding observations on the role of the native workforce, along with some suggestions on health care. He expected skilled labour would be a product of time and patience from the native raw material, but that good results would be obtained, judging from what he saw of the natives round the New Kikori Pastoral Station. However he believed unskilled labour could be a problem because of its instability. He noted that voluntary labour rarely extended beyond two or three years and rarely did it return after periods of 'ease and indolence'. The Papuan pay for unskilled labour at the time was 10 shillings a month plus a return to the village of recruitment on termination of the contract. He went on to say that experience in Sumatra and Borneo indicated that clearing the jungle along the line of wells and around residential quarters, native quarters and pumping stations lessened the incidence of malaria.

> If commercial oil wells be proved on the Vailala and Purari Rivers in Papua, similar precautions should yield like results to the advantage of employees and the locality generally.[44]

In April 1912, when Carne had resumed the coal expedition, Evan Stanley was appointed as the first Government Geologist of Papua. He spent the rest of the year and on into 1913 visiting all the seeps known at that time as well as traversing many parts of the country east of the Purari River. During this period four shallow wells were drilled by percussion methods at Upoia and Orevi on the Vailala River close to the oil seeps. But, apart from the locations and the fact that the first ones were drilled for the British New Guinea Development Company and the later ones for the Papuan Government under the supervision of the Department of Public Works, no details are known. During 1913 the Australian Government decided that it would conduct the search for oil and exclude private enterprise completely. To this end the British geologist Dr Arthur Wade was engaged to take charge and he arrived in Papua in September of that year.

His work included much travel by boat and even more traversing on foot. Measurements were taken using compass, Abney level and chain, while the more important traverses were plotted using a theodolite and stadia. Wade's lengthy report offers advice on weather and climate (which he describes as variable), native labour (good workmen) as well as on potential development of oil. He also mentions potential drilling difficulties in the surface formations of soft clay.[45] After examining the area in the Purari, Vailala, Kerema and Yule Island regions Wade noted that the seepages at Upoia and Orevi were reasonably accessible and he recommended a drilling programme there. Five wells were subsequently drilled between 1914 and 1919, the deepest of which was drilled by cable tools to 540 m. No significant production of oil resulted from these efforts.

At the end of the First World War the Australian Government reconsidered its position and, after some deliberation, was joined in 1919 by the British Government on the understanding that each would provide £50 000 as exploration 'seed' money. The British company Anglo Persian (now British Petroleum) became associated with the enterprise as exploration agent for the two governments and it supplied the technical staff. However there was a strict stipulation that the company's expenditure in Papua should not exceed £100 000 without special approval in writing.[46] From this point, exploration in the country began to fall into two distinct, but complementary, lines. The first was to outline and test by drilling geological structures in the seep areas that might form suitable traps for the oil and gas escaping to the surface elsewhere. The second was the broader task of unravelling the geological history of the region as a whole to try to predict where commercial accumulations of oil might lie, what kind of traps they might be and at what depth they might be buried.

Anglo Persian's Lister James took charge of the operation, arriving in March 1920 and heading a team of six geologists and two surveyors. At the time the last Upoia well was still drilling and James directed that the work cease in favour of exploration elsewhere. It was his opinion that the Upoia region was too highly faulted to justify the hope of commercial oil. Nevertheless Wade's efforts were recognized as sound, based on the information at hand when he had recommended the drilling programme. James pointed out that it was only through the drilling results that a later adverse opinion of the area could be formed. The extensive field work round the coastal regions of the Papuan Gulf which followed led to the selection of the Popo anticline near Yule Island, well to the east of the Vailala–Purari region, as a valid structure worthy of test drilling. However, by that time the joint government money had been exhausted and the British Government decided to withdraw. After a good deal of discussion the Australian Government took over the British Government's share for £25 000 and Anglo Persian was authorized to carry on as agents for the Australian Government alone in 1922.

About this time too, a party was sent to the Mandated Territory of New Guinea — the vast area north of Papua which had been taken from Germany by the Treaty of Versailles at the end of the war and placed under Australian administration. The earliest recorded work in this region was a 1907 survey led by a Roman Catholic missionary in the Aitape district. Rumours of an oil discovery in 1912 were confirmed by the German mining director in Rabaul, Dr Schlenzig, and samples of oil from seepages in the Matapau area were apparently sent to Berlin for examination. In March 1914 the German Royal Geological Institute reported that the analyses indicated it was a heavy oil suitable for making lubricants. The sum of 0.5 million marks was set aside in the German Budget for further investigations, but the war intervened.[47] The subsequent 1922 Anglo Persian party into the region was led by H.T Mayo, a pith-helmeted British geologist who was often accompanied on geological expeditions by his wife. Mayo set out from Rabaul with his group and conducted a rambling survey covering the coastal Aitape, Matapau and Madang districts as far as the Dutch New Guinea (now Indonesian) border in the west.

In the meantime, drilling had begun at Popo after bulky steam-driven drilling equipment had been hauled into place. However the percussion rig was unable to penetrate a mudstone breccia formation below 560 m and the hole was abandoned. It was decided to bring in a rotary rig to take over the drilling at 460 m in Popo 2. Unfortunately in this attempt the well casing was bent and it too was abandoned without reaching the target. Mechanical and downhole problems continued to dog the operation and three more wells at Popo failed during the years up till 1929. By this time the Australian Government had reopened Papua to private enterprise under a new Petroleum Ordinance (1924), retaining only a small block from Yule Island to Popo as a government reserve. A number of small companies like Papuan Petroleum Company, New Guinea Oil Company and Oriomo Oil began work in the country, some making extensive commitments. New Guinea Oil Company, for instance, drilled seven scout bores near Hohoro and one deeper test in 1926 that reported oil and gas shows. Oriomo established a base at Dagua on the mainland opposite Daru Island and within a short time it was fitted out with seven bungalows, a wireless station, provisions store, engineering shop, casing shed, native dwelling area, plus native hospital and a separate hospital for the Europeans. Everything was built from local materials.[48]

In the Mandated Territory, Pacific Islands Investment Company (later called Aitape Oilfields Ltd) and Mandated Development Ltd carried out some surveys during the mid-1920s,[49] while Ormildah Oil Development Company and its consultant — the Australian-educated Danish geologist Dr Harold Ingermann Jensen — drilled an abortive well at Marienberg near the mouth of the Sepik River between 1925 and 1928.[50] But it was not until a few months after Marienberg was abandoned that Anglo Persian sent out another team to the region. This second wave of exploration was again at the behest of the Australian Government and it began with an agreement drawn up in 1927 leaving technical organization entirely to the British company.

In the light of earlier experience Anglo Persian decided that more detail and less haste might bring better results and that this would entail a long geological process including laboratory work to furnish descriptions of fossils and rock types and make accurate correlations. Previously the drillers had followed too quickly on the heels of the geological field parties and there had been too much pressure on geologists to find a drilling location using meagre data. For instance, doubts were raised about the geological validity of the Popo structure and for two years there was a concerted effort to return to areas previously examined and gather more detailed information. There was also a push to visit areas in Papua not previously examined and to widen the extent of knowledge of the Mandated Territory. The whole operation began with a sea plane survey of the entire coastline and included coordinated ground surveys in the Cape Vogel, Madang, Aitape, Vanimo, Yule Island and even Port Moresby areas. However the work was not accomplished without a toll, particularly in the Sepik region where at least two of the geologists caught blackwater fever and were evacuated back to hospitals in Australia.

Valuable as the work might be, however, the expense of the operation meant that it could not be sustained. By late 1929 the world was sliding into an economic recession and the Popo programme had just recorded its fifth failure. In view of this, Anglo Persian's Papua and New Guinea work was halted between the years 1930 and 1935. Its place in those intervening years was taken by a new company called Oil Search Ltd — an organization which was to have the longest continuous presence in the country and, at times, the most significant, of any oil explorer. Oil Search was founded by an Englishman named E.L. Walter who has been described as a great visionary and a dealer in tea, and who had come to Australasia via South Africa. He incorporated the company in Papua during 1929, but it was not until 1932 that work in the field really began. This took the form of a geological survey in the Aitape and Matapau regions of the Mandated Territory and it was carried out by a man destined to become one of the best known of New Guinea's many characters in pre-Independence days — G.A.V. Stanley. Stanley, an Australian (and not to be confused with Evan Stanley, the first Government Geologist in Papua), arrived in the country during 1928 as a cadet to take part in Anglo Persian's detailed two-year quest under its contract to the Australian Government. When the British company left, Stanley refused to depart with the rest of the team. Instead he joined Oil Search. Eric Rudd, another early Oil Search employee, explains the circumstances.

> The trouble was, Anglo Persian treated Papua as a sort of Siberia — an outpost. Much of the company's reason for being there (albeit for the Australian Government) was to deny access to other oil companies. Management in London used to feed

> people through the place, but only allowed them to stay a certain time so that individuals never really became familiar with all aspects. If they subsequently left the company they couldn't tell a new employer very much. All the data was collated in London.
>
> When Stanley refused to leave Papua it upset Anglo Persian, but eventually they had to give him a retainer of 100 pounds.[51]

Rudd remembers Stanley as an eccentric bachelor who was devoted to the country and eventually married a Papuan girl. He became an expert bushman and was meticulous in his work. All his maps were done by hand and he kept detailed notebooks of his expeditions. Another early Oil Search employee, Sam Carey (later Professor of Geology at the University of Tasmania and a world figure in structural geology famous for his theories of Continental Drift and Plate Tectonics), remembers Stanley as the man who first hired him for work in New Guinea.

> At Oil Search Stanley teamed up with a surveyor named Harry 'Dinkum' Eve and together they did the preliminary survey in the Mandated Territory. Both men were good with the natives and learned how to work with them. They had a high sense of morality and they were referred to by the natives as *Masita Mak 1* and *Masita Mak 2*. During that first survey they even ventured outside the government patrolled area without mishap.[52]

On completion of this work Stanley suggested Oil Search assemble a bigger team and Carey was one of several recruits employed for a second survey which began in 1934. The chief geologist by this time was J.N. Montgomery, another ex-Anglo Persian employee, who directed a programme of random and wide-ranging traverses along rivers and mountain ridges throughout Papua and the Mandated Territory. Carey, then fresh out of university in Sydney, still vividly remembers his part in this effort — much of it on his own in the Sepik region in charge of up to 40 native carriers and assistants. Any glamour he might have felt associated with the job was dispelled when news came through soon after his arrival that Eve had died of malaria. He was buried at Wewak and his grave was used as one of the initial trig points in Oil Search's survey work.

> I was sent straight up into the Sepik. Jack Fryer, a surveyor, was working in the area for Oil Search too, but we met infrequently. He was surveying the ridge paths while I mapped the geology up the rivers and creeks. There was no radio and I had to use a runner boy (boi) to the coast for anything urgent.

Generally we were out of government controlled areas because the police only patrolled the coastal regions and a little way inland at that time.

I used a plane table and alidade to make maps of the area and I included detail of who owned what in the gardens, marking it 'bush belong so and so'. Unfortunately these were destroyed by the government during the Second World War. They'd be invaluable for working out land claims nowadays.

There was a trade store at Aitape run by a white named Charlie Goff who was speared later on. He supplied tinned food and we shot a lot of fresh food on patrol. In the party I usually had three 'shot bois' who carried shot guns and they used to go after cassowary, pig or wild fowl. But I have also eaten things like wasp, wood grub lavae, spiders, eel, greenback tortoise, crocodile (*puk puk*), pandanas fruit, spinach tree and sago.

I tried to make sure the bois had a balanced diet and insisted that the peas they were issued had sprouted before they were eaten as they had more nutritional value that way. The government regulation for meat was one pound weight per week per boi. They were also given a stick of black 'twister' tobacco a week as part of their rations. I used to administer cod liver oil too to keep them fit, specially when canned food was the only thing available for any lengthy period.

The government system of the time was to sign up bois for a few years as a way of education — apprenticeship. The parents would be paid and the contract would be officially drawn up and signed. Normally there would be 10 or 12 bois from the same village in one line and there'd be three lines to carry all the food and equipment. I'd also have three 'survey bois' — one to carry the instruments and two with a graduated measuring staff, each whom I could direct to places where I needed to take a reading. Then there were three 'cutters' who went ahead and cleared the lines, and I had one personal boi trained a bit like a theatre nurse to hand me whichever piece of equipment I needed.[53]

When on patrol Carey's day began at dawn with a cooked breakfast. Then he would leave camp with his survey and shot bois to begin the day's traverse. The boss boi would break camp, estimate how far Carey would get by lunch time, pass him on the trail and have lunch set up at that spot when he arrived.

When I got there lunch would be set up waiting — folding card table, knives and forks, table cloth... the works. While I was eating lunch the cook boi would begin making soup for the evening meal.[54]

Carey would work on through the afternoon while the boss boi repeated his performance to set up the night camp. Carey would arrive, have a cup of tea, then bathe in a shallow canvas-sided 'tub' out in the open and dress in his pyjamas before holding sick parade for the bois (and the locals if they were near a village). Then he would have dinner, plot up the maps for the day and finally read by a pressure lamp until bedtime. He did not use a tent, preferring a fly to cover a canvas bed on poles and an A-frame holding up a mosquito net.

> This routine was seven days a week on patrol and I'd make about four miles on an average day. I could go for weeks without seeing another white person. The whole bush allowed a sort of 'go forth into the wilderness' aspect that you can't get today. Sometimes I saw no local natives or gardens either, and then I had to be careful because I was in no man's land and could be coming to an area from the same direction as an enemy.
>
> I had to be self-reliant in everything — including making my own thin sections in the field to study rocks and fossils, and building my own canoe and paddles because the native ones did not have a suitable centre of gravity. I even had to make stadia cross-wires for the theodolite out of spider web across a hole cut in cardboard because the glass and ordinary wires quickly gummed up with fungus.[55]

The nearest health centre was at Madang on the central north coast so Carey had to be medic for himself and all those in his charge, plus any locals who needed attention. With a basic medical kit and a few reference books he tackled problems like a punctured rectum, arrow wounds, stitching up deep cuts and dosing fevers. He nearly died of tropical typhus himself during one patrol. But it was not all constant work and Carey particularly recalls one camp when he and Jack Fryer had one of their rare meetings. On Saturday evenings the bois often had a 'sing sing' and on this particular occasion Carey noticed in his diary that there would be an eclipse of the moon. The two patrol leaders decided to have a bit of fun.

> We told the bois that we would join the sing sing and 'mak im die moon'. Fryer had a mauser pistol which he fired behind me at fifteen minutes to total eclipse time. I had been hamming it up a bit before that and, sure enough, the eclipse began. I explained that it would take a little while. Some of the natives were amazed, but I could see the older ones were upset, so I promised to get it back again. My alarm clock had an intermittent ring and I used that as the voice of the man-in-the-moon while I 'coaxed' him back again.

> The trouble was news of this performance got around and a week later the local villagers wanted to see it done again. I had to think fast and said OK, but that I was not sure I could bring him back again this time. Luckily the elders got wind of it and pleaded with me not to do it after all.[56]

Another natural phenomenon with less pleasant effects was the severe earthquake of September 1935.[57] Luckily Carey and his party were on alluvial flats at the time, away from falling trees and landslides. Nevertheless, waves developed in the river and broke on shore like surf, while the traverse station he had just left was buried under a crumbling cliff. There were a number of after-shocks and Carey had trouble calming the natives. He was concerned that the nearest town, Aitape, might be destroyed, cutting him off from civilization with less than a month's rations. But a district patrol officer arrived to assure him the town was intact and he turned to the more immediate danger of log jams damming the mountain streams. When a dam broke it would send a wall of water down through the ravines and wipe out the entire party. He decided to reverse his traverses and begin at the top of each water course so he would always know what was above him.

Carey's traverses were typical of the random work done by Oil Search during 1934 and 1935. But when the Australian Government replaced the 1924 Oil Ordinances with new ones in 1936, the company took up specific permits covering 33 670 km^2 in the Territory of New Guinea and over 31 080 km^2 in Papua. Wide-scale mapping was then carried out, particularly in Papua, with the most significant discovery being the Puri anticline and numerous seeps between the Purari and Era Rivers which led into the Papuan Gulf; however Oil Search did not do any drilling. By 1936 a number of other companies had entered the Papua and New Guinea exploration scene, some of them only briefly. Shell, through its subsidiary Papuan Oil Development Pty Ltd, began surveys in the western delta and central regions of Papua under the direction of English geologist Dr C.W. Creek. The company drilled a number of shallow wells — the deepest reaching 760 m — before leaving again at the end of 1938 to try its luck in Queensland. The Vacuum Oil Company (now Mobil Oil), through its subsidiary Island Exploration Company Pty Ltd, also entered Papua around 1936, followed by Papuan Apinaipi (which took its name from a village in eastern Papua near its first exploration work) in 1937.

In 1938 Oil Search began to wind down its operations in Australia and New Zealand to concentrate on its New Guinea holdings. But, at the same time, management realized it would have to struggle to maintain effective programmes in the tortuous jungles and swamps. With this in mind it struck up a partnership with Island Exploration and D'Arcy Exploration (a subsidiary of Anglo Iranian — formerly Anglo Persian and soon to become British Petroleum). That year negotiations were completed in London and New York such that all three companies joined to form Australasian Petroleum Company Pty Ltd (APC). This alliance finally marked the start of large-scale, coordinated oil exploration in the country after more than 25 years of scattered surveys. Not that the countryside was any more forgiving, or the methods any less arduous from that time. The only immediate advance was the flying of aerial surveys over the bulk of the permit areas, and the subsequent preparation of photo-geological maps on a scale of 1:40 000. These maps formed the basis of extensive regional surveys between 1939 and the end of 1941.[58]

One young geologist joining the APC team for the first time was Leo Stach, who later went on to become a member of the United Nations technical mission in South-East Asia and later still a consultant geologist based in Australia. Stach took ship from Melbourne bound for Port Moresby, and from there an uncomfortable four days across the Gulf on the 100-tonne steamer *Papuan Chief* to Kikori. A launch then took him to his field camp some miles up the Amarti River.

> I was soon introduced to the routine. The general rule was that a field assistant stayed at this base camp to monitor and send radio communications to Daru, APC's main base for Western Papua back on the Gulf. The geologist (me) went off alone with a line of bois on traverses inland, usually staying out for about 15 days at a time. I'd come back in to the field camp for about two days to reprovision and then out again. It would be three to five months before I went to Daru to write up reports, refit, and generally have a break for 10 days or so before starting a new 'tour' in the field.
>
> It took terrific organization because we had to be completely independent in the field and conditions were far from ideal. It rained 24 hours a day during the southeasterly monsoon season and we had to have special paraffin wax impregnated tent flys. It was impossible to use a normal paper field notebook during the day, so we had to use sheets of sanded celluloid and write in pencil.[59]

Many of the early surveys in the foreland area south of the central highlands were pace and compass traverses back and forth between the main rivers and, for the first few weeks, Stach used the accessible native trails. After that the party had to start cutting its own lines through the bush using big 400 mm bladed knives.

> I'd go by canoe about 70 km up river from the field camp, then be picked up on the other river to go a few kilometres downstream and start back again. Sometimes I'd traverse down river and if it was a bit rough I'd take a day off to make rafts. It made life interesting if we encountered some unexpected rapids.

> The standard party was 60 bois to support one geologist for 10 days. We would all be up at dawn and break camp straight after breakfast to be on the traverse by 8.00 a.m. There'd be a half hour stop for lunch and I'd think about making the evening camp about 4.00 p.m. At night I would transcribe the day's work from the celluloid sheets to a note book which hopefully was still dry in the middle of my bedroll.[60]

A lot of the mapping was done over a formation known as the Darai Limestone, where millions of years of weathering has produced a tortuous surface terrain called karst topography. It consists of sharp ridges and deep sink holes stretching for miles in any direction.

> One time I set out specifically to get across. It took 10 days to go four kilometres. My first attempt was to go round the ridges, but I'd soon come to a knife edge where it was literally impossible to put one foot beside the other. Luckily there was vegetation clinging to the rock so we could use that as support.
>
> But, in the end, I had to go up and down the sink holes using vine ladders. The annoying thing was that although it was raining hard all the time, there was no water in the sink holes at all. We had to cart it daily from the river a number of ridges back — in four-gallon cans![61]

Although the work was arduous and the conditions far from ideal, life for the young geologists taking part in the early surveys was treated as a real adventure — almost like stories from Jungle Jim, Biggles and Indiana Jones. Crocodiles were a danger, especially if a camp or a river crossing was used for any length of time. Sometimes, if the patrol was to be out for a long time, supplies would be air dropped into a gorge or lake side, although it was rare to recover everything sent out. Usually relations with the local villagers were good and there would be a lively barter session for fresh food — two coconuts for a stick of tobacco, for instance. But sometimes there could be trouble, more often in the isolated highland regions than in the Gulf or the Sepik provinces. This might take the form of stealthy thieving or a show of spear-throwing strength at unwanted intrusion. Often police bois, and sometimes the *kiap* (patrol officer) himself, would accompany the parties, particularly if they were penetrating new or rarely patrolled areas. Eugene O'Driscoll, who went into the field for APC and then for Papuan Apinaipi during this pre-World War II period, recalls that the native police were easily identified in the line.

> Each man had a blue loin cloth fringed in red and a bandolier across his chest containing 10 bullets. He carried a .303 rifle,

> but under no circumstances was he supposed to fire any bullets and he had to account for them when the patrol returned. They all treated the rifle better than themselves and I can remember an occasion when our boat hit a snag and tipped one of them out. He insisted his rifle come back on board first.[62]

Melbourne geologist Alan Condon, who traversed the Sepik region for APC in the late 1930s and later joined the Bureau of Mineral Resources in Canberra, recalls a few tense minutes on one Sepik patrol. At the time he was surveying with Queenslander Dick Mott, who post-war went on to join the Queensland Geological Survey and to found the monthly *Australasian Oil and Gas Journal.*

> This particular area was not controlled so we kept police bois at either end of our line. Dick and I were at the front, going single file along a narrow track through one-metre high kuni grass, when we came to a fence and a type of stile with a low arch. It looked a bit ominous so we stopped and the police bois began calling out: 'good fellas... good fellas'. Nothing happened, so the bois continued to talk in loud voices for 15 minutes. Then, just as we were about to continue, 30 natives armed with spears, bows and arrows and some done up in paint, stood up on each side of the track. It would have been a complete ambush if they'd wanted. Some of them had never seen a white man before. But the head police boi shook hands and, in the end, it turned out to be the best evening of the year with a sing sing, a good clean rest hut and plenty of food exchanged for our trade goods, including a pig for the bois.[63]

Naturally enough there was always a bawdy humour in the camps. Sam Carey remembers one incident in the Sepik region where a Hungarian-born surveyor suffered the discomfort of a leech in his urethra. He could do nothing but bear the pain until it came out of its own accord. Afterwards the man refused to go bush again without condoms. When the order was sent to BP in London the stores lady was apparently mortified that the 'wicked men' would want a case of 1 000 condoms sent to New Guinea. She would not believe they were for medical purposes.[64]

As for the geological work itself, results were tentative at first because many of the areas traversed contained few outcrops. Stach says that the surveys did pick up some structural trends, but a lot of imagination was needed. It was hard to tell whether they were measuring true dips and strikes of the rocks or merely false bedding planes. A little later however, a paleontological laboratory was set up in Port Moresby by

Martin Glaessner who came up from Adelaide. Stach says this produced some very detailed work.

> A couple of natives were trained to select the forams [minute marine fossils] from sand grains after chemical washing and they became expert at the job. The idea was to use the fossils combined with the rock descriptions from field notes to draw up an idea of the stratigraphic column — that is, the sequence of different rock types and ages present — concentrating on the Papuan Basin in particular.[65]

The culmination of all this work came in the early 1940s when APC selected its first drilling site — the Kariava anticline — about 100 km upstream from the mouth of the Vailala River. The rig was brought in by barge and then hauled five miles inland by big caterpillar-tracked vehicles along a specially bulldozed road. Roy Richter, who first went to Papua to follow gold mining, became involved with the drilling of this inaugural APC well.

> We were working in hilly country and mudstone plains round the Gulf of Papua. The site was well up river and we used 100-ton barges to get the rig near the site, then built a road for the rest of the way. There was no igneous rock about so we used shale and mudstone on the road surface. During drilling the supplies would come up by the same route. One time there was a dry spell and the river was well down. We had to use canoes fitted with 22-horsepower Johnson motors carrying only 5 or 10 tons each to keep the supply line open.
>
> In those days the rig had a standard derrick which meant it had to be built piece by piece. The draw works [hoisting winch for hauling pipe and casing in the derrick] only had a dog clutch then too. Pneumatics didn't come in till after the war.
>
> The natives worked mostly as roughnecks and roustabouts [labourers round the rig]. They were good workers and the best of them came from the Papuan Gulf region. They were physically strong and there was always a healthy competition between the various crews. We had three shifts of eight hours working round the clock.
>
> The camp itself was built by carpenters about half a mile from the rig. The hut frameworks were of sawn timber, the walls were woven sisal craft and the roofs were thatched. There was a communal mess, a club with bar and billiard table, a tennis court, a small hospital with a doctor on staff, and even a newspaper. The native camp was of bush timber — a long communal hut — about the same distance away

> on the other side of the rig. There was no real problem with grog. It was a close-knit community.
>
> The white camp had about 40 people and included some wives. The married couples had separate huts (dongas). The humidity was high and there were plenty of mosquitoes and sandflies. We took quinine in those early days. It was before atabrin tablets came in. I did get a bad dose of blackwater fever, but recovered and, strangely enough, I must have built up some sort of immunity because I didn't need preventatives during the rest of that stay.[66]

Even so malaria took its toll on others. Leo Stach 'sat' the Kariava well as one of his last jobs for APC.

> My view of personal health was that as long as I was active there would be no problem. During my time in the field I was very fit, but at the rig site there was no real activity — just core logging and sampling. I used to try to wander a bit during the day and specially at weekends, but I did get a bad dose of malaria. So much so that the doctor at the hospital had already arranged a grave just before the fever finally broke. I was told I had a temperature of 105 and the sweats for five days.[67]

The Kariava well itself was a difficult and drawn out affair, made even longer by the Japanese invasion of New Guinea in 1942 and the evacuation of all personnel. Drilling began in March 1941, was suspended at the end of January 1942 at a depth of 1 647 m and not resumed until October 1946. Numerous cave-ins dogged progress and, in the end, it was abandoned as a dry hole on 25 March 1948 at a total depth of 3 850 m. By that time post-war activity had ensured Papua and New Guinea exploration was being tackled on a much broader front.

NORTHERN TERRITORY, TASMANIA and NEW SOUTH WALES

Although they are geographically as well as geologically separate, these three regions are lumped together in the negative sense. They have received the least attention from explorers over the years. Whether this lack of interest stemmed from no success, or vice versa, is hard to judge. Certainly, in the cases of New South Wales and Tasmania, nothing has yet been notched up in the form of commercial discoveries. And if it were not for a quirk of offshore political boundaries which puts the 1980s Timor Sea discoveries off the northwest tip of Western Australia under its jurisdiction, the Northern Territory would have little commercial production either. Nevertheless, at the beginning of the century the forays into adjoining States did spill over into the three regions in question.

Northern Territory

Due to the finds of Price and Okes, the early twentieth-century search to follow up Commander Stokes's original oil find on the Victoria River was concentrated in the Kimberleys to the west. Most travellers in the Territory, particularly during the early years of the 1900s, stuck close to the life-line of the overland telegraph. Adventurers who did penetrate the wilds of Arnhem Land, the lonely wastes of the Tanami and Simpson Deserts or the brooding McDonnell Ranges were after buffalo, crocodiles or gold — not oil. In any case the rocks they did see were hard, weathered, ancient and quite unlike oil-bearing strata they had heard about — if they had heard at all. The one documented exception was a sworn statement by a prospector named William Archibald Campbell who

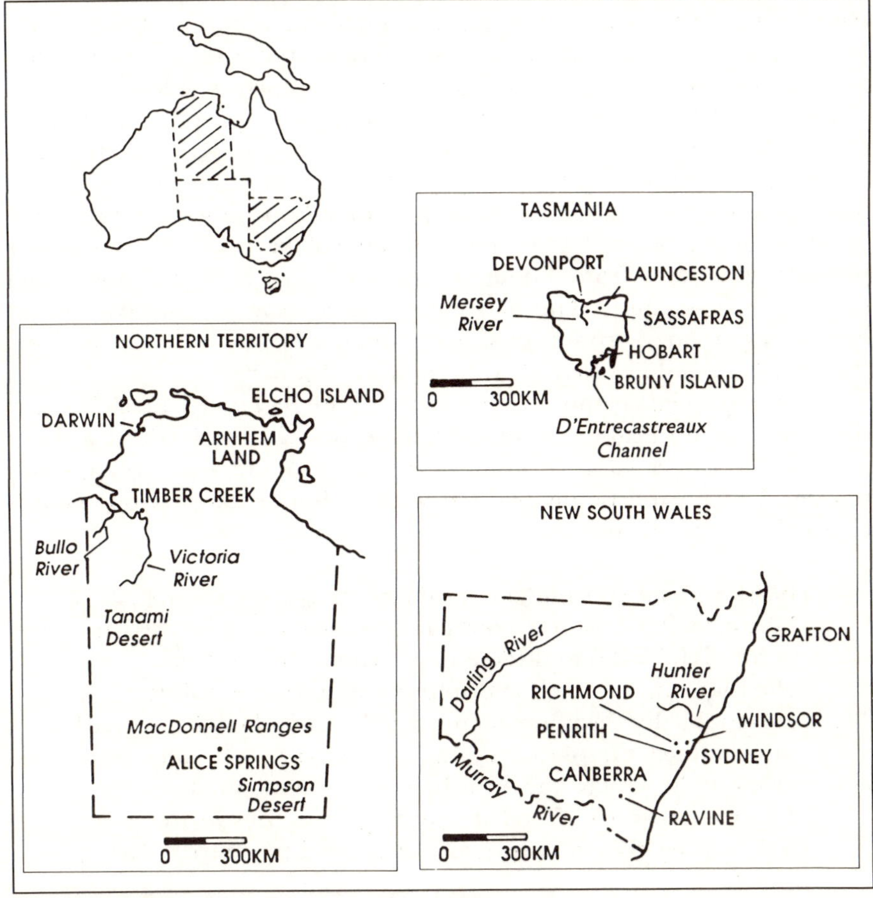

Northern Territory/Tasmania/New South Wales — location of key places mentioned in text.

worked in the vicinity of the Bullo River, a tributary of the Victoria River, in 1904. From his description the region must have been close to Stokes's discovery 65 years earlier.

> Prospecting the adjoining ranges, I found... a very conspicuous mountain peak with noticeable black rocks on its top. While examining in a gorge distant about three miles [4.8 km] bearing 30 degrees west from the said mountain, I saw oil oozing from between the strata of the sandstone rock on the side of the gorge. I examined some, and brought samples of it back to my boat on the river and thence me to Melbourne, and produced me at the time of making this declaration and marked numbers ONE and TWO are bottles containing samples of the said oil strained by me.[68]

When analysed, bottle No.1 contained oils said to resemble 'Borneo petroleum'. Bottle No.2 contained an amorphous wax resembling paraffin. It was not a wildly exciting result, but it did have a sequel, albeit a frustrating one. In November 1907 the government resident officer in Darwin, Charles E. Herbert, wrote to the police constable at Timber Creek whose post was in the locality of the Victoria River asking him to investigate reports he (Herbert) had been given of an oil find in the area. He enclosed tracings of the approximate position of the reported 'springs' and requested that the constable make a patrol as soon as possible. The reply from Timber Creek was dated May 1908 and stated that police Constable Astant had started out to examine the alleged find but, owing to boggy conditions, he reported a failure to reach the area. 'No doubt it would be easy of access by way of water.'[69]

Events obviously moved at a leisurely pace in Timber Creek because the next communication from the police post in the form of a diary is dated December 1910. Mounted constables White and Holland and trackers Charlie and Bobby had left with 15 horses the previous month to search for the oil springs.

> [19th] Sighted conspicuous peak mentioned by Campbell... Owing to flood waters could not cross river. Followed her up for eight miles and fixed depot camp. Swam river... travelled four miles north and found gorge. In my opinion this is undoubtedly the gorge mentioned by Campbell. Prospected and obtained sample of strata, also a bottle of water from a rock hole which had bluish, greasy scum on the surface of water. Owing to flooded waters prospecting in the gorge could not be done to advantage. Returned to camp... Heavy rain.[70]

On the way back the four men were obviously hampered by

floodwaters and were forced to wait on several occasions. On the 28th they were at the Victoria River itself.

> Victoria River a banker. Horse Creek banked up. Had to construct raft to carry packs etc across and swam horses. Difficulty with raft owing to rotten condition of the flys supplied to police station by the Department and in consequence lost private rations and property . . . also the sample bottle of water from the supposed spring.[71]

This stunning anticlimax seems to have deterred any further serious attempt at oil exploration in the Northern Territory for the next 50 years, although there is a fleeting reference to 'some boring on Elcho Island [off Gove Peninsula] during 1924–1926';[72] however no details have survived.

Tasmania

This State has been, and still is, an enigma as far as oil exploration is concerned. Bitumen strandings have been reported along the western, northern and southern coasts since the earliest settlement, as has the presence of intermittent oil 'seeps' in the Bruny Island–D'Entrecasteaux Channel region south of Hobart. But the search for their place of origin has been fruitless, hampered in part by the layer of volcanic rocks overlying, and thus hiding from view, the postulated source and reservoir sediments. Tasmanian Government Geologist W.H. Twelvetrees gave another explanation in a report dated 1917.

> As far as the asphaltum occurrences are concerned, failure has been owing to a misinterpretation of the data. It not rarely happens that prospectors and other intelligent observers, either by accident or assiduous search, bring to light some signs or indications of oil, past or present, but fail to recognise the true inwardness and bearing of the facts. The geological structure of the ground, which should govern all tests, is unhappily only discovered after an expenditure unpleasant to all concerned.[73]

Twelvetrees goes on to cite the case of the Bruny Island Petroleum Company which offered a prospectus during 1915 in which directors made the extraordinary statement that 'liquid bitumen has been known to exude from the ground there [Bruny Island] at two distinct points for over 50 years'. British geologist Dr Arthur Wade was called in by the Mines Department to verify this, but after a detailed examination of the region he could not do so. In fact he went as far as adversely reporting on the chances of finding oil. Bruny Island Petroleum went ahead anyway and drilled a 130 m well on the island penetrating what appeared to be rocks of marine origin. The well was dry and the company ran out of funds soon afterwards. Another attempt to drill for oil in southern Tasmania was apparently made during the early 1920s in what is referred to as Johnsons well on North Bruny Island, where oil was reportedly seen escaping from rock fissures. Eyewitness reports say a light oil was encountered at about 25 m and collected in drums.[74]

However the most active search took place in the northern half of the State, in particular the Mersey Valley region close to Launceston. Here sediments of Permian–Carboniferous age were recognized as having oil potential and one of the earliest figures on the scene was an extraordinary character by the name of George Dick Meudell. He could variously be described as banker, promoter, pioneer, stockbroker, speculator, patriot, traveller and general enthusiast with a zest for living life to the full. He is saved from the additional title of charleton because of his works for charity and his obvious wish to find and develop oil and mining resources for the good and independence of Australia as well as the good of his own pocket. However he was closely associated with the Victorian 'Landboomers' during their reign in the late 1800s and early 1900s. The title of his own book, *The Pleasant Career of a Spendthrift*, published in London in 1929, probably sums him up as neatly as any description.[75] Born in Bendigo in 1860, he based himself in Melbourne, although he seems to have spent a great deal of time on ships and trains round the world.

> I have travelled in 40 countries covering over 400 000 miles [644 000 km] by land and sea, on over 400 steamships and through about 600 hotels... and have probably travelled further than most Australians.[76]

Much of his involvement in the oil scene was in Victoria and South Australia, but he was interested in any potential deposits throughout Australia and New Guinea, including Tasmania. A measure of his promotional 'talents' is seen in the prospectus of Tasmanian Oil Wells Company, dated 1915, which he wrote himself.

> Fortunes are made from oil. What others can do, you can do. Without exception, oil has produced the biggest combination of wealth ever known to man. Oil has made more people rich, and at less risk and smaller outlay to the investor, than possibly any other investment known. Here is a record of what 20 pounds invested in oil stocks made for some of the fortunate investors in:
>
> New York Oil Co. .... 780 pounds.
> Hansford Oil Co. .... 800 pounds.
> San Joaquin Oil Co. .... 1 140 pounds.

> Peerless Oil Co. . . . . 1 250 pounds.
> Kern Oil Co. . . . . 1 900 pounds.
> Central Oil Co. . . . . 6 000 pounds.
> Alcade Oil Co. . . . . 8 000 pounds.
>
> Twenty pounds will not buy much land will it? Yet you see for yourself what profits that same amount made in oil. Forty-four Californian oil companies, whose stock is listed on the Californian stock exchange and Oil Exchange, out of a total of 59, paid 19 million dollars in dividends last year.

The prospectus offered 230 000 shares, priced at a shilling each, to raise £11 500. Meudell's aim was to exploit the oil shale deposits in the Mersey Valley area and also drill for natural oil, which he believed had sourced the shale in the first place. However his unabashed comparisons of Tasmanian prospects with the USA failed to impress would-be investors. Attempts to obtain Tasmanian Government support in raising funds in London also failed, although part of the reason at that time was the stringent conditions imposed by the Federal Government during the First World War.

Another of Meudell's promotions around that time was the Melbourne Oil Exploration Syndicate. Few details survive, but it apparently also included an attempt to spark interest in the Mersey Valley region of Tasmania. On the back of a map of the Devonport region among a collection of Meudell's papers is a series of hand-drawn 'geological' sketches showing the conditions responsible for the phenomenon of reported gas emanations in the region.[77] According to these drawings there were *domal caldera*-type features in the area, one of which supposedly caught fire and burnt for four years. The diagrammatic explanation shows how natural gas might have originated in an underground reservoir and risen to the surface vent by way of fractures in the rock. A graphic red flame and billowing smoke have been drawn to add dramatic effect. [Author's note: The Mersey region does have limestone karst features of the type indicated in the diagrams, including sink holes. The Tasmanian Mines Department says that some gases have been reported coming from them, but this is related more to underground drainage than any trapped oil or gas accumulations. There is no knowledge of a fire burning in the vent for four years as the annotations on Meudell's drawings suggest, nor is there any coal seam in this area which might catch fire and burn for such a long time. The Department does say, however, that if a 'button' grass which grows in northern Tasmania is set alight it can burn or at least smoulder for long periods.]

Two other companies — the Mersey Valley Oil Company and the Adelaide Oil Exploration Company — were formed in the 1920s to drill in the region. The two companies sank a total of 21 wells between them at a cost of £60 000 before becoming involved with Tasmania's short-lived oil shale industry.[78] Mersey Valley Oil Company did attempt oil exploration in the nearby deeply buried Tertiary rocks of the Sassafras Valley and also contemplated drilling near Macquarie Harbour about 1924. All these ventures were unsuccessful and the Tasmanian oil search effort petered out before the 1930s.

New South Wales

The rich deposits of shale oil in the Blue Mountains had been an important industry for the State and for Australia in the years leading up to 1900, but the economics of mining and treatment rapidly fell behind those of importing the new 'natural' oil from overseas. With this in mind New South Welshmen began to look around for accumulations closer to hand. Unfortunately there was a lot more hope than promise. The indefatigable Assistant Government Geologist Joseph Carne gave his opinion about the State's potential for natural oil discovery when writing a detailed report on the declining shale oil industry in 1903.

> The absence of any of the usual superficial indications of petroleum, such as oily exudations, or impregnations of its more or less solid residues in the form of maltha or asphaltum, is not a hopeful augury for this State, for it is doubtful whether any oilfields have ever been developed without one or all of these constant surface adjuncts to encourage and promote research.
>
> A noteworthy and suggestive fact... is the disconnection of petroleum and coal. Productive oil deposits are entirely absent from the Coal Measures [of New South Wales], in which minute quantities only have been found.[79]

Mindful of the gas found in the Roma town water bore in Queensland, Carne did concede that there was a chance of similar deposits being found in the deeply covered Mesozoic-age basins in the State's northwest, in areas not then pierced by water bores. But he cautioned that, with a few exceptions, there was an absence of structural anticlines which could form traps for either oil or gas. The Mesozoic strata of New South Wales, he said, had not been subjected to even moderate folding stress and occupied, undisturbed, their original flat-lying or gently inclined position. Nevertheless, in the same report reference is made to what was seen, initially at least, as a mirror image of the events at Roma. A water bore drilled on the racecourse at Grafton on the northern coast of New South Wales struck gas at a depth of around 950 m which then flowed intermittently for several years. The bore was originally begun in 1897, sited on the advice of Government Geologist E.F. Pittman and drilled under the supervision of the State's Department of Public Works using a calyx drill. In December 1901, Acting Road Superintendent in the Public Works Department, H.M. Baldock, described the phenomenon.

> In June 1900, the gas would ignite at the top of the casing and send a flame four feet high of the size of the casing. In May last the gas ignited from the forge, which is fourteen feet away from the casing, and considerable difficulty was experienced in extinguishing it; the plant was in danger of being burnt at the time. Since then every precaution has been taken against fire. At times the gas is very strong and throws water half way up the derrick, and very frequently throws a strong stream of water from the top of the five-inch [diameter] casing.
>
> There is a perpetual flow of gas which varies in strength from the five-inch casing. I have partly made arrangements to tap the gas in the five-inch casing, and intend trying it for lighting the plant in place of the kerosene lamp now used.[80]

There is no record of this lighting 'experiment' succeeding, and in any case the well was only deepened another 180 m before finally being abandoned in 1902. Disappointingly for Grafton, the end result of five years labour was neither a gas nor a water well. Commenting on the bore in later years, Carne did say that some oil was seen after drilling ceased, but no scientific sampling substantiated the claim. The gas was generally thought to be associated with coal beds penetrated during the drilling programme.

Following this chance 'discovery', the focus of the search shifted south to the Hunter River, Richmond and Penrith regions 80 km or so northwest of Sydney. Early in 1910 reports came from Redbank, near Richmond, of some oil shows. A hand drilling plant was rented from the Mines Department and boring commenced close to the Kurrajong Road later that year. At a depth of 73.5 m the drill rods were reportedly covered with oily slime. However, when J.B Jacquet, the Chief Inspector of Mines, visited the site he did not find any signs to indicate that oil was likely to be found. A second bore was begun 25 m away and drilled to 267.5 m without revealing any of the heavy oil reported in the first. A third attempt did find traces of 'heavy crude' at a depth of 60 m, but it was concluded in Mines Department circles that if the main oil trap was found, it would be a very small accumulation.[81] A well at Penrith, drilled between 1918 and 1920, found traces of gas but because of poor drilling techniques it could not be adequately tested and was abandoned.

During the 1920s and 1930s a number of companies, syndicates and even private individuals made scattered exploratory attempts to find the elusive liquid hydrocarbon, but all were unsuccessful. The names included Oil Search Ltd and its subsidiaries Kamilaroi Oil Company Ltd (Kulnura Bore) and Gas Drillers Ltd (Mulgoa Bore), Oil and Gas Investigations Ltd (Loder Dome), Belford Dome Ltd (Belford Dome), Yerrinbool Oil Prospecting Syndicate (Yerrinbool Bore) and private individuals named Tyler (Tyler Bore) and W.J. Naskell (Parley Bore).[82] Some well locations were chosen by amateurs and were doomed to failure at the outset because a valid geological structure was not present. One of these was drilled near the Lobb's Hole copper mine at Ravine, about 160 km southwest of Canberra, because gas had been detected in the mine workings. These failures angered entrepreneur and founder of the Petroleum Association of Melbourne, George Dick Meudell, who was a passionate advocate for Australia's need to find its own oil reserves. In an address to the Historical Society in mid-1924 he mentioned his own theories about oil prospects in New South Wales, and blamed the government for not hosting a more responsible search.

> Very little search for oil has been done in New South Wales, and the Government is strangely negligent, because the river system of the state must have made enormous deposits of terrestrial vegetation favourable for the deposition of petroleum throughout the ages, especially near the junction of the Murray and Darling Rivers. Shallow boreholes were sunk near Penrith, Windsor and Richmond with ridiculous appliances by ignorant people. That kind of search for oil is most damaging.[83]

Some bore holes were sited on well-defined geological structures like the Loder Dome and the Belford Dome in the Hunter River region, but although several encountered traces of gas, there was not enough to justify appraisal work. One interesting attempt in the mid-1930s was made by Natural Gas and Oil Corporation Ltd from the bottom of the Balmain coal shaft on the outskirts of Sydney. The shaft itself was 896 m deep and the bore was drilled on down to a depth of 1 506 m. Gas was encountered below the 1 220 m mark and a small flow of about 3 360 m^3 a week was drawn from the sealed colliery to use in cylinders and bags as a petrol substitute in motor vehicles during the Second World War.[84]

Fittingly, in a way, the last serious exploration attempt in New South Wales prior to the Australia-wide 1950s revival was back in the Grafton region. A company called Clarence River Oil Prospecting Company began drilling about 30 km southeast of the town at a place called Halfway Creek in 1933. The well was drilled to 787 m and was cased throughout. Even cores were cut through some sections although, inexplicably, the company did not bother to carry out any mapping or field investigations. No gas was found during drilling, but it was discovered later that if the hole was plugged for a few days, gas containing 5 percent ethane would accumulate and burn for one or two minutes. Needless to say this was not enough to sustain a gas development and the well was abandoned.[85] There were reported finds of oil in various parts of the eastern half of the State during the 1920s and 1930s, most of which were investigated by Mines Department officers; however all were found to be entirely baseless. In no case was there any evidence of the existence of oil and, on that note, exploration in New South Wales reverted to a revival of shale mining as the 1940s and the wartime emergency approached.

SOUTH AUSTRALIA

The early oil search throughout Australia was studded with shysters, cranks, and those who simply applied misguided theories in their attempts to find the elusive liquid. But it seems that during the first half of the 1900s or, more accurately, the years up until the outbreak of World War II, a good many of them practised in South Australia. Certainly their efforts were more publicized in that State than any other and, understandably, the results drew no praise from later chroniclers. In his 1943 summary of Australian oil drilling up to that time, Commonwealth Government geologist and later founder of the Bureau of Mineral Resources, Dr Harold Raggatt, was scornful when listing the South Australian wells.

> Most of them call for little comment as they were put down at locations which even the most optimistic geologist would regard as hopeless.[86]

Much of the reason for this can be attributed to the fact that the 'coorongite' controversy was still in full cry, and it was fuelled from time to time by the appearance of bitumen deposits washed up on coastal beaches from the Eyre Peninsula right along the southeastern shores to Victoria. Helmut (Hele) Wopfner, a noted post-Second World War government geologist in the State, aptly defined this period of exploration as one which was entirely based on surface indications. The deposits of coorongite and coastal bitumen were taken as tangible evidence of liquid riches below. One only needed to drill at these surface occurrences to 'tap' the mother lode. Early exploration took place in the relatively accessible regions in order to to be near established settlements for the purchase of supplies and equipment, and for the carrying out of any necessary repairs. At first it was also preferable to be near the coast and a suitable port for shipment of any oil found. But later efforts moved the search out into the plains and highlands. With the benefit of 70 or 80 years hindsight it is hard now to believe that some of the statements and exploration programmes were ever taken seriously. Wopfner quotes a choice example from the prospectus of Lakes and South Eastern Oil Syndicate NL dated January 1931.

> It has been reported to the writer [of the prospectus] by quite a number of local residents that a number of these salt lagoons on Kangaroo Island during several months of the year (summer

> time) on and off at night are like a city illuminated with large balls of fire and beams of light, some vertical, others horizontal. There are very few oil fields of the world that are favoured by such splendid evidence of the presence of oil from below.[87]

The prospectus author's explanation of this phenomenon strains the imagination even further.

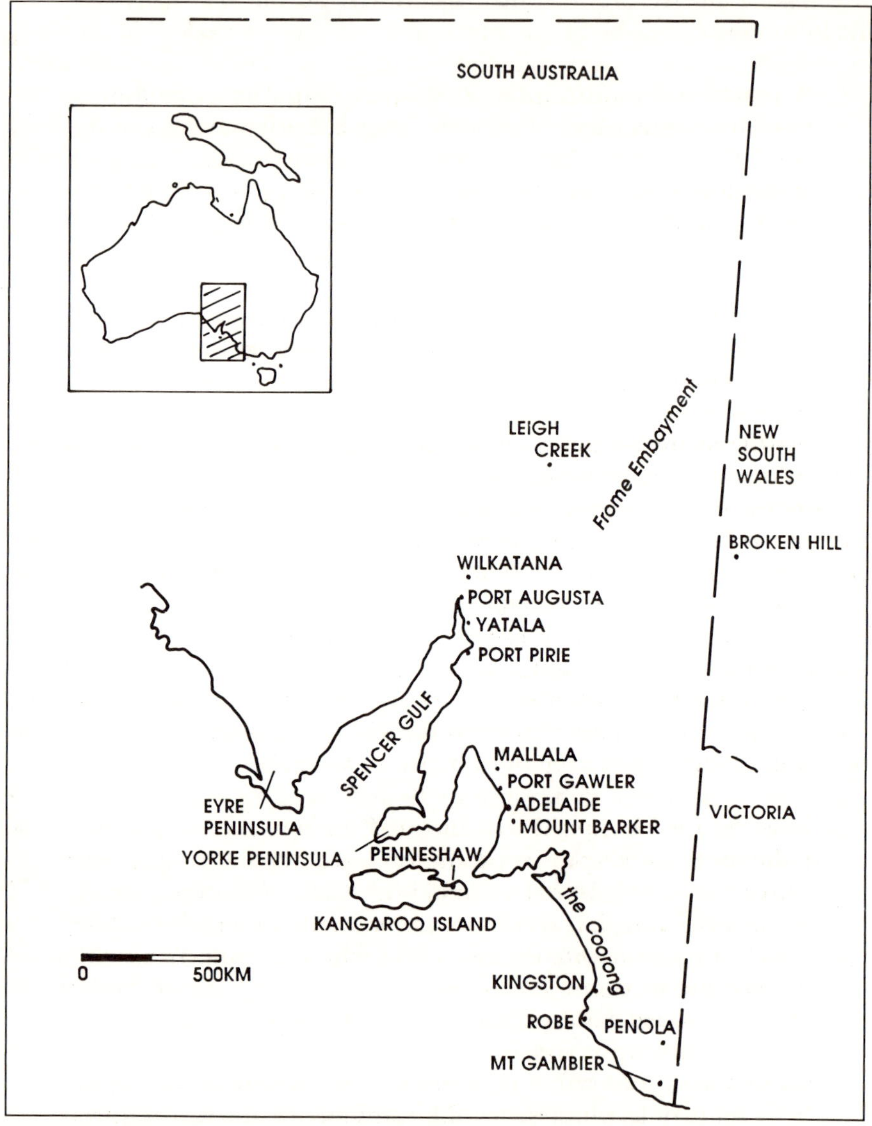

South Australia — location of key places mentioned in text.

> The great pressure of the confined gases at shallow depths below the surface rends the rocks open; this friction causes a spark that ignites the light volatiles of the petroleum oil, and unites with the chlorine gas from the limestone rocks, thus forming sulphur sublimate; this particular form of gas is very dangerous to inhale.

Wopfner is left wondering how much of the gas the writer himself inhaled. Another form of persuasion took the form of 'practical demonstrations' by diviners, and the *Adelaide News* of 8 February 1927 ran the following story.

> A practical demonstration of prospecting for water and minerals by means of an electric divining machine was given by Mr J.P. Spry, of Glenunga, in various parts of the hills yesterday afternoon. Mr Spry produced entirely by his own handiwork a machine which he claims will accurately locate silver, gold, copper, coal, tin, platinum, water or oil.[88]

The machine had a number of 'stops' allowing the operator to set it for the mineral he was seeking, and shutting off the others. Two steel bars were swept over the surrounding countryside and they swung forward when a prospect was located.

Southeastern Australia was also the haunt of the Russian confidence man, promoter and sometime writer, Eugene de Hautpick. Apart from his 'studies' of coorongite, one of his early speculations (oil shale in the Tasmanian Mersey Valley region) is said to have persuaded Prime Minister Billy Hughes to part with £5 000, which was promptly lost. *Smith's Weekly* of 10 December 1927 comments that Hughes's remarks to the Russian 'expert' when he discussed his loss were supposed to have been almost worth the money. De Hautpick left Australia soon after that episode, but he returned in the late 1920s eager to set up another exploration company to try out his 'new invention'. *Smith's Weekly* interviewed him at Fremantle on board the *Ville de Strasbourg* en route to Adelaide.

> Since he [de Hautpick] left Australia, he had invented a new radio metrical method and designed apparatus for prospecting the earth for metals and mineral oils. This method was based on the principle that all metals and mineral oil sent out radioactive emanations into the air. With his new invention it would be possible to detect in the air the presence of mineral oils and valuable metals by testing the air in the locality.
>
> The invention consists of two instruments, which learned savants have told him will associate his name with the twentieth century as wireless has Marconi's with the nineteenth. One shows the intensity of radio emanation present, and the

> other measures the emanation in the air and shows how many ions are present.
>
> Petroleum has its own individual unit. It is his belief that petroleum is liquid radium and that the air of oil fields has a high degree of radio-active emanation.[89]

Looked at now, de Hautpick's 'science' was a quaint portent of today's geophysical tools. A more sinister character of that time was a promoter named Max Steinbuchel. Described as 'a celebrated American geophysicist', he set up Producers Oil Wells during the 1930s to select seven drilling locations in Australia — including one at Mount Gambier. Steinbuchel apparently accomplished this task by means of a gadget he called a 'doodlebug' which was used to detect oil pools. However there was some concern by the South Australian authorities about the company brochure which indicated that all monetary subscriptions should be sent directly to Steinbuchel himself. His unscrupulous and somewhat violent character was called into question further by newspaper reports that he had been committed for trial on a charge of assault of a bookmaker's clerk. After smashing his victim's jaw, he is alleged to have said, 'That's the way we do things in America'.[90]

This sort of report did nothing to soothe the groundswell of public opinion against foreign interests, Americans in particular. Allegations and suspicions of misconduct and malpractice were prevalent right through until the 1940s. Residents of Mount Gambier, for instance, were convinced that a well being drilled for Associated Oil Company in the district had been deliberately sabotaged. In the South Australian Parliament during 1940 it was alleged that the American driller by the name of Ray Fife had actually struck oil at 600 m, but then deliberately mudded the well and returned to the United States. American oil interests were supposed to be behind the scheme, their motive being that they were opposed to the discovery of oil in Australia, as a lot of money might be spent with doubtful results and they had ample supplies elsewhere.[91] Among the oil promoters and business speculators themselves, however, there was recognition that oil geology was in its infancy and expertise was scarce. So men who had worked in or even just passed through oil fields elsewhere were generally much sought after. Enthusiastic Melbourne- based promoter and inveterate traveller George Dick Meudell visited the United States seven times and included numerous oil well tours in his itineries — an experience he used to set up his companies.

Meudell's first syndicate was the Standard Oil of Australia, formed in 1912, which he used to float Australian Oil Wells Company NL, which in turn provided funds to raise capital for a drilling company he called South Australian Oil Wells Company NL. The raising took him 15 months of incessant work in the Australian press and a series of 31 lectures delivered in South Australia and Victoria illustrated by a 'box of lantern slides of gushers and oil wells'. Finally he secured over 2 000 shareholders. In the course of his exploration ventures, Meudell became friendly with a Californian driller named Tom J. Whaley, taking him to likely localities in western Victoria and southeastern South Australia.

> Encouraged by an excellent work on the geology of South Australia by Father Julian Tenison Woods, parish priest at Penola, published in 1862, I took a fancy to the region between Mount Gambier and Kingston, as a likely oil-bearing locality, and so dreamt another dream of Alnaschar. Whaley took charge of a hole near Kingston [for Southern Ocean Oil Company] and drilled to 1 100 ft [335 m] without success. I became part owner of oil licenses over 60 square miles [155 km^2] near Whaley's bore.[92]

Meudell himself admitted to a policy of 'blind stabbing' when it came to selection of drilling locations, but Whaley seems to have at least attempted to apply some structural theory. He selected a number of drill sites on top of fossil beach ridges, mistaking them for anticlinal structures.[93] The deepest of these was Robe No.1 (also known as the Richmond Park Bore) drilled to 1 374 m for the South Australian Oil Wells Company in 1915. Consultant for this well was Herbert Basedow, one of the few Australians with any technical training willing to lend his expertise and support to the early South Australian oil search, and a man prominent in Adelaide during the late 1800s and early 1900s. Caught up in the coorongite debate, for a long time he advocated the mineral origin theory, only to concede the truth of vegetable origin when he succeeded in growing the algae himself in his laboratory. At various times a medical doctor, politician, anthropologist and entomologist, Basedow's main work was as a geologist. Early in his career he was employed by the South Australian Government Geological Survey, but apparently ran foul of officialdom and turned to consulting. In this capacity he attempted to influence exploration policies and select drilling targets, particularly in the southeast of the State. Despite his views on coorongite, he did spend large amounts of money in exploration, reasoning that any knowledge or information gathered about South Australian geology would be beneficial in the long run. He was also relentless in his calls for government assistance, or at least encouragement, in the oil search.[94]

In this quest, however, Basedow and his fellow explorers ran into formidable opposition in the person of Director of Mines for South Australia, L. Keith Ward. A quiet man who was said to be poker-faced even when telling a funny story, Ward was nonetheless an experienced and capable geologist of unimpeachable character. He had little patience with the optimism of the oil companies and the sometimes outright fraudulence of the promoters. A letter to the Minister of Mines in 1913 clearly outlined his attitude to oil potential in South Australia and it was a view he held until his retirement in 1944.

> Nothing has been shown to me in the field which may be considered to be sufficient justification for boring and that as far as I can tell, the geological conditions are not favourable for the occurrence of oil in appreciable amounts.[95]

Ward went further and accused some promoters of unscrupulous 'salting' of wells. Eric Rudd, a student and young geologist in South Australia during the 1930s, can remember Ward commenting acidly that 'if explorers stop putting oil down their wells they will stop getting it back up again'.[96] Although Ward's accusation had some truth, it was (and still is) a fact that drilling fluid, lube and hydraulic oils could be inadvertently squeezed out of machinery and down the hole to reappear as part of the returning formation fluids. Perhaps the prospectors could be accused of sloppy work practices, but many did not deserve the government geologist's scathing scepticism. Consequently, companies became reticent about officials examining their exploration efforts because the reports would invariably be pessimistic. The antagonism intensified when industry accused Ward and his assistants of having no oil experience.

Ward defended himself by saying he had read relevant literature, and he went on the offensive by denigrating the old argument that South Australia, or anywhere else in Australia for that matter, would have oil just because surface appearances were similar to those in other oil-bearing areas like America.[97] He was backed up in this view by the ubiquitous English geologist Dr Arthur Wade who, in 1914, was called in to make a study of South Australia and comment on its oil potential. Wade began his report by deploring the loose way in which the terms 'bitumen' and 'petroleum' were used in South Australian localities he visited.

> ... it must not be forgotten that the bitumen family has a very wide occurrence on the earth's surface. Traces of bituminous matter occur in every country and in every geological formation, but these traces can rarely be taken to indicate the presence of large and payable bodies of petroleum.[98]

In the main Wade's conclusions were pessimistic, although he did concede that much of the material washed up on the beaches round the coast was apparently bituminous. He added, however, that none of the rocks in the vicinity were capable of producing the substance and he concluded that the origins lay offshore. How far offshore was uncertain, but Wade did note that similar occurrences had been reported from Tasmania to Western Australia and also on the coast of South Africa. He speculated that the intermittent timing of the depositions may have something to do with their release from the sea floor after earthquakes.

He dismissed coorongite and supposed deposits in foetid limestone as red herrings, noting in matter-of-fact terms that one sample sent to him turned out to be a 'refined oil of the brand 600W Gargoyle Cylinder Oil' which had its origins in a boiler used in drainage excavation works near Robe. Reports of calm seas off Kingston supposedly caused by oil on the water were, he found, in fact caused by masses of seaweed floating just below the surface.

Wade's study was denigrated by Basedow and the explorationists as no more than a compilation of other previous reports. Basedow in particular reacted with letters to newspapers and journal articles, and continued his opposition to Wade and Ward in parliament when elected as member for Barossa in 1927. His main claim against Wade was that as the Englishman had covered 60 000 km^2 of the State in a mere three weeks, the report was based on a very cursory examination.[99] With the benefit of hindsight, the main criticism that can be made of Wade's work is that he dismissed the vast region in the northeast of South Australia under the heading 'The Rest of the State'. His rationale was that none of the numerous water bores in these parts had revealed indications of petroleum or petroleum gases; however he did mention two exceptions. The first was a report that gas was discharged during the drilling of a water bore at Coonanna about 800 km northeast from Adelaide on the New South Wales border. The second was the occurrence of traces of free oil in core samples taken from the No.2 bore near Leigh Creek. He cited these two examples as the only authentic occurrences of petroleum in the State.

Despite this official condemnation, exploration continued. The difficulty was that many of the efforts were made by small syndicates with meagre finances — often there was only enough money to drill one hole. Eyre Peninsula Petroleum Oil Company took its name from the locality where it drilled a shallow hole during 1914 in an attempt to find the source of 'beach bitumen' in the area. In 1921 an explorer called American Beach Oil Company drilled near Penneshaw on Kangaroo Island. Various attempts were made during the 1920s to locate oil pools by divination at Yatala near Port Augusta and closer to Adelaide at Mallala, Mount Barker and Port Gawler. There was even drilling by the Largs Bay Oil Company on the beach between Port Adelaide and Brighton on the supposed evidence of gas escaping from the sand in the area.[100] Diviners also led the search on Yorke Peninsula and a well was drilled (unsuccessfully) just outside the town of Maitland in 1930. In 1931 a water bore at Minlaton allegedly carried traces of oil and this sparked an enthusiastic search by the Minlaton Oil Boring Syndicate and the Peninsula Oil Syndicate, both companies apparently having the wholehearted support of the local community.[101] In fact the oil traces at Minlaton originated in ancient Cambrian-age rocks, as did globules of oil recovered from an artesian bore at Wilkatana near Port Augusta at the head of Spencer Gulf in 1934. At first these 'finds' were dismissed by Ward as impossible phenomena for rocks of that age and he made accusations of well 'salting'. Interestingly, the possibility of oil from Cambrian rocks was authenticated in later years and the Wilkatana shows provided the initial impetus for the formation of SANTOS in the 1950s.[102]

Most of the exploration work up to the Second World War years was conducted in the southeast of the State, where a total of six companies were involved, mostly along the coastal strip and inland as far as Mount Gambier. This was the area favoured by Basedow in his consulting work, but it was not until 1930, with the arrival of Oil Search Ltd, that any geological or geophysical mapping was attempted as a prelude to selection of a drilling site. That company mapped the Burnda area near Mount Gambier and located what became known as Knight's Dome. A magnetometer survey confirmed the structure and a percussion rig was brought into the area. Eric Rudd remembers that in his early work with the company the well was dry, but a more amusing anecdote sticks in his mind.

> An old driller named Sol was in charge. He was an ex-Papua man and a little deaf from taking quinine over a number of years. The percussion rig had a big bull wheel for raising the rods in the bore and we used to hide under the drill floor and whistle and squeak till Sol got the oil can out to try and remedy the problem. We'd stop while he did that, then start again when he put it away. He never did catch us at it.[103]

Rudd also remembers a cook at the Mt Gambier wells who made more money out of introducing women to the drillers than he did with his culinary talents!

When Oil Search left the State in the late 1930s to concentrate on Papua and New Guinea, exploration in South Australia had almost ceased. Keith Ward's continual warnings to the public about oil promoters and their wild schemes were emphasized by the failure of the explorationists and diviners to fulfill their promises. Basedow, the industry's 'champion', had died from a rare tropical disease in 1934 leaving Ward the final say in their 20-year battle of words.

> No business in Australia has been so infested with sharks and charlatans as this business of looking for oil. The result of all this futile expenditure has been the creation of a widespread apathy and lack of confidence in any venture connected with the search . . .[104]

Blobs of bitumen were still arriving on coastal beaches, but Ward and others developed theories of Antarctic or South American origin and long distance transport involving surface currents. Few took up the lead given by Dr Wade in 1915 that the material might be the product of seeps located not far offshore with its release prompted by earthquakes along the many faults in the region. Convincing evidence of this truth had to wait for the sophisticated oceanographic technology and equipment of the 1960s and 1970s. The early 1940s passed with little activity in South Australia. Then, in common with the rest of the country, war's end brought new life to the search. Work resumed in the southeast of the State, but more significantly, for the first time explorers began to seriously cast their eyes over the arid regions of the far northeast.

Prompted by a wish to find and tap gas or oil for use in the lead smelter at Port Pirie on Spencer Gulf and in the mines at Broken Hill over the border in New South Wales, the Zinc Corporation took up areas in the Frome Embayment area in 1946. In association with the Vacuum Oil Company the Frome-Broken Hill Company was formed and some of Australia's first geophysical prospecting was carried out in an attempt to define the thickness of sediments and gain an indication of potential structures. Although there was no reward for these initial efforts which continued into the 1950s, some encouragement was taken from the work of Dr Ivan Tschebotarev of the South Australian Mines Department. Reviewing hydrochemical aspects of artesian basin waters in 1949, Tschebotarev drew on experience from his native Russian oil fields and agreed that the northeast held a good deal of promise for the discovery of hydrocarbons.[105] The course towards the State's ultimate success during the 1960s had finally been set.

VICTORIA

During the first half of the twentieth century, oil search activity in Victoria rivalled that of South Australia and Queensland. It certainly caused as much, if not more, excitement because the explorers actually found legitimate (albeit small) accumulations, the first of which (Lake Bunga No.1) lays claim to being the earliest oil discovery in the country. The main focus of attention was in the Gippsland region in the southeast of the State, but there was also interest in the coastal areas stretching west from Melbourne right up to the South Australian border. A few permits were even held at various times in central Victoria round Kilmore and Bendigo, but there is no clear record of any drilling attempts in those locations. In all just over 50 wells were drilled between 1900 and 1950, many of them with the financial and/or technical assistance of the Victorian Mines Department. The earliest work appears to have been a continuation of the rather futile and misguided drilling begun during the last years of the nineteenth century in Gippsland. This was based on 'evidence' of seeps at the surface which are now thought to have been nothing more than iridescent films of iron oxide on the surface of water in marshy ground. Dubbed 'fool's oil', it was a common trap for the inexperienced settlers, and several private enterprise schemes came to grief on this false premise during the first two decades of the 1900s.

Promoters were never far away, and in 1921 a new company with the long-winded title of Boola Boola Petroleum and Natural Gas Company NL began a campaign to solicit subscribers. Advertisements in local papers ran with bold print catch lines which proclaimed that 'Australian Oilfields must become a Reality and every indication points to BOOLA BOOLA being THE FIRST PAYABLE FIELD'.[106] The accompanying screed unashamedly pointed out that there was little doubt oil existed at Boola Boola: the geological formation was identical with the rich fields of Roumania, USA, Mexico, Austria and Galicia. What is more, three world-renowned experts had opined that the oil was there! Boola Boola was located between the Thomson and Latrobe Rivers about 80 km west of the Gippsland town of Sale and the well was spudded during October 1921, much to the dissatisfaction of Mines Department officials who had misgivings about the geological data supplied in the company's prospectus. Not surprisingly the well was found to be dry, but not before it had dragged on for four years to reach a mere 560 m.

By that time, however, oil exploration had finally gained some credibility due to the efforts of two mineral prospecting partners named George Shirrefs and James Duncan. During 1922 the pair were looking for silver-lead deposits in the limestones between Buchan and Nowa Nowa when they decided to try their luck with the search for oil in the younger

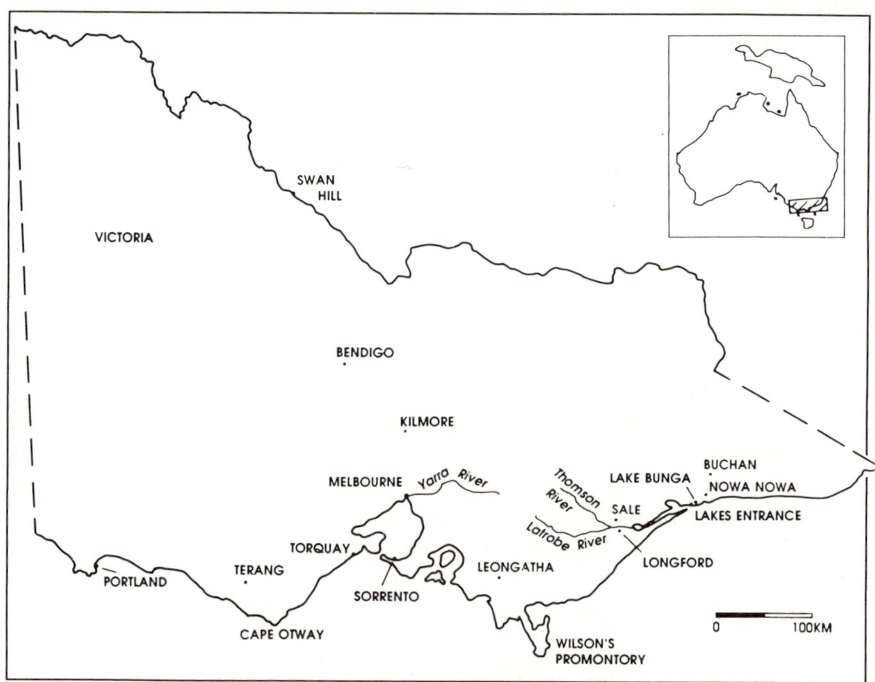

Victoria — location of key places mentioned in text.

Tertiary-age rocks to the south around the coast. Consequently they applied for a lease covering 14 km^2 in the region east of Lakes Entrance. With Mines Department advice a well site was chosen at the marshy, mosquito-ridden head of Lake Bunga about three kilometres east of the town and drilling began using leased Mines Department equipment and crew on 5 January 1924. Writing on Victorian oil history several decades later, Nicholas Boutakoff, himself destined to play a role in post-1950s exploration, described the outcome.

> After about seven months drilling, on 25 July, 1924, at a depth of 1 070 feet [326 m], much gas and some oil was struck in fine glauconitic sandstone. Quantities of water (125 000 gallons [27 500 L] a day) and methane gas escaped, together with a very small amount of oil and oily sediments, forming an emulsion. Analysis of the oil by Mr Watson, Chief Chemist of the Mines Department, showed it to be a mineral base heavy crude.[107]

It was not the spectacular, earth-shattering event of the Spindletop gusher in Texas which set the US oil industry on course twenty years earlier, but it was proclaimed as Australia's first authentic oil discovery. If one overlooks the early bitumen finds in the Kimberleys and those along the South Australian coast, and agrees that the indications of condensate in Queensland in 1900 are more closely defined as natural gas liquids, then the claim is accurate. Today a miniature derrick and painted sign mark the spot — a swampy, still, almost eerie place. Water is still flowing freely from the bore into a copse of ti-tree and there is the strong rotten egg smell of hydrogen sulphide probably caused by a combination of gas in the water and decaying vegetation in the brackish, stagnant lake itself. Back in 1924 though, the discovery was greeted with a great deal of enthusiasm and it prompted a flurry of company flotations bristling with glowing prospectuses about the Gippsland potential.

Soon after the discovery Shirrefs and Duncan formed Lake Bunga NL (in August 1924) with a property value of £25 000 and 100 000 shares of £1 each. Two years later both men became subscribers in Lakes Entrance Development Company which raised £40 000 and drilled a second well 3 km further west, with the idea of lighting Lakes Entrance township with gas from the discovery.[108] This attempt again found the glauconitic sand reservoir, but exhaustive tests indicated that only about 0.5 L of oil a day could be recovered. Gas flow was not recorded and the company apparently ceased to exist soon afterwards.[109] A short time later the Victorian Government decided to sink its own bore and chose a spot on what is known as 'North Arm' close to where today's Princes Highway crosses the bridge into Lakes Entrance township. It produced a similar result. More significant was a correlation of these three holes showing that the bedrock dipped in a southerly direction. To the geologists of the day this indicated that the basin deepened to the south and caused them to make the first tentative guesses that oil was coming from a main accumulation somewhere off the coast.

In the 1920s offshore exploration was out of the question technically and financially. But this did not stop the fertile mind of Melbourne's confirmed optimist and entrepreneur George Dick Meudell. Undoubtedly spurred on by what he saw during visits to the Californian oil fields, he was soon trying to drum up interest in an enterprise called Summerland Oil Company NL which had applied for a lease extending three kilometres along the sea beach at Lakes Entrance. His idea was to drill for oil from derricks erected on jetties, delving on the beach and under the sea.[110] His proposed programme was accompanied by the extraordinary and unsubstantiated claim that the Lake Bunga gas field covered 2 070 km^2. What is more, when the scheme failed through lack of support, he angrily attacked the government for not tapping the supply which he calculated was flowing at a rate of 280 000 m^3 a year from the original well, only to be lost into the air.[111] Regarding the prospects of a company called Port Phillip Oil Company, which actually spudded a well at Williamstown just west of Melbourne in 1924, Meudell was in a more technical, but no less optimistic, mood.

> ... the seeps of oil undoubtedly [there] came from the beach of the ancient River Yarra, and petroleum is invariably found in the deltas of ancient rivers which debauched on old marine beaches.[112]

Not everyone was enthusiastic about the 'boom'. Despite the official government authentication of oil at Lakes Entrance, the Sydney journal *Smith's Weekly* was acidly sceptical.

> ... the published official announcements are entirely non-committal. Boiled down to plain English the whole summary is 'go ahead at your own expense as hard as you like and win the offered prize'. But the fact remains that only an equivalent of 2 ounces [57 gm] of oil to a ton of earth has yet come to light — in other words, the contents of a desert spoon scattered over a dray load of muck — an absolutely unpayable proposition.[113]

Nevertheless, the rash of new companies (joined in some parts by the Mines Department) swarmed to the Lakes Entrance area, drilling in and around the town itself as well as west along the coastal lake system to Wilsons Promontory, Leongatha and Sorrento, past Melbourne to Torquay, Cape Otway and Portland. Torquay, on the coast west of Melbourne, attracted early attention because the Tertiary sediments there were quite visibly folded into a series of anticlines. The Point Addis Company drilled two wells of just under 300 m depth about 11 km from the town in 1924, penetrating brown coal beds, but reporting no oil or gas. A short distance away Torquay Oil Wells Company drilled five wells ranging from 213 m to 460 m deep with much the same result. Other attempts were made further north and west by companies with long tongue-twisting names like Victorian Petroleum Boring Company, Mouta Oil Wells Company, Jennawarra Oil Wells Company and Boonahwah Oil Wells Company. There was even a Lake Boga Oil Wells Syndicate formed for work near Swan Hill, although there is no record of any drilling in the area. One explorer called Victorian Oil Properties, with a capital of £18 000 and a field of operations near Lake Terang in the Western District, attracted particular sarcasm in the columns of *Smith's Weekly*.

> ... [its lease] is a swamp which in wet weather emits methane or marsh gas such as gave rise in medieval times to the legend of the will-o'-the-wisp.[114]

But there was no success so, as the 1930s approached, most of the work contracted back to the Lakes Entrance region where small amounts of oil were being found and, despite *Smith's Weekly*'s earlier scepticism, even produced using bailing or pumping methods. Clive Wallace Smith, whose father was a well-known stockbroker and underwriter of mining and oil companies in Melbourne, remembers the fever, and the frustration, of the time.

> Percussion rigs were used in the main and there was always great excitement when gas was found down the hole, then a little oil was bailed. But suddenly water would break through and ruin it.
>
> All the government officials tried to help exploration as much as possible, particularly the chief geologist and director of the Victorian Geological Survey, W. Baragwanath — a lovely, scruffy little man affectionately known as 'Mr Barry'.[115]

Baragwanath was a prolific report writer as well as an enthusiast, and in one short article published after the Second World War he summarized what was known about the Lakes Entrance oil in the 1930s.

> The existence of free oil in glauconitic strata has been established. The quantity, though small, is appreciable. The oil has a specific gravity which permits it to rise slowly to the surface of the water, but, its high viscosity gives it a tendency to adhere to the sides of the bore until dislodged by drilling tools or water flushing. When drilling stops the sediment in

> the return water rapidly settles and acts as a filter, retaining the oil. Gas pressure exists in the glauconite, and probably in the sand layers.[116]

The exploration operations of the 1930s are best related by Keith Scarce who, although educated at the Mildura Agricultural College, soon followed in his father's footsteps and became a driller, rising to be a drilling supervisor with the contractor W.L. Sides and Son. However his first experience with the Victorian oil search was as a driller's offsider in Gippsland, where he began work for a company called Tanjil Oil No.1 Company in June 1930.

> They were using a steam-driven cable [percussion] rig and I was put on bailing oil out of the hole. The well, like others in the area, would start off at 100 gallons [450 L] or so a day, then taper off as the water cut increased over about a fortnight. Eventually it got too costly for men and bailer so they installed pumps. Because I was young and single I had the job of watching the pumps at night.
>
> The lighting system on the rig was several powerful kero lamps which would need recharging every evening. That was my job too. For a long time I couldn't understand why the meths I used to prime the lamps was always low. Then I caught the cook one day siphoning it off. He mixed it with milk and called the concoction a 'white lady'.
>
> We had two 12-hour shifts in those days and each shift was manned by the driller and his offsider. When the water got too strong we'd move the rig and drill again.
>
> The area was wild scrub then — all bracken and thicket. The company provided you with a one-man hut of galvanized iron, a single bed, small table, enamel plate and pannikin, a knife, fork and spoon, plus a wood stove. For the rest you were on your own. No fridges in those days and we weren't supplied with a Coolgardie safe or ice chest either.
>
> The huts themselves were about 10 feet [3 m] wide and pulled on a four-wheeled trailer by a Chevy truck. The married blokes would push two singles together if the wife was with them at the rig site.
>
> There was no such thing as safety clothing, helmet or boots and there were no safety regulations, although there were regular machinery inspections. You learnt the hard way like 'don't tighten a joint on a steam line when it is under pressure' and 'watch where you put a crow bar if close to the draw works'. I was pretty lucky and didn't get hurt, but if there was an accident the company would pay everything even if you were not insured.

> When we were working round Lakes Entrance a lot of the men were local, but if we moved west around Sale they came from further afield. A 48-hour week was standard. There was overtime, but you had to work the first eight hours of it before penalty rates applied, and then the maximum was time-and-a-half. There was also a shilling a day camping out allowance.[117]

Scarce recalls that Tanjil Oil was always running out of money. The drilling crew would work for a week or a fortnight, and then there would be a call from headquarters to stop. A little later another call would start them again. This was a regular occurrence.

> It was difficult because the firewood carter, petrol and other suppliers had to be paid and they weren't any better off than the rest of us.

Tanjil shareholders kept the company going in this stop-start fashion for a number of years, including a programme round Longford, south of Sale. One ploy was for the crew to burn flares at night and at holiday time to attract investors by giving the impression of success. But eventually the company was taken over by Lake Wellington Oil Company. By that time Scarce had become a driller and was employing a few techniques of his own to facilitate the work.

> There was always a problem with cable drilling in sand. The bit kept 'grabbing'. I'd read about drilling mud in an American magazine so I talked to the owner of a white pottery clay quarry over at Heyfield and organized to buy some. We mixed it up with water and made our own drilling mud. It worked OK for much of the time, but there was one snag. If there was a holdup in drilling for any reason the clay would settle out in the hole and be just as hard to drill through as the ground itself.

Most of the companies had geologists to direct the work, but some of them had fixed ideas that were more a hindrance that a help. Many associated oil with the coal seams and said that once the brown coal layers had been passed through there was no point in continuing. Scarce recalls one occasion when South Australian Oil Wells (a Dick Meudell company) drilled into granite and continued for a long time because the geologist said this was common in the US and sandstone was often found underneath. His erroneous assumption of similarity between Gippsland and the American overthrust belt cost the company dearly, but such comparisons with overseas oil fields were common in those times. The era was also ripe for the entry of oil 'diviners' and Scarce particularly remembers the appearance of Max Steinbuchel in 1938.

Calling himself an 'Oil Field Locater', he was a most imposing man, maybe six feet four inches [200 cm] tall, 21 stone [133 kg] and a little over 40 years old, who quickly obtained the interest of battling oil company directors and even government officials. He claimed major successes in the USA for his divining method as a sure-fire means of locating many large pools.

As far as I could see he used an old metal torch shell, a bit of wire and a marble. Listening to his patter and watching his act, my reaction was that he was a cheap crook. But he collected large sums from gullible people who thought their money was to be used to drill a test well in their area.

The Victorian Mines Department [like its South Australian equivalent] apparently went along with the method of raising funds because any well drilled provided further geological information. But eventually the government became uneasy about the lack of drilling action and probably advised him, for his own good, to quit the country — which he did, leaving quite a few debts unpaid. I was with Sides and Son then and we were drilling a well for Great Eastern Oil Fields — one of his companies (probably his only real attempt at exploration). We quit at 1 700 ft [2 735 m] when the prospects for payment looked dim.

Some time later news spread about that he had been killed in a shoot-out in San Francisco, but I don't know whether that was true or not.

Early settlers in Victoria had quickly judged Gippsland to be good farming land and so, by the 1930s, the tide of oil companies trying their luck in the region continually came in contact with the man on the land in the search for, and with access to, drilling sites. Keith Scarce, as company representative-on-the-spot, had to negotiate and often persuade the farmers to allow the rigs, vehicles and equipment through. Prior to 1940 the pegging of leases also needed consent of the landholder, and companies would agree to pay 'entry' fees for access or perhaps buy a small piece of freehold to cover the access and drill site. Scarce says he was sometimes authorized to offer royalties on production of up to 5 percent as further inducement for their cooperation. But quite often his familiarity and friendship with the locals was persuasion enough. Overall more than 20 companies and syndicates were promoted and floated on the chance of success in the Lakes Entrance and surrounding districts during the 1930s. Apart from those already mentioned some of the more prominent ones were Kalimna Oil, Midfield Oil, Mid West Oil, Lake View Oil, Texland Oil, Signal Hill Oil Exploration, Amalgamated Oil Syndicate, Oil Search, Cobden Lakes Entrance Oil Trust, Mac's Lakes Entrance Oil Wells, Valve Oil Wells and Austral Oil Drilling Syndicate.

Promotions and share hawking methods were many and varied. Rees Withers, an accountant and auditor for several of the early explorers and later the founder of Woodside (Lakes Entrance) Oil, recalls such activities as giving away small bottles of Gippsland oil at the Royal Melbourne Show with an advertising pamphlet seeking subscribers. It was also common to invite government heads and leading dignitaries to 'turn on the tap' of a Lakes Entrance well. The photograph of this event would be included in promotional brochures to lend credibility to the company concerned. Lakes Entrance, even then, was a holiday spot and it became a tourist attraction in the Club Hotel to 'light' water taps in the garden or the bathroom. Methane gas was always present in the artesian water wells which supplied the town.[118] Records show that drilling between 1930 and 1941 delineated an area of at least 21 km^2 within which oil occurred in glauconitic sandstone at a depth of around 366 m. The companies concerned reported either pumping or bailing a total of just over 486 000 L from wells in the district during this period, produced in the form of an emulsion which was then dehydrated to remove the artesian water. Close to 455 000 L of this heavy, dehydrated product was purchased by the firm Ramsay and Treganowan and then sold to the Melbourne Metropolitan Tramways Board for lubricating suburban trams.[119]

As the figures show, this production was quite minor and totally inadequate for supply to anything but a small local market. Yet Lakes Entrance was Australia's only oil-producing area (apart from oil shale in New South Wales) at the end of the 1930s so that, when the Second World War erupted, the Federal Government turned to the Gippsland town to see if production could be increased to alleviate the impending Australian shortage. Officials in Canberra were concerned that the country depended too heavily on the arrival of oil shipments from overseas — an increasingly uncertain and hazardous supply line in the early 1940s. Consequently, in July 1941 two American engineers, Leo Ranney and Charles Fairbank, visited Lakes Entrance at the invitation of the Australian Government. Apparently these men had experimented with a horizontal drilling technique to increase production from shallow oil fields in the United States, and Australia's Geological Adviser, Harold Raggatt, believed it might be applicable to the low-yielding glauconite reservoirs in Gippsland.

Ranney and Fairbank recommended that a vertical shaft be sunk to reach the oil sand, and then horizontal wells be drilled out from its base into the thin reservoir to allow oil seepage from as wide an area as possible back into containers for haulage to the surface. The scheme appealed to Raggatt and, in what has since been called one of his few misjudgements, he recommended it to the government. In 1942 a joint Oil Departmental Executive Committee was formed by the Federal and Victorian Governments to oversee the operation and a lease just outside Lakes Entrance township, held at the time by Austral Oil Drilling

Syndicate, was taken over under National Security Regulations. Finally, A.B. (Bert) Clarke, an experienced mining engineer who learnt his profession at the Ballarat School of Mines, was called in to plan and manage what turned out to be a venture unique in the world — the Ranneywells Project.

> I was living at Bendigo, having just volunteered for the army, when I was called to Melbourne for an interview with the Department of Supply and Shipping. They put the oil shaft proposal to me then and, as I understood it, the idea was that while ordinary bore holes would not get up much oil, a shaft with oil draining into it just might. It sounded OK. My services application was cancelled and I went back to Bendigo for three months to plan the operation.
>
> One problem for me at the time was that I was not familiar with the ground, although there was Mines Department well data from the area. Another was that I knew more about timber shafts than the concrete lining proposed, so I had to have a good concreter as part of the team. But they said I could have anyone and anything I wanted so I took three good miners from Bendigo with me to Lakes Entrance, hired some more blokes there and we got to work.
>
> First thing was setting up the engineering shop and facilities and then came the shaft. The iron poppet head was brought down from Bendigo. The ground was reasonably soft and it was fairly easy going with pneumatic spades. We didn't need any explosives. The practice was to dig out a section up to 12 ft [3.6 m] deep depending how the walls looked, and then put in form boards to pour the concrete lining. We'd leave holes in the sides to let water out while the concrete set and then plug them later.
>
> The shaft was 10 ft [3 m] internal diameter and we kept a drill hole going down in advance of the shaft sinking so we'd know what was coming. There were only two main water flows — one at 230 ft [70 m] and the other at 600 ft [180 m] — but the pumps coped with that OK till we got the lining in. The shaft had two cages and the winder engines were steam driven. We also had an electric winch.[120]

By June 1945 the shaft had reached a depth of just under 366 m, and it was close to the top of the glauconite reservoir. However fears that high pressure artesian water underneath the thin oil sand would break through the base of the shaft and endanger lives caused the project committee to halt operations while H.J. Cook, the government supervisor, went to America to investigate operations there. His report, delivered at the end of July that year, was far from rosy. Cook had found no successful application of the Ranney technique and a prominent US consultant oil engineer named Pemberton had concluded that the Lakes Entrance project would fail simply because there was not enough oil present to make it pay. In view of this pessimism the committee ran new estimates and indeed found that recoverable oil would only be worth between £20 000 and £60 000. Since £150 000 had already been spent on the project and more would be needed to complete the work before production could begin, the scheme was declared nonviable and the two governments handed the lease back to Austral Oil in May 1946.[121] No doubt they were aided in this decision by the fact that the war had ended and a resumption of regular fuel supply lines to Australia was in sight.

No one would have been surprised if Austral had let the lease and the project lapse, but one of the syndicate members, Charles S. Demaine, had other ideas. A consulting engineer himself, he was instrumental in reforming Austral into Lakes Oil Ltd and, as managing director, he pressed on with the Lakes Entrance scheme. Bert Clarke and his team of miners completed the shaft and widened the base into a work chamber about 6 m in diameter. Then, four 610 m horizontal holes were drilled, one in each quadrant, and put on test for about six months. It was found that most oil was coming from the two southern wells, so the next, more delicate, step was to drill shorter holes into the oil sand in that direction.

> It was difficult drilling as the oil sand was only 10 ft [3 m] thick. The drillers would go about 15 or 20 ft [5 m] at a time and then do a clinometer survey to check the angle and direction. If the auger went out of the sand they'd soon know and stop immediately, so some holes were 500 ft [150 m] long and others only 100 ft [30 m]. The reservoir rock itself was relatively strong and each hole was only three inches [76 mm] in diameter so there was no need to line them. All told we must have drilled 40 or 50.[122]

Once drilled, the holes were tapped with a valve, and a four-gallon can was attached. It was a tedious business because the flow from each had to be carefully monitored. Some cans filled during a shift. Others took two or three days. The full cans were hauled to the surface where water was removed chemically in a settling tank. Clean oil was then piped up to the main holding tanks ready for transfer to road transport and market.

The words 'unconventional' and 'slow' would probably best sum up the winning of oil in the Ranneywell project, but this did not seem to worry managing director Demaine.

> I look at it this way. From seeps and dribbles you get cups, then pints and gallons. They build to barrels, drums, then tons.[123]

Demaine's hope was that the shaft scheme would lead to similar developments over the rest of the 1.6 km^2 lease and eventually result in a production of around 32 000 L of oil a day. Alas, it was a dream never to come true. Lakes Oil spent over five years coaxing production into the shaft but, despite the managing director's optimism, the company was fighting a losing battle with economics. The operation was finally suspended in December 1951. The total tally of oil won was 785 000 L. It was of poor quality and valued at no more than £10 000 — a disappointing, if not unexpected, end to an ambitious (some would say foolhardy) scheme.

The post-war wave of exploration has bypassed Lakes Entrance, but there are still a number of reminders of Victoria's, and arguably Australia's, most sustained pre-1950s oil boom. The shaft can still be seen about one kilometre from Lakes Entrance town centre. Although blocked at the top with old car bodies, legacy of a more recent occupant of the land, it is still open for several hundred feet below. The winding house is there, but the winding gear and cable are long gone. A notice board with the shaft cage signal codes is the only reminder of its original purpose. The four concrete storage tanks remain, with their tapped pipe outflows at various levels; also the main sheds, some outhouses and a water tank. Beside these is the incongruous sight of a healthy-looking apricot tree annually bearing a bounteous crop of summer fruit. The surrounding fields, and even vacant lots within the town, contain a number of wellheads, some still oozing oil which enterprising farmers use to grease and lubricate their farm machinery. The vertical, capped pipes rising calf high are fitted with taps minus handles so there is no temptation for unauthorized people to turn them on.

Chapter 4

THE GOVERNMENT IMPETUS

Although a number of individuals and companies were in the field ostensibly looking for oil during the first half of the twentieth century, the private sector lacked credibility overall. Some of the companies were in earnest, if somewhat misguided in their ideas of geology. But others were outright frauds whose only intent was lining their pockets at the expense of a gullible public. The industry quickly developed a bad name and when schemes failed to find the promised bonanza, reactive comments like 'Australia is too old to have oil', and 'There is no oil south of the equator' gained popular credence. Certainly the almost complete lack of genuine surface seeps and the vast amount of featureless sand and alluvium, which hid rock strata from view, did not make matters easy for the handful of geologists working in the country during the first years of the 1900s. In any case, most of them were employed by the various State Government Surveys which had begun in the 1800s and concentrated on gold and the search for other hard rock minerals like copper, silver, lead and tin. A few others took positions at the fledgling universities, but they too focused on igneous geology, or perhaps coal if they looked at sediments, for most of the time.

Yet it is these men, along with the governments and institutions they represented, who provided the first real impetus to understand Australia's geology and to apply that knowledge to the vexed question of the country's petroleum potential. For most, the transition from 'hard rock thinking' was not easy. It was a long time, for instance, before they recognized that significant sedimentary basins existed and that structures with oil-trapping potential were present. Preconceived notions of what oil country should look like often clouded their judgments. Nor were they helped by the barrage of private get-rich-quick schemes and ill-considered bore locations they were asked to examine. Even Sir Edgeworth David, arguably Australia's greatest geologist, was unenthusiastic about the country's oil potential. This was partly due to his having been caught up by an oil promoter during the early 1900s and his subsequent wariness of oil 'cranks'. But, as experienced post-1950s geologist Reg Sprigg points out, David's otherwise quite remarkable

Geological Map of Australia is deficient in that the great internal basins were represented principally by the colour green to denote alluvium. This covered a lot of sins of omission and, thus, ignorance.[1]

Such was David's standing and skill as a teacher in all other aspects of geology that his work had wide influence and his ideas were passed on to following generations. South Australian's Sir Douglas Mawson (Professor of Geology at Adelaide University) and Keith Ward (Director of Mines for South Australia), both students of David, did not believe in the possibility of oil in their State. To be fair to all three men, the evidence at the time — coloured by the coorongite debate — was against petroleum discovery in the region. We now know that this was an erroneous view. On the other side of the coin, it is interesting to note that after a visit to Gippsland in eastern Victoria in the late 1920s, David was insistent that if oil were to be found in that part of Australia in commercial quantities, it would be offshore near Lakes Entrance.[2]

He was, by all accounts, a remarkable man.[3] Born Tannatt William Edgeworth David in 1858 in south Wales, his enthusiasm for geology was kindled at an early age. He studied under Professor Sir Joseph Prestwick at Oxford before leaving Britain for Australia in 1882 to take up the appointment of Assistant Geological Surveyor in New South Wales. Nine years later he became Professor of Geology and Physical Geography at the University of Sydney, and held that position until his retirement in 1924. Described as a frail looking man with keen, blue eyes and a delicate complexion, he had dramatic abilities that could hold an audience in rapt attention long past his allotted lecture time, and rarely needed notes. He often led student parties in the field, setting a strenuous agenda with no thought of his own health. Geology was his hobby and recreation as well as his work, meals being an irritating interruption. Walter George Woolnough, another of David's students who soon carved his own name on the Australian geological scene, told a typical anecdote.

> David, one evening, became absorbed in a piece of work and went ahead with it till his brain began to weary. Thinking it must be nearly midnight he prepared for bed. While doing so he heard a cab drive up to his door. Wondering who could be so late a visitor he peeped out. Day was breaking and the milkman had arrived.[4]

David was always eager to promote cooperation between scientists, particularly geologists, and in the summer of 1908 he and young Mawson joined Ernest Shackleton's attempted dash for the South Pole. During this expedition David led the first party to climb Mt Erebus and he was a member of the sledging expedition which discovered the South Magnetic Pole. During the First World War he was partly responsible for forming the Mining Corps, which he then joined at the age of fifty-seven. He spent the war years at the front lines displaying his usual pluck and physical vigour, alternately tunnelling under enemy lines to set mines and going down old wells to check rock formations, water levels and the type of aggregate the Germans were using in their concrete.

Most of his geological research was in the broad subjects of stratigraphy and glaciology but, when pressed, he did offer opinions on specific topics. On oil matters, for instance, he came down on the side of vegetable origin for coorongite, and this partly explained his lack of enthusiasm for petroleum prospects in South Australia. He also felt the oil exploration work in New Guinea lacked an ordered, scientific approach and he believed a Federal Geological Survey should have been in charge of the programme. The tendency to drill at haphazard places, he said, would lead to costly disappointments and the public would be likely to abandon in disgust the companies involved.[5] His knighthood was conferred in 1920, while his Geological Map of Australia and the first accompanying volume was published in 1932. He died in August 1934.

Although David's connections with the oil industry were fleeting and almost at arm's length, his work more than any other built on the start given by nineteenth-century Australian practitioners like the roving Reverends Clarke and Woods. Certainly there were a number of other excellent geologists at work around the country as members of the various State Government Surveys during this period. Men like Alfred Selwyn, Richard Daintree, H.Y.L. Brown, Logan Jack, Charles Wilkinson, Gibb Maitland, Charles Gould and W.H. Twelvetrees made important contributions in their regions and to the geological literature, while Professor Ralph Tate was an outstanding figure at the University of Adelaide.[6] But it was the efficient, energetic and enthusiastic Professor at Sydney University who lifted the science into the twentieth century, giving it the status of a credible profession in the process.

David's call during the 1920s for a Federal Geological Survey to oversee oil exploration work in New Guinea was in accord with a growing number of concerns expressed after the First World War about the state of Australia's mineral and oil well-being. In terms of petroleum, an article published in the *Times* of London in 1921 put the matter succinctly when it said that in 1910 Australia consumed 126 million L of petrol. In 1914–1915 the figure was 195 million L and in 1918–1919 it was 232 million L.

> For this supply of oil Australia has been almost entirely dependent on outside sources. The problem is there has been little success in attempts in Australia to detect 'free' petroleum.[7]

Melbourne entrepreneur George Dick Meudell struck a similar, if more dramatized, chord in his book *The Pleasant Career of a Spendthrift*. He pointed out that leading sailors and soldiers — Lord Jellicoe, Admiral

Henderson, Generals Birdwood and Fitzpatrick — all said Australia should find an oil supply of its own to be independent. He added that there was bound to be a Pacific war and Australia would be easy prey to a raider or invader if left without domestic supply.

> There is not a reserve of petrol in the Commonwealth which would last more than two months. There is no reserve of fuel at all for the use of the Australian Navy. We have no domestic supplies of lubricants, benzol, or engine kerosene. Our automobile industry is like a subverted pyramid, plenty of cars, trucks and tractors and no petrol to run them. It is farcical and would be laughable if it were not so damnably dangerous.
>
> ... Until a first-class Australian engineer, who understands petroleum mining, is given the right machinery with proper plans and maps, and endowed with enough money, it cannot be said there is no flow-oil in Australia. My experience has made me prejudiced against oil experts who are strangers and outsiders, and I don't trust them.[8]

In fact, there had been attempts to set up a national body as early as 1910 in response to requests from organizations like the Council for Scientific and Industrial Research and the Australasian Institute of Mining and Metallurgy. They had failed on three main grounds: there was a lack of appreciation of the magnitude of the problem of geologically surveying Australia; there was too narrow a conception of the importance of geological studies in mineral (and oil) discovery; and, not least, there was fear among some States that their own survey organizations would languish or be disbanded. But these early requests were made more on philosophical grounds as distinct from the post-war reasons of practical and economic necessity.[9] It is true that the Federal Government had been quick to appoint a government geologist in New Guinea (Evan Stanley in 1911) once the first oil indications were found. It also tried to become more technically involved in the region by sending English geologist Dr Arthur Wade to Papua in 1914 to carry out investigations on its behalf. Nine wells were drilled and the fossiliferous material was examined in Melbourne by Frederick Chapman of the National Museum. However, by 1919, the government in Canberra had reduced its input to a financial arrangement with the British Government and Anglo Persian to carry out further exploration work.

In Australia itself the earliest direct government involvements with the oil search came at individual State level. In Queensland the original Roma water bore was drilled with State Government money, and the State retained an area round the Hospital Hill discoveries for its own drilling programme until well into the 1920s. In South Australia the colonial government there had tried to stimulate the search for oil as far back as 1874 by providing a fund of £1 000. Parcels of £250 were to be awarded for the discovery of each 45 500 L of kerosene oil. As it turned out, the government's coffers were never in any danger of depletion from that promise.[10] Nevertheless, in 1914 the South Australians made a new bonus offer, this time a sum of £5 000

> ... for the person or body corporate which first obtains from a bore or well situated in the State of South Australia 100 000 gallons [454 600 L] of crude petroleum, containing not less than 90% of products obtainable by distillation.[11]

To claim this reward applicants had to fulfill a number of conditions including the supply of a monthly record of work done, a full log of all bores whether successful or not, and samples taken from every 15 m of hole drilled. Notification of a discovery had also to be made within 24 hours. The West Australian Government preferred a reward in kind by offering to grant a claim of 2.6 km^2 and two 0.2 km^2 blocks to the successful oil discoverer. In October 1920 the New South Wales Government offered £10 000 to any person or company who could produce 450 500 L of 'natural mineral oil' within the State and report it to the Mines Minister. Earlier that same year the Federal Government itself had become worried about the increasing import bill for petroleum products, and decided to offer a reward of £10 000 for the discovery of commercial oil fields anywhere in the country. But there were conditions attached.

> No payment of the reward shall be authorised unless it is proved to the satisfaction of the Minister that oil to the extent of 50 000 gallons [227 300 L] has been obtained, and that the bore is still flowing freely and producing oil in commercial quantities.[12]

The Gazette notice was signed by Prime Minister Billy Hughes. A few months later the incentive was increased to £50 000 but, after a five-year trial period without success, it was concluded that the bonus scheme was ineffective and the reward offer was withdrawn. The individual States carried on with their own schemes, Victoria in particular setting up Mines Department drilling programmes to explore the Gippsland region. In hindsight, the failure of financial enticements in the early 1920s was not surprising. The rewards were offered at a time when public perception of the oil industry was at a low ebb. The work of David and others, including those in the State Government Surveys, was drowned out by the shysters who promised a bonanza with every well no matter where it was located. Clear geological thought was also hampered by the quasi-technical outpourings of men like George Dick Meudell.

> The most likely spot to find oil on our big continent is in Central Australia where for ages the rivers draining Queensland

> and New South Wales have been depositing earthy sediments containing foramnifera, the true bacilli of petroleum, along the two ancient beaches made when a sea ran diagonally from the Gulf of Carpentaria to the Great Australian Bight.
>
> ... The Basin of the River Murray must contain more or less petroleum because from the beginning of time it has been draining an extensive basin of 10 000 square miles [26 000 km^2] and carrying the detritus and alluvium towards the sea.[13]

New South Wales Government Geologist E.C. Andrews was actually moved to defend his profession in his report on Australia published in 1924, although in doing so, he put forward a pessimistic view of the country's oil potential.

> The Australian public is told repeatedly that oil in abundance must, and does occur within Australia, but that the Australian geologist does not appear to be alive to his responsibilities. This statement must be taken as criticism of untrained irresponsibles of the 'get-rich-quick' type.
>
> In any case, the work of the prospector and the geologist indicates that oil, if found to exist in Australia, does not occur in the same manner as it does in countries such as America, Asia and the East Indies, for if such were the case, it would have been found easily[14]

Andrews concluded that some of the signs found to that time in the Kimberleys, around Roma, and in Tasmania were possible indications of oil, but they did not necessarily point to commercial accumulations. He ruled out the possibility of Australia having Tertiary-age oil fields and finished his report by advising the general public to keep away from Australian prospecting altogether.

> If we may judge by the experience of oil prospecting outside Australia, the value of the expectation of securing crude petroleum in commercial quantities within the area under consideration is small and is not to be considered as a business proposition by small investors.

After Andrews a succession of geologists made their assessments and comments about Australia's oil potential. Some were discouraging and others were just trying to put the Australian search in perspective. New York-based consultant Frederick Clapp belonged to the former category. Brought out by a private Australian group to assess the country, he confined his visit to Western Australia, although he did study all four basins in the State. His conclusion, published in 1926, could hardly have been more damning, saying that drilling for oil in Western Australia was a complete waste of money and should be discouraged.[15] Dale Condit, who also came to Australia at the request of private companies, was more charitable.

> If there be any criticism of efforts in the past, it is that... they have lacked sound geological guidance. For instance the accidental oil and gas discoveries in Queensland, though of great significance, may be said to have been unfortunate in some respects through having set up a clamour for quick results and the consequent intensive drilling activity.
>
> The bore locations could be little more than random shots, considering the featureless character of the country in which, as shown by subsequent experience, reliable structural results are obtainable only by the tedious and expensive process of scout boring.[16]

Dr H.I. Jensen of Queensland's own Geological Survey pointed out that the delay in finding commercial oil in that State was due to the fact that there were no seeps to mark the way. His view was that oil would be found in sealed basins with the probable oil formations only outcropping in the highlands where seeps would be unlikely because of the elevation and consequent gravitational forces. On the western margins oil formations were covered with a thick mantle of young sandstone and soil.[17] R.H. Cambage, in his presidential address to the Royal Society of New South Wales in 1924, tackled the problem of oil generation.

> It would seem the question which has to be solved is whether the amount of animal and vegetable remains quite apart from coal and shale, in rocks of Australia, is sufficient to have supplied mineral oil in anything approaching commercial quantities, and if so, whether such oil has been retained. Much of the evidence so far obtained during the somewhat limited investigation which has been made is not yet convincing, though the possibility of success still remains, and the question can only be answered by further careful geological survey and boring.[18]

However none of this public debate and learned discussion in the literature was promoting oil exploration, and there seemed little likelihood of anyone coming forward to claim the Federal Government's reward. In an effort to clarify its own idea of the country's prospects, the Nationalist-Country Coalition Government of Prime Minister Stanley Melbourne Bruce, which had taken over from the Hughes administration in 1923, recalled English consultant Arthur Wade to make a thorough examination of the likely petroleum areas. This he did, visiting the Kimberleys in 1924 and five areas of Queensland plus New South Wales in 1925.

Wade himself was an interesting character who had a distinguished geological career. Born in Yorkshire in 1878, he graduated from London University with a Bachelor of Science in the first years of the twentieth century, and was then granted a Doctor of Science degree in 1911. His first involvement with the Australian Government came in 1913 via a nomination from its British counterpart for a study of the prospects in Papua. Consulting jobs in Australia followed, for State and Federal Government and private clients, interspersed with numerous and varied overseas contracts in places like Algeria, Turkey, Arabia, Spain, Portugal, Poland, Greece, Albania, USA, Venezuela, Columbia, Trinidad, Sumatra, Borneo, Timor and Madagascar. In the tumultuous preamble to the First World War he ran blockades, tangled with various rebellious factions in the Middle East during the break-up of the Turkish Empire and was held captive more than once. During the Second World War he was an officer of Allied Intelligence. Though notably equable in temperament for much of the time, he could become the typical dour Yorkshireman when occasion demanded — blunt and tenacious. In his student days he had apparently shown some skill as an amateur boxer. He was widely read and, surprising to some, his spare moments were given to poetry. He himself was the author of a volume called *Vagabond Verse* which was first published in 1917 and went into a third edition 30 years later. Nevertheless he was essentially an outdoor man, happier in the field than in the office. He died while surfing on the Queensland coast in 1951, having spent the last years of his life based in Brisbane.[19]

Wade's findings for the Federal Government in the mid-1920s were optimistic, but not wildly so. He believed the Kimberleys held promise although, strangely, he threw some doubt on the presence of a suitable seal for the potential reservoir rocks. In Queensland he gave preference to the Bowen Basin region north of Roma, while in New South Wales his choice was the Sydney Basin although he believed it would be more gas than oil prone.[20] Taking these reports into consideration the Bruce Government felt that some sort of oil exploration incentive was still justified and it introduced the Petroleum Prospecting Act in 1926. The legislation provided £60 000 as an advance for anyone or any company searching for oil in Australia. The money was given as a pound for pound subsidy on the drilling of deep wells although, naturally enough, the location had to have prior government approval. The Act was extended to Papua and New Guinea the following year along with an increase in subsidy to £160 000. An additional sum of of £50 000 was approved in 1928.[21]

Perhaps more important than the actual incentive to industry, the 1926 Act sowed the first seeds for the formation of the long-awaited Federal Government geological organization — a body that was eventually to become the Bureau of Mineral Resources. The early moves were modest, but they did bring to national prominence the charismatic former student of Sir Edgeworth David — Dr George Walter Woolnough. In 1927

Woolnough was appointed Commonwealth Geological Adviser, a position which required him to guide the government in oil matters, including the arrangement and award of the drilling subsidies. The position also covered the mineral industry, although in the initial years this was seen as a less urgent task. Other appointments were Frederick Chapman, the National Museum in Melbourne's paleontologist, to Commonwealth Paleontologist, while one of the first women on the Australian geological scene, Irene Crespin, won the post of Assistant Paleontologist. The latter two scientists were to make an important contribution to the knowledge of Australian sedimentary geology through the study of fossils taken from bore hole samples required under the Petroleum Prospecting Act.

The organization became known as the Geological Branch of the Department of the Interior. It was destined to remain a small group for some years and for much of that time it was dominated by the boundless energy and bouncing personality of George Woolnough. Most published photos of this man depict him later in life and, from descriptions of his deeds, it is hard to visualize him otherwise. The pictures show a bald head, neatly trimmed moustache and white beard pointed at the chin, with bright blue eyes sparkling out of a cherubic face, and a neck disappearing into a high starched collar. A former acting head of the Geology Department at Sydney University and the first Professor of Geology at the University of Western Australia, this leprechaun of a man was a brilliant scholar. He was awarded the first Doctor of Science degree bestowed by Sydney University, was an excellent mathematician and a linguist competent in 12 languages. He was also an experienced bushman and during his career he travelled widely in the outback, equally at home on bike, car, horse, camel or on foot. He apparently once walked 1 600 km in seven weeks and even drove trains when the engineer made room for him in the cab. He was an enthusiast in all things and had a sense of humour to match, although his tendency to bounce from one thing to the next earned him a reputation of trying to show how much he knew, or thought he knew, and of trying to be a one-man band. Nor was he shy of publicity and this trait did grate on others from time to time.

Woolnough tackled his new position with gusto and was soon putting his views forward. He quickly saw that the subsidy scheme was deficient because, while it encouraged drilling, the work was being carried out in regions where little was known of the geology and where locations were often poorly chosen. There was also a tendency to use the government assistance as a 'stalking horse' in approaching the public for market flotation purposes. Late in the decade the policy was changed to one which provided the subsidy for surveying and geological investigations made prior to the choice of drilling location.[22] The revised subsidy was set at one-third of the expenditure incurred in the preliminary work. But before the regulations could be amended, the available funds had been used up, the Bruce Government had been swept from power by the Labor Party machine and J.H. Scullin had become Prime Minister. Financially they were difficult times. On the edge of the Depression years the new administration did not want to spend more funds on the oil search, but it did decide to send Woolnough to the US to seek out ways of effectively stimulating Australian exploration. With exploration companies throughout Australia desperately trying to raise funds, the change from direct drilling subsidy to one based on preliminary surveys prior to drilling caused heated debate, and much invective was directed at the government geologists at both state and federal level.

> The breasts of these gentlemen glowed with contempt for the 'blind stabber' — that is, the small driller working independently of the big companies with their geological staffs. This contempt is scarcely justified... blind stabbing had found nearly every important field in the United States... It certainly found all the promising fields in Australia... one wonders whether they [the government geological experts] desired to find oil or well-paid geological jobs!... [Professor Woolnough] had to be sent to America at his country's expense, to learn the business on which he was supposed to be an expert... a few score 'blind stabbers' scattered over the country would probably have been even more effective.[23]

Woolnough spent most of 1930 in the United States and Argentina studying the American oil industry. His subsequent report did not offer much in the way of specific incentives for the home scene, although he suggested that help from overseas would not go amiss. This was a controversial comment at the time because feeling was running high in Australia that foreign companies were deliberately sabotaging wells. They were also suspected of the more subtle act of pretending to set up a vigorous search and then contriving a failure that would downgrade the region and discourage further work. The rationale for these suspicions was that foreign companies wanted to retain the market for petroleum products in Australia and a local discovery would spoil this dominance. Woolnough's report dismissed such fears.

> The competition in America is so intense that the US blockage of oil search in Australia would have to be 100 percent perfect before it could be effective at all. It is considered absurdly improbable that such a condition could exist.
>
> Given intelligent, active, honest and up-to-date methods, the prospects for Australia are good. The introduction of overseas capital and technique under stringent safeguards is desirable.[24]

Although not overly optimistic of the outcome, Woolnough considered the most promising places for oil discovery in Australia to be in the Carnarvon Basin of Western Australia and the Hunter Valley-Gunnedah region of northeastern New South Wales. He also suggested that work be carried out in the marine rocks of the Great Artesian Basin of western Queensland and that 'buried ridge' structures be looked for in the Gippsland Basin of eastern Victoria. In addition to his liaison with the government, the Commonwealth Geological Adviser saw his job as one of educating the public about oil and trying to guide any local company or individual who might want to join the search. To this end he wrote a number of articles for papers and magazines in the 1930s explaining the industry in layman's terms. Probably his most notable effort was the contribution of a chapter on oil in the fourth edition of the popular book *Prospecting For Gold, Other Minerals and Precious Stones* at the invitation of its best-selling author, Ion Idriess.[25]

In these 20 pages (which are just as relevant today) Woolnough attempted to correct some misconceptions about oil occurrence and then explained clearly the types of oil-bearing formations. He pointed out that, whereas gold prospecting was often the province of the individual needing little more than strong arms and common sense, oil exploration was more the domain of the large companies because of the sophisticated equipment and the geological training needed. However he went on to outline the methods used to collect uncontaminated samples of any surface evidence of oil that might be found by the prospector, and concluded by saying that government officials would not unnecessarily discourage anyone who had real indications to submit. But undoubtedly Woolnough's most outstanding contribution to exploration was his pioneering work in the technique of aerial photography. His overseas trip heightened an interest he had in the use of aircraft for general mapping and geological reconnaissance and, on his return, he began experimenting with technique. In Woolnough's obituary, written in 1959, Harold Raggatt includes an amusing snapshot of the man and his experiments as seen by a colleague of the early 1930s.

> A picture emerges of a somewhat elderly geologist, camera in hand and safety belt loosened, leaning far out of the Gipsy Moth in uncomfortable proximity to the propeller in order to secure the best results.[26]

His first trip was in a modified single-seater where the photographer lay full length on the floor and Woolnough directed operations seated next to him on a drum of a well-known brand of petrol. The intrepid geological 'aviator' was himself tickled by the humorous side. 'For months after, every time I sat down I could feel the word "Plume" and the imprint of two feathers.'[27]

Following the success of this initial trip he flew long distances over northern, central and western Australia using two RAAF Wappity aircraft and logging over 1 000 hours in the air. The technique of air photography was pioneered during the First World War when maps were scarce, and by 1918 the Allies had taken over 215 000 negatives and made 10 million prints of the Western Front alone. Apart from their use in cartography and for confirming the success or otherwise of bombing raids, it was found the photos could also pick out camouflaged installations where turf from another meadow was used as the cover. This latter feature sparked the beginning of their use in peacetime geological and archeological work.[28] However the art of air survey was not so much in the taking of the photos, but rather, in flying the aircraft straight and level. Lack of instrument aids hindered the early work. Yet even when this was remedied in the more modern planes of the 1920s and 1930s, pilot skill was still a big factor in the success of the operation. As Woolnough himself remarked, what passed for straight and level flying in ordinary commercial aerial operations was quite inadequate for the special purposes of air survey where tilt, drift, tail winds, head winds and roll all had to be taken into account. He added that it was not straightforward for the geological operator either because shadows, mists and the strangeness of flying all gave wrong impressions until read properly.

> ... it takes something like 100 hours flying before one can sufficiently disassociate from the act of flying to concentrate on systematic scientific observation.[29]

Woolnough was also quick to point out that air photography should not absolve the geologist from ground survey. There was no substitute for investigation on the spot and the surveys needed to have accurate ground control points to be of any use. Studying the photos themselves called for special skills too because oblique angle shots could mask features or cause misleading assessments. He cautioned that the whole operation was a costly exercise because of the need for a great deal of aeroplane time to produce the photo overlap needed for an accurate mosaic. Nevertheless, when employed properly, Woolnough was convinced aerial photography would be a boon for Australian explorers and he set out to prove it. His 'Report on Aerial Operations in Australia' during 1932, particularly over northwest Australia, illustrated the results that could be obtained.

> Striking as are the distinguishing characters of the individual horizons which build up the hill masses when viewed from the ground, these distinctions become even more remarkable when viewed from the air. Poole Range Beds and Grant Range Beds [sedimentary strata in the Canning Basin of Western Australia] are recognisable at a glance even by the non-geologist, and can be delineated by photography with admirable clearness.

> The folding and faulting which have affected these rocks is sharply defined, and can be detected even at great distances. In addition there are other effects, obscure and unrecognisable at close quarters, which stand out with remarkable clearness when the ensemble effect gained from the bird's eye aspect is obtained.
>
> Thus individual marker beds, the continuity of which is not apparent when one attempts to follow them on the ground, stand out like ruled lines. Scarps and dip slopes... are seen in their entirety, and are not lost in the confusion of detail caused by erosion and the accumulation of detritus, as is the case when they are viewed... from ground level.
>
> The remarkable and complex jointing which is characteristic of the Grant Range Beds, which makes areas occupied by these beds almost impassable even on foot, and which entirely dominates all indications of structure, falls back into its proper perspective, and becomes a striking criterion by which to distinguish this horizon from others.[30]

It was during this 1932 survey that Woolnough identified two large domal structures in the Exmouth Gulf region which he enthusiastically believed were deserving of 'the most searching investigation'. Harold Raggatt, then a junior geologist, was to remember this recommendation and later directed William Walkley of Ampol to the area where the two domes were identified as the Cape Range and Giralia Anticlines.

Air photography and ground geological work were not the only initiatives carried out by the Federal Government during the late 1920s and early 1930s. In June 1928 the Geophysical Survey Act was passed by parliament to give birth to a proposal for running geophysical surveys in Australia for oil, minerals and also water. The idea came from an Imperial Conference decision taken in London after the First World War that experiments in geophysical surveying should be carried out in some part of the Empire as it then stood. Provision was made for the work to be financed by the Empire Marketing Board and the Australian Government on a pound for pound basis (£16 000 each). Geophysical prospecting methods for mineral deposits using electrical techniques had been developed in Europe during the years prior to the First World War. During the war itself geophysics was used successfully in Germany in the search for new oil deposits. Gravity torsion balances indicated the presence of buried salt domes and hence led explorers to the oil reservoirs associated with them. At about the same time South Australian Government Astronomer G.F. Dodwell recorded magnetic anomalies during a geomagnetic survey in the Musgrave Ranges in the northwest of the State, making this the first recorded use of applied geophysics in Australia.

Immediately after the war, advances in application and technique were made in Europe and America although they were mostly in the mining industry. Aware of the good results, two prominent scientists, Sir Edgeworth David and H.W. (later Sir Herbert) Gepp, advocated its use as a search tool to revive flagging Australian mineral exploration during the early to mid-1920s. Gepp's influence in particular was instrumental in Australia being selected as the location for the Imperial field tests. The investigation was known as the Imperial Geophysical Experimental Survey (IGES) and a major condition of participation was that detailed descriptions of methods and equipment used would be published along with the results obtained. Initially a number of private geophysical contractors made offers of cooperation, but such was the competition in the fledgling industry that the requirement to publish details of their procedures and equipment was not acceptable and they withdrew.

An English geophysicist who had worked in Spain, Portugal and Africa, Professor A. Broughton Edge, was appointed Director of the IGES, while the Assistant Professor of Physics at McGill University in Montreal, Canada, Dr E.S. Beiler, was appointed to the deputy position. Unfortunately Beiler died after contracting pneumonia in Western Australia during 1929 and his work was completed by Australian geophysicist J.M. (Jack) Rayner, who was seconded to the survey from the New South Wales Geological Survey. Rayner later went on to play an important early role in the Bureau of Mineral Resources. During the course of its 18-month charter the IGES worked in all States and experimented with a number of geophysical tools and methods. Much of its work was in the mineral industry, but a notable oil application was run in the Lakes Entrance region of southeastern Victoria. This was in the form of a gravity survey to try to locate a supposed granite ridge which might have trapped oil in the sediments along its flanks or over its crest.

Although basically a test programme not intended to specifically search for new mineral or oil deposits, the IGES is recognized as having laid the foundations of applied geophysics in Australia. It was not long after completion and publication of the survey report (1931) that private companies began to look more closely at the potential role of this prospecting method in the country. Some surveys were even run in the Roma area as early as 1928 and 1929 while the IGES was still in existence. The applications gained more prominence, however, in the early 1930s, particularly via the work of Rayner, who was again seconded from the New South Wales Geological Survey, this time to run magnetometer surveys for Oil Search Ltd in South Australia and Victoria.[31]

In 1934, heartened by the success of the IGES, the Federal Government (by this time led by the United Australia Party Prime Minister, J.S. Lyons, who had replaced Labor leader Scullin in 1932) came to an agreement with the Western Australian and Queensland State Governments which resulted in the formation of the Northern Australia Survey. As before, it was Sir Herbert Gepp, as development consultant to the Federal Government, who influenced events. He conceived the idea to stimulate the interest of Australian mining companies in the less known and less accessible parts of the country north of the 22nd parallel. The Northern Australian Survey Act, passed through parliament in December 1934, was significant in that it was the first time there had been cooperation and pooling of the resources of State and Federal Governments in the search for mineral deposits.

The arrangement called for the Federal Government to provide £75 000 and the two States £37 500 each to finance the survey for a three-year period. This was later extended to five years although Western Australia dropped out before the final year was completed. Jack Rayner was appointed as Geophysical Consultant and the first party assembled at field headquarters in Cloncurry, western Queensland, during May 1935 under geologist P.B. (Percival) Nye, seconded from the Tasmanian Geological Survey, and geophysicist R.F. (Robert) Thyer. Although exclusively aimed at mineral exploration, particularly gold, the survey embraced aerial photography (with the help of the RAAF) and ground geological parties as well as the fast-developing procedures in geophysics. As such there was liaison with the work of Woolnough, and the considerable number of maps and reports produced helped to increase the overall knowledge of Australia's geology and mineral/oil potential.[32] The work was not without its drama. On several occasions forced aircraft landings in isolated areas put party members in some danger, as noted by Irene Crespin in her memoirs.

> Dr Woolnough, when flying in the Northern Territory with medical doctor Clyde Fenton, made a forced landing near Victoria River. Being without a wireless (apparently damaged) he had to walk to Victoria River Downs Station homestead for assistance.
>
> Geologist Nye was involved with two forced landings during the Northern Australian Survey. One was near the Granites north of Alice Springs in June 1937 during which he was lost for 10 days. Water was collected through condensation in apparatus devised by the party on the plane and using two-gallon petrol tins as a boiler.
>
> The second time was in 1938, although only 20 miles [32 km] from Alice Springs. However wireless contact was made with Oodnadatta and rescue cars were sent out from Alice Springs.[33]

Other flights turned out to be more historic than dangerous. Crespin noted that a flight taken by Woolnough, Nye and the heads of the Western Australian and Queensland Government Surveys at the time (F.G. Forman and L.C. Ball) was made in the *Southern Cross*. It was the last long flight of the famous aircraft, its next and final journey being the unsuccessful expedition piloted by Charles Kingsford Smith in an attempt to cross the Tasman Sea to New Zealand. Field work for the Northern

Australia Survey ceased in 1940 and the organization was disbanded in 1942. But by that time the administration of petroleum exploration in the country had taken a new turn.

During the mid-1930s exploration round the country had sunk to a low ebb. Although the Federal Government's subsidy scheme had been geared to preliminary survey work since the beginning of the decade, funds were provided to help individual bona fide drilling programmes, such as the Freney Kimberley work in the north of Western Australia and Oil Search in a number of States. But there were no cries of 'favouritism' because few companies were in the field during these years except in the Gippsland area of Victoria. In that region the Victorian State Government itself helped companies as much as it could, both financially and technically. With equal financial participation from the Federal Government it also ran its own drilling programmes, many of which were aimed purely at obtaining stratigraphic information to help piece the State's geology together.

But clearly there was not enough being done and, as the decade advanced, the concerns about defence and strategic supplies began to emerge once more. Once again, at the instigation of Sir Herbert Gepp, the Lyons Government in Canberra introduced the Petroleum Oil Search Act in May 1936 and parliament passed the Bill later the same year. Although Gepp had called for a much larger sum, the Act appropriated £250 000 to encourage drilling operations. Particular attention was to be paid to the supply of suitable drilling machinery, and part of the money was earmarked for advances by way of loans to approved companies or persons on a pound for pound basis. It was a return to the 1926 system, and to deal with applications under the scheme the government set up a three-man Oil Advisory Committee staffed by the three prominent 'W's of Australian geology at the time: Dr Woolnough as Chairman; Dr Keith Ward, the South Australian Director of Mines who still disbelieved in the likelihood of finding oil in his own State; and Dr Arthur Wade, the British consultant who had already worked extensively for the Federal Government and State Governments throughout Papua and New Guinea and Australia.

The Committee did help to improve technical standards by insisting that detailed geological mapping and scout drilling were carried out as a necessary precursor to drilling, and that competent drillers were employed to do the work. Nevertheless, the response was still slow, Oil Search Ltd and its subsidiaries being virtually the only companies to seriously attempt exploration on a large scale. Towards the end of the decade an American driller named Earle Koyden was engaged to supervise the drilling machinery purchased and a specially equipped truck was obtained to transport this equipment. Most of it went to Queensland, although the Freney Kimberley Oil company spudded the Nerrima 1 well in Western Australia during 1939 using a government provided rig.[34]

Apart from the work of Woolnough in particular, but also of Wade and Ward, plus Chapman and Crespin providing the paleontological support, the main thrust of Australian geological advancement right through the 1930s was provided by the Geological Surveys and Mines Departments in each State. Hard rock geology and mining was still a principal activity, but there was also an increasing amount of work done in sedimentary mapping and stratigraphic correlation. Some of it, like Lockhart Jack's surveys of the large anticlines in the northeast of South Australia, went unrecognized for a number of years. Other efforts, like W. Baragwanath's support for oil exploration in Victoria, were widely acknowledged at the time. A number of other names were prominent during the period — I.H.C. Croll in Victoria, Leo Jones in New South Wales, L.C. Ball in Queensland, P.B. Nye in Tasmania and Frank Forman in Western Australia. But the man destined to make the greatest impact was Harold Raggatt. He had finished a First World War-interrupted geology degree at the University of Sydney in 1922 and immediately joined the staff of the New South Wales Geological Survey. A modest man of infinite good humour and patience combined with a great geological talent, he was often likened to Professor Sir Edgeworth David.[35]

Although he worked in several short secondments elsewhere, Raggatt stayed with the New South Wales Survey until 1940. In April that year he was appointed Assistant Commonwealth Geological Adviser. Six months later he was appointed to the top job when Dr Woolnough and Dr Ward retired and the Oil Advisory Committee was disbanded. By that time the emergency measures of the Second World War were in force and, asked by the Federal Government to recommend likely sources of indigenous oil supply, Raggatt recommended the Ranney scheme, which resulted in the Lakes Entrance oil shaft project. In addition, the Federal Government and the State of Victoria combined to drill a well close to the Victoria-South Australia border called Nelson's Bore. Both projects were failures — the shaft because oil could not be recovered at an economic rate, and the bore because no oil was found at all despite the great thickness of sediment penetrated.

Putting these disappointments aside, during the war years Raggatt also saw the need for a comprehensive minerals inventory of Australia. He gathered together a group of geologists and geophysicists, most of whom had worked in the recently disbanded Northern Australia Survey, and others who had been withdrawn from the Government Geological Services in Papua and New Guinea during the Japanese invasion. The group became known as the Mineral Resources Survey, with Raggatt as Director and P.B. Nye as Assistant Director. Towards the end of the war (September 1944) the Mining Industry Advisory Panel was formed, drawing its membership from mining employers, trade unions and the Australasian Institute of Mining and Metallurgy, with Raggatt as Chairman. Its function was to advise on Australia's post-war mining policy.

Events began to move quickly. A month later the Aus.I.M.M sent a pamphlet to Prime Minister John Curtin renewing its request of several decades earlier to establish a Federal Geological Survey. Then, in 1945, State and Territory Mines Directors were added to the membership list and the enlarged panel recommended that Raggatt and geophysicist Jack Raynor visit North America to study the organization of the US and Canadian Geological Surveys. Finally, in May 1946, the Chifley Labor Government created the Bureau of Mineral Resources, Geology and Geophysics — the BMR as it became known — and Harold Raggatt was appointed its foundation Director.[36] The stage was set for the post-war oil and minerals revival. Certainly the sceptics were still vocal, particularly those against the likelihood of commercial oil. However there was no denying the vital contribution that government scientists had made, under sometimes difficult conditions in the field and in the laboratory, to the advancement of knowledge of the country's geology and resources potential during the first half of the twentieth century. The state of pre-1950s oil exploration in Australia was best described by Victorian geologist I.H.C. Croll.

> [oil exploration]... has called every branch of science and engineering to its aid, and is still thinking for new and more modern methods and laboratory saving devices. Methods of exploration have had to keep pace with the advancement, and have risen from mere 'wildcatting' to highly technical operations in which the surveyor, airman, geologist, engineer, physicist, chemist and driller all play important parts.
>
> One feature has not been superseded, and that is the necessity for a good topographical and geological map of the area.[37]

Although written in 1938, eight years before the formation of the BMR, the sentiments could not have been more aptly expressed as Raggatt began to build on the foundations he and a long line of predecessors had helped to lay.

PART TWO

THE 1950s

A BULL BY THE HORNS

Chapter 5
BMR LEADS THE WAY

The Bureau of Mineral Resources was founded on two main aims — to promote the search for oil in Australia, and to assess the nation's mineral wealth. Harold Raggatt, the BMR's first director, felt that the best way to bring these rather lofty ideals to earth was for the new organization to take the lead and provide technical data of a standard that would encourage the private sector to follow. In practical terms that meant setting up detailed geological and geophysical surveys over a wide area of unknown or poorly explored country, and it called for great skills in coordination. Raggatt was the man for the job. In the course of his earlier work with the Geological Survey of New South Wales he had carried out a number of field assignments which included systematic geological mapping in several parts of that State. The maps he produced were of a high standard and Raggatt quickly developed a love of field work along with an appreciation of the importance of systematic surveys. It was an attitude which greatly influenced his approach to resources development on a national scale.[1]

A modest man, noted for his good humour and patience, he was also a person of great integrity with a deep sense of the national interest. Raggatt was convinced that, despite all the opinions to the contrary which persisted into the 1950s, Australia was prospective for oil. His belief was based on the fact that there was a large volume of marine sediments onshore Australia: the Great Artesian Basin; the Bowen Basin of Queensland where Shell had carried out extensive work in the 1940s; the Sydney Basin where he had done work of his own in previous years; and the massive limestone reefs outcropping in the Kimberleys of Western Australia. He also remembered the huge structures in the Carnarvon Basin discernible in the pioneering air photos of Dr George Woolnough.

In 1947 the task in front of him was to prove his belief, or at least set in motion the initial work towards a proof. The Chifley Labor Cabinet had already given approval for the appointment of a staff of around 55 technologists including 17 geologists, 24 geophysicists, a petroleum technologist and a mining engineer. The initial problem facing Raggatt was finding people to fill the positions. Some were still available after the break-up of the Northern Australia Survey and the subsequent Mineral Resources Survey, but in the immediate post-war period there were few newly trained geologists and geophysicists. Undaunted, he organized the Bureau into three main divisions and carefully chose his principle officers: P.B. (Percival) Nye became Deputy Director; Norman Fisher was appointed Chief Geologist; J.M. (Jack) Rayner was Chief Geophysicist; Harry Temple-Watts was Petroleum Technologist; while H.J. Cook became Mining Engineer.[2] Nye and Rayner had been involved with the Northern Australia Survey, Temple-Watts and Cook had overseen the government's part in the Lakes Entrance oil shaft project during the war, and Fisher had been Government Geologist in Papua and New Guinea. Mineral economist J.A. Dunn was added to the team shortly afterwards.

One other key member of the early BMR should not be forgotten. Dr Irene Crespin, the Melbourne-born geologist who had been Assistant Commonwealth Paleontologist under Frederick Chapman during the 1920s and 1930s, moved to the top position when Chapman retired in 1936. She and Raggatt formed a close working relationship when he became Government Geological Adviser and they made a number of field trips together. Her work with fossil assemblages, particularly microscopic foramnifera, was instrumental in identifying and correlating rock formations in many of the sedimentary basins where oil exploration was being carried out. Always enthusiastic and dedicated, Crespin was the

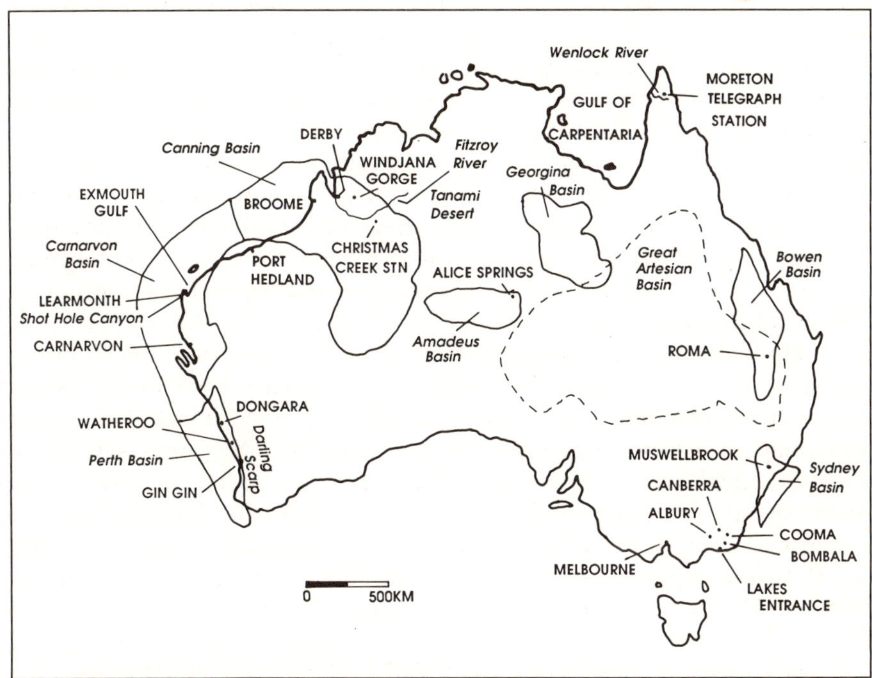

Location of key places mentioned in text.

only woman to descend the Lakes Entrance oil shaft during its construction, an experience she related many times in later years.

> I visited the site every four weeks and usually descended in a kibble [a large jar-shaped bucket] which I had difficulty clambering in and out of. However the shaft provided a unique opportunity to study the Tertiary sediments of the area because there was no outcrop anywhere [on the surface].[3]

The route from Canberra to Lakes Entrance was 450 km of very poor unsealed road via Cooma, and Crespin often had to contend with mud and fog in winter and sometimes bushfires in summer. On one later trip with Raggatt she drove via Bombala in an old Chevrolet truck, a relic of the Northern Australia Survey affectionately known as 'Leaping Lena'.

> We stayed the night at Bombala and it was very cold. Next morning ink in the hotel register inkwell was frozen and icicles were hanging from the trees alongside the highway.

Gippsland was plagued with mosquitoes, even in the cooler months, and Crespin took to wrapping newspaper round her legs for protection. She worked as often as possible in sheds where the drillers had fires burning. If anything, her contributions to the knowledge of Australia's sedimentary geology intensified when the BMR was formed because the survey parties sent a steady stream of fossiliferous material in from the field. Crespin continued to make her own excursions to the various operations, although most of the analytical work was done in Canberra. Her output was prolific and she authored or co-authored 110 written papers during her working life. Never content with her state of knowledge Crespin went on to gain a Doctor of Science from Melbourne University in 1960 at the age of 64, and was honoured with an OBE in 1969. Probably the best summation of her attitude to geology and the quest for knowledge was contained on a small plaque on her desk: 'Label today — Tomorrow you will have forgotten'.[4]

Having assembled the basic BMR team, the next step was to gather together all the equipment that would be needed in the field, including a fleet of vehicles. The only ones available at the time were army disposal jeeps and in the immediate post-war days of 1946 they were parked in their hundreds at Bonegilla, near Albury on the Victoria–New South Wales border. The Federal Government's Department of Supply had the job of disposal and, as the BMR came under the control of this department, the fledgling organization had the inside running on choice of vehicles. Norman Fisher remembers assembling the first BMR fleet.

> I collected as many people in the BMR who had driver's licences as I could and we went down to Albury to pick some out.

Admiral John Lort Stokes. As a former lieutenant on the British Sloop HMS *Beagle*, he made the first recorded discovery of petroleum in Australia in 1839 on the Victoria River now in Northern Territory. Courtesy La Trobe Collection, State Library of Victoria.

Reverend William Branwhite Clarke, the 'Father of Australian Geology'. Courtesy Geological & Mining Museum, Department of Mineral Resources, NSW.

Reverend Julian Tenison Woods, horse-riding geologist, naturalist and clergyman who won acclaim for his treatises, particularly on the geology of southeastern South Australia when based in Penola. Courtesy Department of Mines and Energy, South Australia.

H.Y.L. (Henry) Brown, Government Geologist in Western Australia and South Australia. Courtesy Department of Mines and Energy, South Australia.

Joseph Carne, Assistant Government Geologist in NSW around 1900. Noted chronicler of oil shale and petroleum industry in Australia and Papua New Guinea. Courtesy Geological & Mining Museum, Department of Mineral Resources, NSW.

Below: William Fell, a member of the Fell family company which took over shale oil production at Newnes, NSW, from Commonwealth Oil Corporation in 1914. The Fells also owned a shale oil plant in Sydney which became Shell's Clyde oil refinery in 1927. Courtesy AIP Library, Melbourne.

Dr Herbert Basedow, geologist, botanist, anthropologist, doctor of medicine and politician who championed the petroleum-origin theories for coorongite pre-1900 before reverting to an algal-origin theory post-1900. Courtesy AIP Library, Melbourne.

All that remains of the Alfred Flat bore drilled to 8 m depth by Coorong Kerosene Company in 1866 in the Coorong district of southeast South Australia. Courtesy South Australian Department of Mines and Energy.

The Joadja shale oil plant near Mittagong, NSW, about 1903. Courtesy AIP Library, Melbourne.

Battery of retorts under construction in 1938 for the shale oil works of National Oil Pty Ltd at Glen Davis, NSW. Courtesy AIP Library, Melbourne.

The Okes-Durack bore drilling during the early 1920s near the junction of the Ord and Negri Rivers in northeast Western Australia. Courtesy Geological Survey of Western Australia.

The Freney Kimberley Oil Company rig drilling at Mt Wynne in the Kimberley district of Western Australia in 1921. Courtesy Geological Survey of Western Australia.

Michael Patrick ('M.P.') Durack, pastoralist, politician and founder of the Okes-Durack Kimberley Oil Company NL. Courtesy Dame Mary Durack Miller.

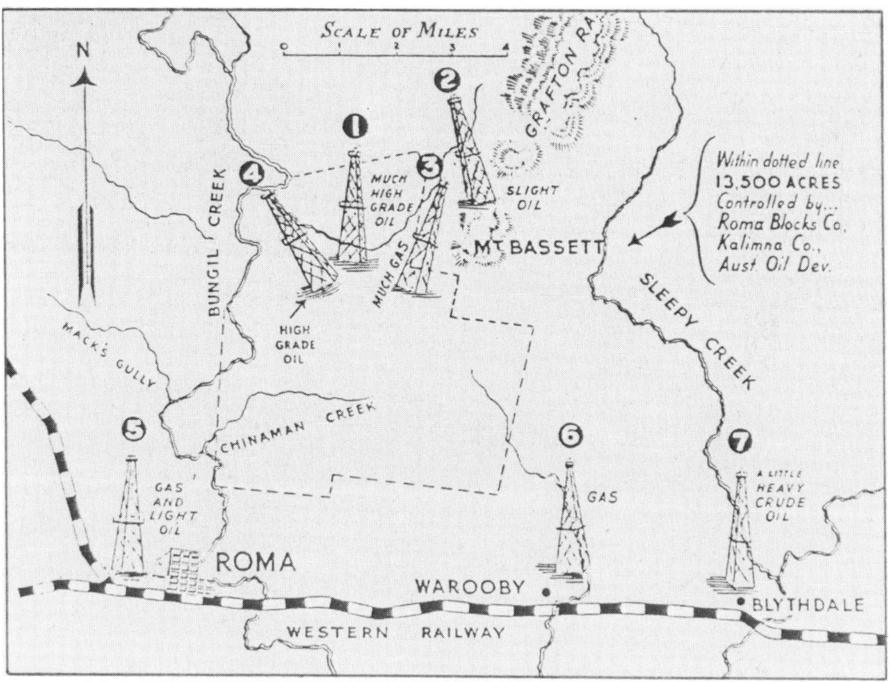

Map from a prospectus pooling the resources of Roma Blocks Oil Company NL, Kalimna Oil Company NL and Australian Oil Development NL in 1948. Roma North Oil Company NL joined the pool shortly afterwards and eventually the group was reformed as Australian Associated Oilfields NL in 1951. Map notations indicate the persuasion used to elicit subscriptions in the immediate post-war years:

1. Here — Roma Blocks No.1 well produced over 1 000 gallons high grade petroleum.
2. Here — Roma Blocks No.2 well tapped slight evidence of oil.
3. Here — Roma Blocks No.3 well tapped large volumes of gas.
4. Here — Roma Blocks No.4 penetrated high grade oil sands. The preliminary shut-off disclosed quantities of oil impregnated mud.
5. Here — Roma Oil Corp produced light oil direct and thousands of gallons of petrol from gas.
6. Here — Oil Search Ltd tapped rich petroliferous gas sands.
7. Here — Australian Roma Oil Co penetrated heavy crude oil sands.

The Freney Kimberley Oil Company's Ford truck crossing the Fitzroy River in northern Western Australia during the 1920s. Courtesy Geological Survey of Western Australia.

A bogged camel after two hours of digging during a survey journey through Western Australia. Courtesy Geological Survey of Western Australia.

Oil Search Ltd personnel outside George Street office in Sydney during the 1930s. From left, Dee (manager), McKillop (driller), Rudd (geologist) and Atkinson (surveyor). Courtesy Eric Rudd.

T.W.H. Dee, manager of Oil Search Ltd, during a field trip to Wooramel River area of northwest Western Australia about 1932. Courtesy Eric Rudd.

Stuck in the mud flats near King Sound during Wade's Kimberley survey in the 1930s. Courtesy Eugene O'Driscoll.

Camp typical of the 1930s Wade surveys in the Kimberley region of Western Australia.
Courtesy Eugene O'Driscoll.

Plane tabling during Wade's 1930s Kimberley surveys in northern Western Australia. Courtesy Eugene O'Driscoll.

Posing with samples of oil (condensate) from the ROC No.1 well at Roma. Queensland Minister of Mines, A.J. Evans, is on the far left. Government Engineer and early analyst of the Roma oil and gas potential, J.B. Henderson, is standing second from right. Courtesy AIP Library, Melbourne.

Inspecting ROC No.1 drill site in 1927 are (from left) L.C. Ball (Queensland Government Geologist), Professor H.B. Richards (University of Queensland) and Dr W.G. Woolnough (Commonwealth Government Geological Adviser). Courtesy AIP Library, Melbourne.

The Roma Oil Corporation's absorption plant was finally installed and operational in 1928. Courtesy AIP Library, Melbourne.

The Roma Blocks Oil Company's Mt Bassett well in 1929, north of Roma township. It took six months to transport the rig to site and set up the derrick prior to drilling.

Shell Company's Queensland exploration programme during the 1940s and early 1950s used this mobile Failing (truck-mounted) rig.

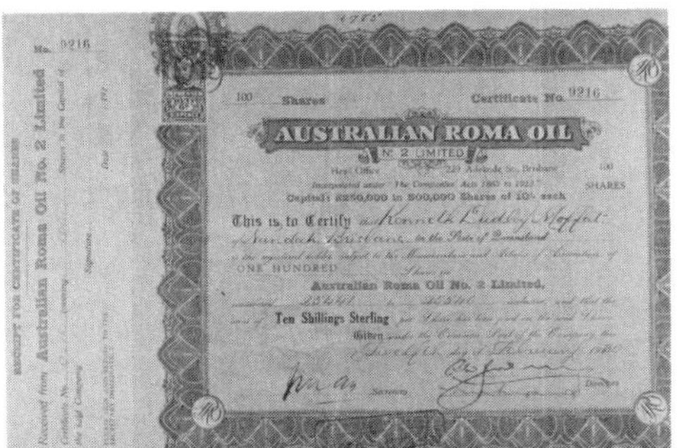

No. 9216
Receipt for Certificate of Shares
Received from Australian Roma Oil No. 2 Limited

100 Shares Certificate No. 9216

AUSTRALIAN ROMA OIL
No. 2 LIMITED

100 SHARES

This is to Certify
ONE HUNDRED
Australian Roma Oil No. 2 Limited,
Ten Shillings Sterling

Share certificate for Australian Roma Oil, one of the many companies set up during the 1920s-1930s Roma oil boom. Courtesy Doug Traves.

Professor S.W. (Sam) Carey, an early employee of the APC Group in Papua and New Guinea in the 1930s. A pioneer in the theories of global plate tectonics and head of the Geological Department at the University of Tasmania. Photo taken in 1983, courtesy University of Tasmania.

Henry (Dinkum) Eve, surveyor for Oil Search Ltd in New Guinea during the 1930s. His grave at Wewak was used as one of the trig points in the company's survey work. Courtesy Eric Rudd.

Rafts made to ferry survey party equipment down or up river, Wuroi, Papua New Guinea, 1930s. Courtesy Eugene O'Driscoll.

Anglo Iranian (BP) field party office in Papua New Guinea during the 1940s. Courtesy BP Australia, Melbourne.

Dr Martin Glaessner, Chief Paleontologist for APC in Port Moresby, Papua New Guinea laboratory, 1939. Courtesy BP Australia, Melbourne.

A young G.A.V. (George) Stanley, geologist, botanist, expert bushman and, later, coast watcher and New Guinean 'personality'. He joined Anglo Persian (BP), then went to APC and, finally, Papuan Apinaipi NL. Courtesy BP Australia, Melbourne.

Australasian Petroleum Company's (APC) headquarters at Port Moresby, Papua, in the 1940s. Courtesy BP Australia, Melbourne.

Early cable tool drillers in Gippsland, Victoria. Courtesy Keith Scarce.

Cable tool rig in Gippsland during the 1930s. Courtesy Keith Scarce.

Steam-driven rig at Tanjil Oil Company's Tanjil No.1 Glenco Bore at Longford in Gippsland, Victoria, 1932. Courtesy Keith Scarce.

W. Baragwanath Jnr ('Mr Barry'), Director of Victorian Geological Survey 1920-43.

Keith Scarce (left) and Jim Kilgour — drillers at Tanjil No.1 well in Gippsland, Victoria, 1930. Courtesy Keith Scarce.

The shaft headframe at Lakes Entrance oil project in 1950. Courtesy Victorian Department of Industry, Technology and Resources.

The official spudding of the Lake Bunga No.1 well near Lakes Entrance, Victoria, in 1924. Courtesy Keith Scarce.

A survey party from the Victorian Geological Survey about 1900. A young W. Baragwanath is on the far right. Courtesy Victorian Department of Industry, Technology and Resources.

The remains of the Lakes Entrance oil shaft project today: Winding house (far right), cylindrical storage tanks (left) and an old exploratory well (foreground). Photo: Sharron Wilkinson.

Flaring gas from Midwest Oil No.1 bore, Lakes Entrance, Victoria, 1931. Such events attracted much local attention. Courtesy Keith Scarce.

Lakes Entrance today has remnants of the early oil search. This old well — still oozing oil — is located in a vacant block within the township. Photo: Sharron Wilkinson.

Sir Edgeworth David, Professor of Geology and Physical Geography at University of Sydney, geological consultant and Antarctic explorer.

A caricature of George Dick Meudell — banker, pioneer stockbroker, oil and mineral promoter, traveller and Australian patriot. Drawn about 1925.

Arthur Wade, British geologist who consulted for commercial oil companies as well as Australian State and Federal Governments in Australia and Papua New Guinea during the first half of the 1900s. Also a member of the three-man Federal Oil Advisory Committee in the 1930s.

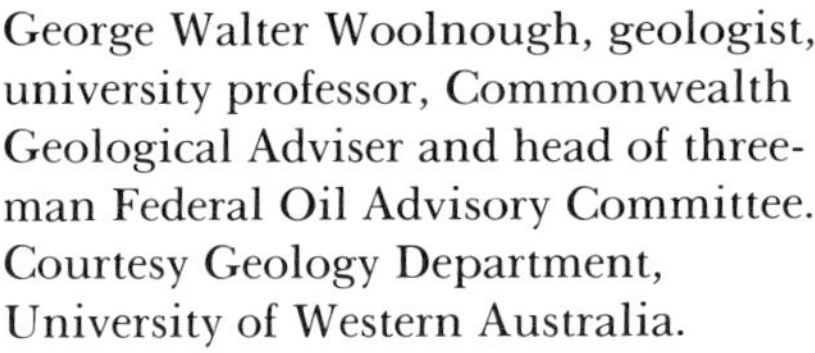

George Walter Woolnough, geologist, university professor, Commonwealth Geological Adviser and head of three-man Federal Oil Advisory Committee. Courtesy Geology Department, University of Western Australia.

L. Keith Ward, Director of Mines for South Australia during the first half of the 1900s, and member of the three-man Federal Oil Advisory Committee in the 1930s. Courtesy South Australian Department of Mines and Energy.

T.W.H. Dee and Eric Rudd going for a 'spin' in the RAAF Wappity aircraft used by Dr George Woolnough during early airphoto surveys in northwest Australia — about 1932. Courtesy Eric Rudd.

> We ended up with about a dozen and drove them back to Canberra where they were given an overhaul and sent out to the field. Many of them functioned satisfactorily for a number of years.[5]

Later, when in London for a conference during 1948, Fisher recalls seeing the post-war revolution in rough-road motor vehicles in a show room on Piccadilly — the first Land Rover.

> After the war we were forbidden to purchase from the United States so I recommended to the Australian Government that we buy Land Rovers, and eventually we did take delivery of some. They were the first to reach Australia and the BMR was the first to try them out. The early ones were accident prone and in the bush they had a maximum life of three years, some of them having a 50 percent downtime.
>
> In the northern conditions they were also dust traps so the parties would often take the hoods and windshields right off. At one stage I did write a long list of suggested improvements and sent it off to Land Rover in the UK. I never had a reply, but later models of the vehicle performed better.[6]

Influenced to some extent by the previous work of Woolnough, Wade and the Northern Australian Survey, Raggatt directed the first BMR field parties to Western Australia — the Desert Basin (soon renamed the Canning Basin) and the Northwest Basin (renamed the Carnarvon Basin) in particular. Before work started on the ground, the RAAF's 87 squadron operating from Port Hedland began a systematic air photo project over Australia using Mosquito bombers. It was the first time vertical shots had been taken, and these provided a marked improvement on the less accurate and more distortion-prone oblique photos obtained during the war years. The flights over northwestern Australia were coordinated so that the runs were made three months before a BMR ground survey party traversed the same area. Prints were then ready for the geologists to take into the field and, largely discarding the old inaccurate military maps, the parties worked directly from the air photos to orientate themselves. Much of the geological survey was carried out in trackless country, and the only pre-war references were basic half mile to the inch maps round mine sites. In most cases there had not even been an attempt to join the mine maps together.

As it had always been Raggatt's intention to set up a comprehensive geological and geophysical mapping programme, when the BMR ground surveys began, the National Mapping section also became involved, and the project to cover Australia on a scale of four miles to the inch was initiated using the air photos and the field work data.[7] To work properly the scheme needed to have national uniformity, and Raggatt was instrumental in setting up agreements between the Federal Government and the individual States on such things as layout, map symbols and colour scheme. He also influenced the adoption of the US system of stratigraphic nomenclature — that of using the 'rock unit' scheme rather than the European system which was based on fossil units, which he thought could be misleading in Australia.[8]

The BMR's first field exercise began in 1948 in the Fitzroy sub basin (a part of the Canning Basin centred on the Kimberley region of Western Australia). Party leader was Doug Guppy, a graduate from Adelaide University who had recently returned from a year in Iran working for British Petroleum. Back in Australia, Guppy had found there were few oil companies working in the country and so he joined the BMR. In the Kimberleys, a field season ran from May to about August or September to avoid the monsoons which swept the north at the end and beginning of each year. Guppy was to find that the Fitzroy project took five seasons in the field and a similar time back in Canberra preparing maps and reports.

> I liked the party to travel as light as possible so it was quick and easy to move. The group consisted of four geologists, a mechanic and a cook. Our base camp was set up around a four-wheel drive truck, but we also had four jeeps. Two of these were used in the field and two were kept in camp as back-up in the event of breakdown or emergency.
>
> The cook and the mechanic usually stayed at base camp, while the geos worked in pairs, taking the vehicles out along traverses for a week at a time without returning to base. The four geologists were always in contact with each other in case of trouble and generally we made one-night camps together during the week, returning to base for the weekend. The base camp was moved maybe once a month depending on our progress.
>
> Our work consisted of mapping rock outcrops along traverse lines using the air photos as a base. We'd first make a rough working map and plan our routes using the photos, then go along the traverse filling in the geological details as we came to them.[9]

Geologically, there was early success when the Fitzroy party found the first Ordovician rock outcrop seen in Western Australia. Workers of high standing, including Arthur Wade, had missed it previously and Guppy believes the air photos were instrumental in the find.

> Sitting in the jeep one day we were checking the photos to see what we'd tackle the next morning when we noticed a small patch on the print that appeared to be different to the

> rest of the countryside. Such a change in rock strata usually meant a fault or an unconformity. It was pretty close so we drove over to have a look and suddenly found ourselves in an area where we picked up graptolite fossils which could only be in Silurian- or Ordovician-age rocks, much older than the strata we had been mapping.
>
> The rock smelt petroliferous too and everyone recalled the small amount of oil recorded from Price's bore way back in the early 1920s. Although the government's initial idea was to use the BMR as an oil search group, this was modified so that the BMR's job was not to directly look for oil itself. It became Raggatt's objective to upgrade an area so the oil companies might be interested enough to come and start exploring.[10]

The party camps themselves were harmonious, but there was never any alcohol. On that point there was no argument and any breach of the rule meant being fired on the spot. Wives and families of the geologists were allowed to live in the field if they paid their way, although they stayed in the main base camps and did not go on the weekly traverse mapping. A hot water service was installed in each camp fairly quickly, consisting of a four-gallon drum in a tree connected to another drum set in a clay fire place and fitted with a relief valve. A third drum was filled with cold water. The whole lot was then connected to nozzles down the hill. The drum in the tree provided the head, and the water flowed down through the heated drum to the showers below.[11] Usually the cook had a fly sheet rigged up for the kitchen, but there were no personal tents in the early days — just a stretcher and a mosquito net. If it rained the party members took shelter in the vehicles. According to Guppy, that was a rare occurrence during the clear warm days of the mid-year months.

> The dry season generally meant just that. But one time we did have a three-day downpour — a 15-year freak. On the Fitzroy River plain the water rose around us and the jeeps bogged in no time. We had to walk 50 km knee deep in slush to Gogo Station to get out.[12]

Although party members sometimes shot wild game for the pot, there was little need to live off the land. Often arrangements could be made with the station people for fresh meat after they had killed a beast for their own home consumption. The base camps had kerosene fridges and usually the party leader went into town once a month for provisions — mostly tinned and dry goods.

> I was issued with a Commonwealth cheque book which had an account total of around £300 to pay for things. The receipt

> was then sent to the Department of Supply in Perth and the money put back into the account so the amount came back to the 300 level.
>
> Looking back, the trust was interesting. There was no need for a counter signature and there were no questions. The locals preferred this too because the alternative of putting goods 'on the slate' often meant them waiting months for accounts to be paid.[13]

The vehicles, particularly the jeeps used constantly during the week, were worked hard and a good mechanic was indispensable, particularly for major repairs. But the station people were also very helpful in those times. Doug Guppy remembers one occasion when a clutch gave out during work near Christmas Creek Station. The mechanic came out from base camp and the vehicle was towed to the station homestead where the manager happened to have a block and tackle rigged up for his own use. The BMR mechanic was able to lift out the engine and remove the clutch plate. A new one was sent up via the Department of Supply in Perth and the vehicle was made bushworthy in much quicker time than the alternative of taking the vehicle back to Derby or Broome for repairs. Arthur Lindner, one of the geologists in Guppy's party, recalls that it was not always up to the mechanic to keep the jeeps working.

> One thing we always had to cope with was tyres continually being staked on fire-hardened roots and branches. We carried three extra wheels per vehicle. We'd start each morning with seven good tyres and hopefully they would last till midday. Then we'd stop round lunch time to mend them all and hope they'd last till the evening. At night we'd mend them again for the next day's work.[14]

Another problem was the possibility of the jeeps suddenly catching fire. When scrub bashing off the tracks, spinifex grass would mass up round the engine manifold and the exhaust and if it was not cleared out at least twice a day it would suddenly ignite. One jeep was lost in this way before the parties realized the danger. Fire also caused the demise of one of the base camp's Chevvy four-wheel drive trucks during the third field season of the Fitzroy survey. Arthur Lindner remembers everyone was away from camp at the time when the truck somehow caught alight — possibly from a camp stove. As no one was hurt, the incident had its humorous side, even though they had to spend time restocking the provisions.

> We had the base camp right at Windjana Gorge in the limestone outcrops a couple of hundred miles east of Derby. We were all in the field and the cook had gone a little way down

the track himself for some reason when the truck suddenly exploded in flames. It burnt with incredible heat and left nothing to salvage. The cook came racing back and saw tins of pineapple slices exploding, with bits of the fruit flying round like Catherine wheels.[15]

The burnt-out truck shell was still at the gorge when Doug Guppy went back on a private trip in 1986.

Despite the fact that the active life kept everyone fairly fit, medical attention was always a concern for the isolated field parties. Radios were carried so that help could be summoned via the flying doctor if needed. Guppy took an additional, and not altogether popular, precaution with his party.

> I never liked anyone to go to the annual race meetings because Perth bookies and other visitors often came up specially for these events and they usually brought 'southern germs' with them. Often the Kimberley people would become sick afterwards and it wasn't just from hangovers. The BMR parties were on a tight schedule to fit in as much as possible in the field season and we couldn't afford to carry a 'sickie'.[16]

BMR operations in the Carnarvon Basin during this period were similar. Work consisted of traverses across isolated country using an ex-airforce jeep and an old army-style trailer which carried three months supply of tinned and dry food. Most of the time was spent mapping outcrops, measuring rock sections and collecting fossil and rock samples for later identification back in Perth or Canberra. In 1949 the Carnarvon party moved to the Cape Range area for a five-month season of field mapping. Daryl Johnstone, a geology/geochemistry graduate from the University of Queensland, was among the group which took part.

> We had a base camp at Learmonth and lived in huts virtually on the old wartime aerodrome. The place was looked after by a superintendent, his wife and a mechanic employed by Ampol. The company had just taken up permits in the area, but didn't have technical staff at that stage so the BMR made a deal to share the field data we obtained in return for accommodation facilities.
>
> We used to do five-day trips into the range. It was pretty rugged, but we managed to get the jeeps right up along the ridges. Then we'd trek up and down the canyons and limestone potholes mapping the formations. It was strenuous work and quite hair-raising at times, but no one ever fell or was hurt.[17]

During the 1950s other BMR geological parties moved out across northern Australia — the Bonaparte Gulf region, into the Gulf of Carpentaria region of Queensland and further south into the Georgina Basin and the Amadeus Basin of inland Australia. One notable trek was made in 1954 across the forbidding southern Canning Desert of Western Australia east from Port Hedland to the old Canning stock route, using three modified Land Rovers operating from a three-tonne, four-wheel drive Commer truck. The party, led by BMR geologist Doug Traves, included two of his geological colleagues, a mechanic, a cook, two field assistants and two Western Australian State Lands Department surveyors. The group found the going very slow and tedious but, navigating by air photos, the men successfully completed the journey and returned without mishap. Valuable geological data was obtained during the journey as well as a clearer understanding of the geography of the southern Canning Basin and some pointers on how to tackle the logistics of isolated inland surveys.[18]

Although the work was long and tiring for much of the time, a camaraderie grew up among the BMR field parties and the camps did have their lighter moments. Max Reynolds, a South Australian geologist who joined the technical division before going on to the Geological Basin Studies Group, recalls two humorous incidents.

> An English chap just off the boat signed up with one of the field teams as an assistant. A few days after setting up camp the cook came to me puzzled that the four-gallon drum of fly-spray was rapidly diminishing. I couldn't offer an explanation either till one night I spotted this new bloke preparing for bed. We were using stretchers at the time that were maybe 15 cm or so off the ground and he was sprinkling the dirt round his bed with copious amounts of fly spray in the fond hope it would keep the snakes away.
>
> Another time the amusement stemmed from our very public communications system. We had morning and afternoon 'scheds' with the flying doctor radio network, and there'd be station news and chit chat in between our drilling reports. Everything went over the air, even telegrams. Then one time, a member of the crew received a telegram from his girlfriend, read out for everyone across Queensland to hear: 'Come back quickly. I have told mother'. Poor bloke didn't know where to put himself.[19]

The BMR's geophysics division differed from the geological section in two main ways. Firstly, in conjunction with the technical division (drilling), the geophysical work needed laboratory space and so it was housed in an old ammunition factory in the Melbourne suburb of Footscray. The geologists, on the other hand, were based in Canberra and it was not until the 1960s that all sections of the Bureau were joined under one roof in the nation's capital. Secondly, while much of the BMR's geological survey and mapping work was initiated under Raggatt's direction, the geophysical programme was carried out more as a response to oil company requests. In the immediate post-war days and into the early 1950s, few oil explorers had their own geophysical equipment, and none of the overseas service companies had set up in Australia at that stage. Soon after formation, the Bureau bought and equipped a DC3 aircraft to carry out regional aeromagnetic surveys. A second DC3, along with a Cessna, were purchased and equipped for work in Papua and New Guinea in the early 1950s. Bob Smith, a physics graduate from Melbourne University, was one of the first to join the BMR's geophysics team on the ground.

> The first work was at the request of four small companies round Roma who soon afterwards amalgamated to become the Associated Oilfields Group. I was involved with gravimeter and magnetometer surveys in the region, but around the same time the two head geophysicists, Rayner and Thyer, were in the US purchasing seismic equipment. This was brought to Australia and used at Roma, so I took part in the country's first reflection seismic survey at the end of 1949.
>
> We had no geophone cables in with the equipment so we had to make our own. Many a long hour was spent paring wires, connecting and taping them up again while sitting in the hot sun. We ended up with two cables, each about quarter of a mile long.
>
> Trouble was, we had no auto-gain control in those days and we had to judge the force or strength of the dynamite-induced shock wave beforehand, and set the instruments accordingly. Consequently there was no tamping the shot hole with dirt. We used water so if the setting was incorrect we could easily recharge the hole and shoot again.
>
> The seismic team consisted of 11 or 12 people — three geophysicists, a shooter [explosives handler], four field hands for laying cable, and three or four more manning the shot hole drilling rig.[20]

The next major exercise came at the request of Ampol in the first years of the 1950s. The company had just taken up leases in the Carnarvon Basin and wanted some help with seismic surveys in the Exmouth Gulf region. Smith was with the BMR party which began work in the Cape Range area. The only access was a canyon towards the north end of the rugged terrain but, as it turned out, the reflection survey disappointingly gained very little data. The narrow entry into the midst of the range was christened Shot Hole Canyon on account of the seismic work done. Jesse Haddock, who went on to become West Australian Petroleum's field manager in later years, joined the BMR's Exmouth geophysical crew as a field assistant. The base was a tent camp on Burrara Station.

> There were two men to a tent and I can remember living out of a suitcase which I had resting on a case of dynamite. Often it was up to two hours drive to the work place. We only shifted camp when we had to survey a new area.
>
> In that period we worked a 10-hour day plus travel time to and from camp, seven days a week. My job was placing the geophones, loading the holes with dynamite and carting water from the ocean or a bore to tamp the charge.[21]

The BMR party then went to the north end of the Giralia Range (one of the features shown on Dr Woolnough's pre-war aerial photo runs) and obtained better results technically. Unfortunately, from a geological point of view, they indicated a syncline lying beneath the surface anticline and hence the potential for oil accumulation was downgraded. Bob Smith recalls that these surveys were being run at a time when the US majors Chevron and Texaco (a partnership known as Caltex in Australia) were considering an offer to take a percentage interest in the Ampol areas, and two Caltex geologists were with the BMR crew as observers. There were traps for those who were not familiar with the region.

> The two Americans were concerned that the surveys were not tied in to a sea level datum, so they borrowed an Ampol jeep and did some barometer traverses from the work sites to the coast about 30 km away. At the time there were exceptionally low tides and the wind had been blowing strongly from the south for a week which was also keeping the water back.
>
> These two drove out onto the exposed mud flats, got bogged and had to walk back to camp. Unfortunately the wind abated and the tide came in further than it had for a number of days. By the time they went back to rescue the jeep it was well under water.[22]

Unlike the small geological party field camps, the BMR's geophysical groups did allow alcohol, although any abuses of the system ended in dismissal. Smith says it did not happen often, although one occasion stays in his mind.

> During 1951 in the Exmouth area working with Ampol we had a particularly bad start to the season. The first weekend one chap badly gashed a foot and had to be sent to Carnarvon Hospital. Then in Shot Hole Canyon someone flicked a match

> and the spinifex blazed out of control towards the equipment. That caused a panic, but luckily not much damage.
>
> Then, a weekend soon after, the crew were yarning before tea when they heard a rifle crack. The only bloke not sitting with them had blown his brains out with a .303. What with the heat, isolation and some personal problems back in Perth, it had all got a bit much for him.
>
> Anyway with all these things on top of one another the camp was a bit depressed and we decided to have a party. The cook was excellent and he rustled up a good meal. But then he got onto the rum and eventually fell off his stool. He was never the same again — must have been a reformed alcoholic.
>
> After that we had to lock up the grog, but then he took the meths used to prime the tilley lamps. We had to let him go after that.[23]

Long periods of relative isolation did not suit everyone and there was some turnover in the crew personnel. As the 1950s advanced the BMR acquired newer equipment and the surveys and camps became more sophisticated generally. Smith's group spent the 1953, 1954 and 1955 seasons in the Canning Basin at the request of West Australian Petroleum (WAPET). The senior crew often had wives and kids (even babies) along, and often the married quarters were a bit away from the main camp. The women looked after themselves and their families. The newer equipment helped speed up the survey lines and usually a main camp was only in one spot for about six weeks.

> We'd start the day early and finish early because of the heat. Saturday was usually vehicle maintenance and general chores day. Sunday we wrote up reports if they were pressing, or took some time off for fishing and shooting in the Kimberley water holes. Fresh meat was supplied by the local station.
>
> Sometimes unseasonal rain was a problem because it cut the supply lines. Then we might go out shooting for a feed. The BMR supplied a .303 and a .22 rifle, but discouraged private firearms.[24]

During the second half of the 1950s the geophysical crews had spread out over most of the country — particularly eastern Australia. As the seismic contractors became more numerous the Bureau did less work directly for the oil companies and concentrated more on regional work to help fill in the gaps in the broad geological picture. Smith says there was one notable BMR contribution to geology, and that was in the Perth Basin.

> The Darling Scarp was known to be a major fault, but it was always thought the sediments of the sand plain in front of it consisted of only 600 m of recent sediments. On his way to the Kimberleys, one of the party decided to calibrate some of his instruments at the government observatory at Watheroo using the known details of the area.
>
> He was puzzled by an anomaly shown up by his instruments that did not fit the accepted data for the fault scarp region. The only explanation was that there were thousands of metres of sediments under the sand plain — not a mere 600 metres.
>
> There was disbelief in Perth geological circles at first but, sure enough, when doing the first seismic line in the Gin Gin area between Perth and Dongara shortly afterwards, it was found a great thickness of sediments existed. This really opened the way for companies (WAPET in particular) to look more closely at the Perth Basin for oil potential.[25]

The other main arm of the BMR, the technical division, took the field a little later than the geological and geophysical teams and concentrated at first on drilling for coal, not oil. The Chifley Labor Government had bought a brand-new National 100 rig capable of drilling to 5 000 m, but when the Liberals under Robert Menzies came to power at the end of 1949, all thought of the government itself drilling for oil was abandoned. The big rig was never uncrated by the BMR and was eventually sold intact to WAPET. Instead, the drilling teams used shallow capacity, truck-mounted Failing rotary units to delineate coal reserves in the Muswellbrook region of New South Wales, in the hope of proving up sites for open cut developments. That work continued from 1951 through to 1954, although teams from the technical division also supervized shot hole drilling for the geophysical surveys during this period.

Then, in about the mid-1950s, the geological surveys were expanded to include stratigraphic drilling. This took the form of continuous coring to augment the surface outcrop maps with data from the subsurface. The holes did not have to be as specifically located as commercial oil wells because they were only used to give a picture of regional stratigraphy. Harry Taylor-Rogers, an English geologist who came to the BMR in 1953 after working for Shell in North and South America, was put in charge of the Bureau's Australia-wide stratigraphic drilling programme.

> A drilling party consisted of a driller (who headed the individual operation), two assistants, a cook, a mechanic and a driver. Among other things, the driver's job was to maintain supplies to the party, and this could be quite arduous.
>
> One site in the Tanami Desert of Northern Territory, for instance, was 800 km out of Alice Springs and it took

> three days to drive into town. They made the journey on their own and on that job two drivers were hired so they could alternate the six-day round trip.[26]

The drilling programme for each year was based on the results of the previous year's geological surveys. Planning was finalized during December in readiness for the field season which, like the other northern Australian operations, ran from April to about October. Taylor-Rogers says the work consisted of a daylight shift only, running five days a week.

> Wells were drilled to basement, or to the capacity of the rig. Usually this was a six- or seven-week operation. Cores were logged on the spot and samples sent off to Canberra for reference and storage. Full reports and stratigraphic correlations were compiled at the end of each season.
>
> The biggest problem came when drilling salt dome regions because water could not be used as a lubricant. It would dissolve the salt formations and make coring difficult. So the technique of drilling with air was tried — probably for the first time in Australia.[27]

The camps themselves were pretty good compared to those of the survey parties. They were fully equipped with freezers and kerosene fridges. Each man had a tent of his own and all the vehicles had transceivers. Nevertheless, an eye had to be kept on the weather, especially towards the end of a season. Taylor-Rogers recalls one Gulf of Carpentaria party almost caught in the 'wet'.

> The last hole of the programme, located south of the Moreton telegraph station near the Wenlock River, was running late because of rig equipment breakdown. I was in the area at the time and witnessed the most eerie sunset I had ever seen — blood red and quite frightening. Locals said it could be an early monsoon, so I told the driller to work till the Friday shift of November 20 and then get out.
>
> The crew actually finished the hole by the 19th and packed up ready to leave next morning as arranged. That night it rained four inches [100 mm] and they couldn't move. There was only a week's food left in camp so the only thing to do was neutralize the trucks, tie everything down and try and get to Moreton telegraph station. They managed to cross the Wenlock and make Moreton and we airlifted them out by helicopter. But the equipment had to stay put until April the following year.[28]

Throughout the 1950s the various BMR programmes spread out over Australia and also began to geologically map Papua and New Guinea. In all States except South Australia there was cooperation in the form of personnel and data swaps with the relevant Geological Surveys and Lands Departments. The South Australians, however, felt that any BMR work within their boundaries would be a duplication of effort. There was also a setback in 1953 when a fire in the Canberra office destroyed all the records and samples, and the work had to be redone. Nevertheless, as the 1960s approached, there was little doubt the programmes had successfully advanced the country's geological data bank. The instigator, Harold Raggatt, had left the Bureau in 1951 to become Secretary of the new Department of National Development set up by the Menzies Government. But, as this portfolio embraced the country's natural resources, he was still enthusiastic and hence influential in the BMR's work. He was succeeded as BMR Director by Nye, and then by Rayner when Nye retired in 1958.

It had always been Raggatt's aim to use the national mapping and survey programme to entice oil explorers to Australia by showing them that, geologically, the country had untouched potential for discoveries. With this in mind, perhaps his greatest success was in finding a kindred entrepreneurial spirit willing to take the commercial gamble. That spirit was New Zealander William Gaston Walkley.

Chapter 6
WALKING WITH WALKLEY

William Walkley and a colleague, William O'Callaghan, arrived in Australia from New Zealand in the mid-1930s bent on breaking into the oil business. They founded the retail firm Australian Motorist's Petrol Company (Ampol) in 1936 and battled hard to keep their noses above water through the supply-troubled war years. Walkley in particular had great drive and initiative, and it was he who conceived the idea that the company would be on a much sounder footing if it had its own oil field and its own refinery. He decided that discovery of an oil field was first priority and, in that mood, approached Harold Raggatt in Canberra for advice on the best place to begin the search in Australia. Raggatt pointed the Ampol managing director at his own favourite stamping ground — the Carnarvon Basin. This liaison of kindred spirits in 1946 was to have the most profound and lasting effect on the country's oil exploration effort during the 1950s, but it would be wrong to say that Walkley was the only man with an oil 'mission' in the country at that time. Another was Maurice Mawby at the Zinc Corporation in Melbourne.

It was largely at Mawby's instigation that the company Frome-Broken Hill Pty Ltd was formed in 1947 as an amalgam of Standard-Vacuum Oil (Esso and Mobil), D'Arcy Exploration (BP) and the Zinc Corporation. Initially, the aim was to search for gas in the Frome Embayment of the Great Artesian Basin to fuel the lead/zinc smelter furnaces at Port Pirie on Spencer Gulf. But Mawby's enthusiasm for the search convinced Frome-Broken Hill also to take up permits in Queensland and in the Canning and Carnarvon Basins of Western Australia. With the technical and financial resources of major oil companies behind him, Mawby was able to put men in the field before Walkley had organized the Ampol push. To evaluate the Canning permit, Frome-Broken Hill brought out noted American geologist Frank Reeves for his second visit to Australia — the first having been for Oil Search Ltd in Queensland during the 1930s. It was Reeves, along with Zinc Corporation's own geologists Harry Evans and his stepbrother Bill Patterson, who conducted the country's first post-war commercial company field surveys for oil.

Probably the best positioned people to observe Reeves and the Canning Basin survey in action were the Zinc Corporation pilots. Apart from having to land their aircraft on anything from a regular town airstrip to a claypan surrounded by sandhills, they were able to see and hear the preparations for a trip as well as the post-mortems after it, and they could also catch the various moods of the players. Alan Polkinghorne, who flew a number of sorties for both the Canning and the Great Artesian Basin work programmes, was one of those involved in a typical Reeves-led survey.

> He was a shortish, tough old bloke about 60 years old at that stage, wearing glasses and a pith helmet. He disliked civilization

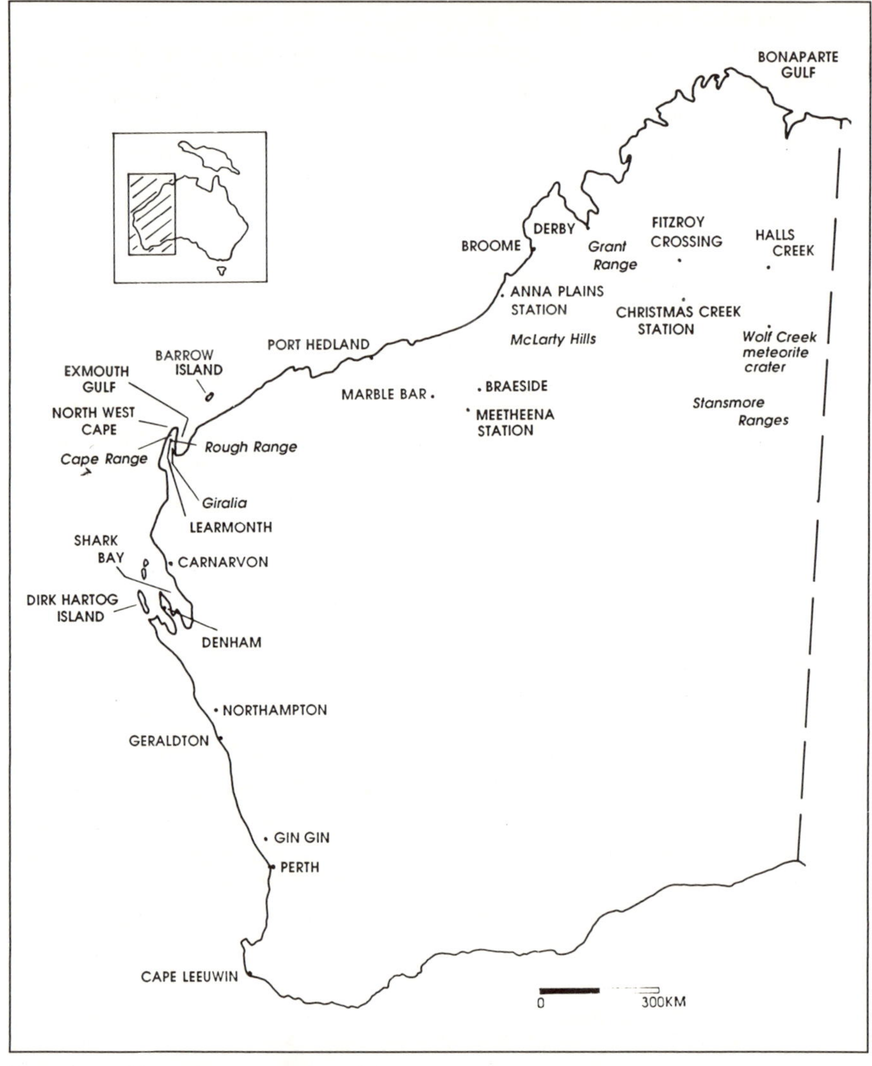

Location of key places mentioned in text.

and hard rock [mining] geology, preferring the study of sedimentary rocks and the search for oil. He also believed geology should be done on foot or on horseback.

Anyway, early in the piece he set out on horseback with 12 pack animals and spare mounts in tow, plus a young full-blood Aborigine and a half-caste who was head man from Meetheena Station near Marble Bar. His idea was to travel round the southeast of the Canning — it was still called the Desert Basin then — to Lake Tobin. It was planned as a three-week trip and I was to pick him up at the end of it. In the meantime we did three parachute drops of food and mail to him on the way. They had to find their own water and for that Reeves relied on his Aboriginal companions.

I set out at the appointed time from Broome in our Norseman aeroplane, an eight-passenger craft which had a high single wing and could fly at 120 miles [200 km] an hour. On board with me were Harry Evans and Bill Patterson and a couple of BMR geologists, Doug Guppy and Jack Cuthbert. We stopped a few times on the way out to the salt lake to look at outcrops, and that was OK. But then when we had Reeves on board on the way back, he wanted to stop a few more times.

I was watching the fuel gauges by this time. The main tanks were finished, but I still had two 50-gallon [230 L] tanks in the wings. Trouble was, they were measured by old boiler gauges and were a bit unreliable. Anyway Reeves insisted so I said 'the hell with it', landed at the spot he wanted and then made for Braeside Station. We could ring up Marble Bar from there and get some fuel trucked out.

It meant a wait of two days, but Reeves was as happy as Larry sitting in the plane on the ground writing up his report. Next thing, he wanted to restock with provisions and go back out into the bush. Marble Bar had told me I couldn't have a full load of fuel because they were short and needed some in reserve for the flying doctor.

When I told Reeves we were going back to Broome he was as niggly as hell. The others warned me he'd want to stop at every claypan on the way home. Sure enough, he sat in the back and kept passing notes up to the front: 'Can we land here ?'. I said no and pressed on for Broome. That night he came up to me, puffing on his pipe, and said quietly, 'I think you put one over me today. I'm goddam sure we could've landed on some of those spots'.[1]

Others remember Reeves as a wiry man, determined to the point of stubbornness, who travelled long and frequently, often pushing planes, jeeps, horses and men close to the limits of safety. On one occasion, although Reeves was not present, a tragedy was narrowly averted when a three-member Frome-Broken Hill field party was caught in the Canning Desert without water or transport. Doug Guppy, party leader of the BMR's 1947 Fitzroy survey, recalls the incident.

> Reeves was in Perth at the time when Evans, Patterson and one of our geologists seconded to Frome-Broken Hill, Jack Cuthbert, set out in one vehicle from Broome bound for the McLarty Hills to study the outcrop. The trip was so rough that the jeep completely cracked up and the water tank was holed. They were in parallel sand dune country and walked only at night to avoid the sun's heat. Even so they were in a pretty bad way, with no water and Evans suffering from a centipede bite. They only just made it the 70 km or so to one of the few permanent springs in the whole region.
>
> We all learnt lessons from that. Always travel with at least two vehicles and never go anywhere without a radio.[2]

Despite geologists' suspicions that pilots always flew by the seat of their pants, the men who supported these survey expeditions from the air never took unnecessary risks and were prepared for most eventualities. The Zinc Corporation pilots were all certified in basic mechanical work so they were able to carry out many of the running repairs in the bush without calling in more qualified help. Alan Polkinghorne confirmed that pilots subjected their planes to constant checks.

> Every time we landed to let the geologists out to check an outcrop or to drop off supplies, we would leave the cockpit and walk around the plane checking everything. We looked for telltale signs like oil leaks, loose bolts, cut tyres — anything that might be an early warning of trouble.
>
> We carried spare bits and pieces like washers, self-tapping screws, hoses and wire plus some engine parts and certainly a good tool kit, so we were able to fix most things. If not we radioed the nearest main base and a mechanic would be sent out.[3]

Polkinghorne says navigation round the inland was a bit difficult for a pilot new to an area, but after a few flights he became familiar with it.

> Some of the charts were pretty rough, but we soon picked out landmarks. We did use compass headings for any aerial geophysical work like a magnetometer survey where the grid lines had to be accurate. I'd never allow any survey instruments

or meters in the cockpit because they threw the navigation aids out. On a longer trip we'd go up to 3 000 ft [900 m] and on a clear day we could see 60 or 100 miles [100–160 km] ahead anyway.

Landings in the bush could be a bit tricky, though. If it was in sand hill country we'd 'slow' fly over them and scan the ground. Often the dunes were separated by claypans. If not, we could land in hard sand. A lot of the camps, particularly in the Great Artesian Basin area where occupation was longer, had graded strips.[4]

Reeves's survey of northwestern Australia included the eastern Canning (where he and Evans discovered the spectacular Wolf Creek meteorite crater) and the Bonaparte Gulf region further north. But he barely had time to more than glance at the Carnarvon or Great Artesian Basins before he returned to the US in 1948. His final report was not enthusiastic and the oil majors in the Frome-Broken Hill company decided to leave Australia for Sumatra where oil prospects were deemed much brighter. Then, in London during 1949, Zinc Corporation merged with Imperial Smelting Corporation under the new name Consolidated Zinc (later to merge again and form Conzinc Riotinto) and Maurice Mawby received word that the Australian operations were to concentrate on lead and zinc. Somewhat reluctantly he obeyed and the company's emphasis moved away from oil for the time being. Although Reeves' report to Frome-Broken Hill was confidential, his lukewarm attitude became clear when he published a report on Australia's oil prospectivity in the prestigious Association of American Petroleum Geologists' *AAPG Bulletin* in the US during 1951. The introductory sentence set the tone in no uncertain terms.

> Australia has no commercial oil fields and, because of unfavourable geological conditions, probably will never produce enough oil to supply its domestic needs.[5]

Basing his conclusions on his own field work for Oil Search in the 1930s and the 1940s Frome-Broken Hill survey, plus a review of the existing literature, Reeves dismissed many surface indications of oil as unreliable, saying that elsewhere in the world surface oil showings commonly occurred in sharply folded strata, whereas Australian basins were only slightly deformed. He said that oil had not been encountered in the thousands of water wells drilled throughout the Great Artesian Basin and this was fairly conclusive proof that hydrocarbons would not be found in Mesozoic-age rocks within that region.

This latter observation prompted a defensive reply from Harold Raggatt.

> The lack of oil in artesian bore holes is no reason to write off the potential for petroleum because most artesian wells

> were not drilled on structures favourable to accumulation of oil. One or two were, but they were so long ago with unsuitable rigs by men untrained to note signs of oil. Also, once the hole is filled with water, the oil would have to overcome water pressure of the whole column.[6]

Reeves's article went on to say that the failure to find oil in Australia, a continent with an area the size of the United States, could not be attributed to lack of enterprise or technical ability because a number of large companies had tried and been unsuccessful.

> The most obvious deficiency is the scarcity of sedimentary basins with a fair thickness of unmetamorphosed marine sedimentary rocks. Two or three basins on the west coast which are exceptions to this generalization may yield some oil when they are adequately tested.[7]

Reeves added that the Tertiary-age sediments onshore Australia were too thin to have generated hydrocarbons, although in the Gippsland Basin there might be some prospect of discovering commercial volumes of oil offshore. He was as prophetic on that point as others had been before him. Nevertheless the article as a whole could hardly have been classified as a glowing endorsement of Australia's oil potential and was a disappointing result for those, like Raggatt, who were trying hard to guide scientific opinion away from the old attitude that Australia was not prospective. The upshot was that Frome-Broken Hill moved out, leaving the field wide open for Ampol and the surprisingly undiminished enthusiasm of William Walkley. The New Zealander, who now called Australia home, had not been idle since his initial interview with Raggatt in 1946, although his activities during the late 1940s were more concerned with administrative matters like setting up an operations base in Western Australia and trying to interest a major oil company in joining Ampol for the mammoth exploration task that obviously lay ahead. There was also a short-lived diversion into Papua New Guinea when Walkley took a substantial shareholding in Papuan Apinaipi Petroleum Company Ltd and tried to revive its flagging exploration programme.

Soon after speaking to Raggatt, Walkley used his connections with the big US company Atlantic Richfield Corporation (at that time Ampol's major petrol supplier) to invite the vice-president in charge of exploration, Frank Morgan, to visit Australia and the Carnarvon Basin (still called the North West Basin at that time) in particular. The rugged terrain in the Exmouth region immediately caught Morgan's attention as it had Woolnough's and Raggatt's before him, and he pronounced the ranges text-book oil structures. Walkley moved swiftly. In 1947 he secured exploration permits over the Exmouth area and out into the Carnarvon Basin beyond. During the next four years Ampol steadily increased its holdings in the State to include all the coastal areas from Cape Leeuwin in the far south to the Bonaparte Gulf in the north, extending inland up to 800 km and including the islands off the coast. This amounted to a huge 1 million km^2 of largely unknown and uninhabited country taking in most of the four major sedimentary basins — Perth, Carnarvon, Canning and Bonaparte.

In acreage terms the permits presented a major challenge to the fledgling explorer, although financially the burden was not onerous. Western Australian Government charges had been set at £3 000 per permit per year and there were only six permits in all. The whole of the Canning Basin, for instance, was covered by a single holding. Influenced by Raggatt and Morgan, Walkley concentrated his initial attention on the Exmouth region and, in 1948, Ampol leased the Federal Government buildings and airstrip comprising the former wartime RAAF base at Learmonth. The idea was to use the facilities as field headquarters for the exploration effort. Charles Knife, administration assistant to Walkley and later a director of Ampol, was given the job of overseeing the company's exploration programmes and made his first visit to Learmonth in March 1949.

> The buildings had been partially rebuilt after being hit by a cyclone and tidal wave soon after the RAAF had vacated the base. They were just habitable and we used the old officers' mess and kitchen complex as a nucleus to develop our regional headquarters.
>
> We had a staff of four in the base at that time — a superintendent, a cook, a mechanic and a general hand. It was a lonely place in those days, the nearest neighbours being a family at the North West Cape lighthouse 56 km north, and the Lefroy family at Exmouth Gulf Station homestead about 22 km southeast. Carnarvon was 320 km south and Perth another 1 000 km further south again.
>
> In the first months there was no radio equipment so telegrams had to be sent to and fro via the Lefroys. Then I discovered that in the RAAF days there had been telephone links between Learmonth, the lighthouse and the homestead. I went out and found some of the old wires still strung along fence posts and, with the help of the base staff, we traced and repaired them all. It meant the operations were not quite so isolated. A little later pedal radio equipment was installed and the Learmonth base joined the flying doctor network.
>
> It was a condition of the lease from the government that we keep the airstrip airworthy. We needed it to bring in supplies and personnel anyway, so there was no quarrel. The trouble was it was 2 745 m long and formed of thin bitumen over sandy soil. Cracks all over the place. Cement was hard to get during the immediate post-war years and all we could

> do was patch it with bitumen emulsion and pull the weeds out.[8]

Ampol began in Western Australia with no technical staff of its own and much of the early geological field work was done with the help of the BMR teams already in the area. During the late 1940s the Bureau was still philosophically geared up to be an oil explorer, the role originally intended for it by the Chifley Government when the BMR was formed in 1946. Although Raggatt directed that the BMR should upgrade areas for later oil company work, the notion of a State-operated oil search was carried on for a short time in political circles after the Menzies Liberal Government won office in 1949. In a cooperative deal, BMR geologists worked out of Ampol's Learmonth base while the first mapping and seismic surveys were carried out on Cape Range, and then nearby Rough Range. During this time geologists from a succession of potential Ampol partners — Richfield, Signal Oil and Gas, Sun Oil and finally Caltex — joined the teams to make their own evaluations of the prospects. Most of them were Americans making first visits to Australia and they were amazed that anyone could find their way round in the monotonous terrain miles from anywhere. Murray Johnstone, who began work in the area for the BMR, thought the same thing during his first trip to Exmouth in 1949.

> There were two of us, and we started out from Perth in an ex-army jeep. At that time the road north contained a very short bitumen section and a bit more of graded gravel. By the time we got to Northampton, just north of Geraldton, we were following two wheel-tracks.
>
> We encountered 11 gates in the rabbit-proof fence between Northampton and Carnarvon and another 26 gates from Carnarvon to Learmonth. Once on the Exmouth Peninsula it seemed that everyone in the district knew where we were but us. We passed one homestead, which was some miles off the road, early one morning. It was slow going for us as new chums feeling our way, and when night fell we camped by the side of the track.
>
> Next morning we pushed on and came to a homestead that was right by the road. A woman came out saying she'd expected us the night before and had kept a dinner ready, not realizing we were unfamiliar with the area. The thing was, we hadn't told anyone we were on the way, nor had we seen anyone. It was my first experience of the 'bush telegraph'.
>
> A couple of years later I had the job of taking the Caltex geologists to the area. They were amazed that I knew where to go. There were no maps and the road signs taken down

during the war to hinder any Japanese invasion force still had not been replaced.[9]

Athough the focus of their work was in the Exmouth Gulf region, the overseas experts also inspected further afield in the Carnarvon as well as the Canning Basin, again with the BMR parties as guides. Most liked what they saw, but eventually it was Caltex, already with some knowledge of the northern Canning region via its own brief reconnaissance exploration during the early 1940s, that agreed to join Ampol in exploring the permits on a commercial basis. A preliminary agreement drawn up in April 1951 and formalized in March 1952 called for all six permits to be transferred to a new company called West Australian Petroleum Pty Ltd (WAPET) in which Caltex held 80 percent and Ampol 20 percent. With Caltex supplying the initial $US3 million in exploration funds, WAPET lost no time in hiring staff and organizing field parties. Because there were few other exploration groups in Australia at the time, most of the new company's geologists and geophysicists were recruited from the BMR. Some others were hired from the universities as consultants.

Although the teams were set up to examine all the areas under licence, everyone agreed that first priority should be given to evaluation of the massive structures in the Exmouth Gulf region that had been enticing geologists for nearly two decades. However there was some hesitation in choosing a place to drill the first well. The BMR's early seismic work along Shot Hole Canyon in Cape Range had done little to clarify the subsurface picture there and, even though it was the biggest structure in the region, it was equally obvious that hauling a rig onto the crest of the rugged mountain would be extremely difficult. On the other hand, the Bureau's survey on the nearby Giralia Anticline had shown that the low hills in the area known as Rough Range presented far fewer access problems. WAPET decided to make Rough Range its first target, and so began the biggest coordinated oil exploration programme Australia had ever seen.

In the autumn of 1952 the company purchased a National 130 drilling rig capable of drilling to 2 400 m — at that time one of the largest and most powerful land drilling units in the world. Built in the US, it was shipped directly to the Exmouth Gulf in the freighter *Ellen Maersk* and landed along with hundreds of tonnes of ancillary equipment and drilling materials. Code-named 'Operation Potshot' after the strip of beach opposite Rough Range that acquired the name from wartime activities at Learmonth, this massive unloading task took 18 days to complete. Preparations onshore had included the building of a road from the beach through the sand dunes to the drilling camp site and, as there was no pier or jetty, the construction of a concrete ramp to bridge the soft mud at the water's edge. The *Ellen Maersk*'s cargo included three landing craft which, along with a hired Australian

Navy lighter, were used to ferry the rig, two big Mack trucks and the rest of the equipment from the vessel's anchorage to shore.

Unusually heavy swells, unpredictable tides and strong currents plagued the operation, making it difficult for the craft to move off the beach after disgorging their cargo. More than once they were caught on sand banks at low water. Powerful lights were rigged up on the beach to allow the unloading to go on through the night. At the conclusion it was counted a success as there was no damage to equipment and only two minor items were lost overboard.[10] However preparations to erect the rig on the chosen well site were interrupted when results from the BMR's Giralia seismic survey suggested the structure did not continue at depth. In an attempt to clarify the position, WAPET contracted the British company Seismograph Services of London (SSL) to run new surveys over the whole area but, after nearly 12 months work, the seismic data still did not tally exactly with the surface observations and, as a result, Rough Range No.1 was a compromise location chosen to best fit the data from both sources.[11]

Having battled to overcome the physical difficulties presented by Operation Potshot and then wrestled with the problems of geological interpretation, WAPET management and staff would have been pardoned for thinking they had seen the worst of the difficulties involved in exploring the isolated west coast region. But there was more to come. No sooner had the site decision been made in March 1953 than a wild cyclone tore in from the Indian Ocean, flattening the Learmonth base and the seismic tent camp. Winds reaching 200 km per hour and a rainfall of 356 mm destroyed large quantities of food, cement and drilling materials. Dry creek beds became raging torrents, roads were made impassable and field parties were marooned for several days. WAPET estimated that the cyclone had added $US150 000 to the cost of the well, and it had not even begun![12] As in most adverse circumstances someone found an amusing side. Jesse Haddock, who rose in the ranks to become WAPET camp manager at Barrow Island, was then a general hand at Learmonth.

> We'd just finished putting up some new buildings at Learmonth base, including a toilet block, when the cyclone struck. Emerging from cover after the blow we could see the toilet bowls lined up in a row on their concrete bases — bare to the world. There were sheets of new corrugated iron winking in the sun all over Cape Range for months afterwards.[13]

Despite the setback, the spudding of Rough Range No.1 on 5 September 1953 was a gala occasion. Present were the West Australian Premier, Albert Hawke, Harold Raggatt and other government dignitories, heads of Caltex and senior executives of Ampol, and a specially chartered plane-load of journalists from all the Australian capital cities. Centre stage during the ceremony was William Walkley, the man who had worked tirelessly for seven years towards the moment when he could spit for luck on the shiny new drill bit just before it began the first well on the Exmouth location. During the previous evening the press contingent had presented Walkley with a cherry-red 10-gallon sombrero in mock protest at the 'inefficient management' of their trip. Accepting this gift, the Ampol managing director declared he would wear it down Sydney's Pitt Street the day WAPET struck oil. Not even Walkley was prepared for the fact that he would be called on to do exactly that just three months later. Rough Range No.1 struck oil in a Cretaceous-age sand at a depth of about 1 100 m on Sunday 1 November 1953. But the cold facts belie the amazement which turned to excitement at the rig site. Owen Evans, drill camp clerk at the time, will not forget the events of that day.

> Normally Sunday was little different from any other day of the week at the camp because drilling was a continuous 24-hour-a-day operation. In the morning I had the usual amount of rush paper work to do to catch the midday plane at Learmonth and I had also attended to several wireless 'scheds' with the flying doctor base at Meekatharra, our main link with the outside world.
>
> However the first of November began with a mild air of expectancy because a drill stem test was to be run late in the morning to evaluate a porous zone penetrated by the drill almost two weeks earlier. Several previous attempts to run the test had failed because of mechanical problems. In fact the 11.00 a.m. test that morning failed too and we went to lunch a bit disappointed.
>
> On returning to the rig at 1.00 p.m. the drill pipe was being pulled from the hole and the word was that the test this time had been mechanically successful. Although working steadily in the 100° F heat [38° C] it took another hour for the rig crew to raise, disconnect and stack each 90 ft [30 m] stand of pipe to reach the test tool. Then suddenly, the drill floor was awash with hot, kerosene-smelling, green-brown oil, mixed with grey drilling mud.
>
> The excitement was tremendous. Most Americans expressed amazement at the luck of finding oil in the first hole they had drilled in a country the size of Australia. The drilling superintendent kept telling everyone not to walk around the surrounding ranges with high-heeled boots on or there'd be gushers everywhere.
>
> One of the geologists was trying to catch the hot oil in an Emu brand beer bottle, oblivious to the soaking he was getting. Needless to say the camp was also very wet in other ways from late afternoon till the early hours of next day.[14]

Jesse Haddock also remembers the day Rough Range struck oil.

> I arrived at the well site with a load of barytes [used to add weight to the drilling mud] and I could see they were pulling pipe out of the hole. It was one of those hot days when the drill crews stripped to the waist. But I couldn't tell if they had clothes on or not because they were covered in a green waxy stuff that smelt like kerosene.
>
> As each stand of pipe was broken out, an oily liquid came out and flowed over the rig floor and everything on it. In spite of the heat, the liquid set quickly on contact with the air. It was impossible to get a firm footing on the floor, but nobody seemed to care, they were so excited. It was the first time most of us had seen crude oil flowing from the earth.[15]

Naturally enough, it was impossible to stop rumours of a find reaching the outside world despite the isolation of Learmonth, but the Caltex management at WAPET decided not to make an official announcement until samples were positively identified. Because the oil solidified so quickly there were fears it might be a worthless wax known as Okerite, and a sample sent to a laboratory in Perth was pronounced to be just that — the analysts having little experience in petroleum. This prompted WAPET to deny that oil had been found, a statement which caused anger in press circles and embarrassment to Ampol when a month later the results of a second sample, which had been sent to Caltex headquarters in the US, were declared to be a fine light crude. Walkley himself was particularly peeved to hear the news via a press cable from America instead of being informed directly by his US partners.

After the fulsome press coverage of the spudding ceremony and the continued Australian journalistic interest in the well's progress, the newspapers felt they had been deliberately misled. West Australian Newspapers in Perth felt strongly enough to dispatch a reporter and a photographer to the rig site where they maintained a presence for the next 18 months, changing shift every six weeks or so. At one point suspicions were voiced that WAPET was withholding information and this led to the journalists being banished from the camp. They then hired a woolshed from the Lefroys on Exmouth Gulf Station to use as their reporting headquarters. Despite these apparent misunderstandings, newspaper articles trumpeted the virtues of the find, while on the stock exchange Ampol's shares rocketed up the scale amid opinions that the company would soon be awash with oil.

WAPET itself grew in confidence and purchased two smaller T32 drilling rigs (1 500 m depth capacity) from the US, plus the Federal Government's big National 100 still lying uncrated in Melbourne. The latter unit was sent to the Canning Basin, and the two US rigs arrived at Rough Range during 1954 to commence an appraisal programme after being road convoyed up from Perth. In the meantime the discovery well drilled on to evaluate deeper horizons — a process that turned out to be long and tedious. Murray Johnstone, a geologist with WAPET, 'sat' the well for much of this time.

> The Rough Range No.1 rig was then the ultimate drilling machine, but knowledge of local conditions was still lacking. The drillers used bits that were quite suitable in the US but, in the hard Rough Range limestone, some came back out of the hole with teeth completely worn off and some without the drilling cones at all. That meant a lot of fishing jobs [retrieving tools and pieces of metal from down the hole].
>
> Put with the time needed to frequently haul up all the drill pipe to change the worn bits and the slow drilling progress anyway [at times a mere one metre in seven hours], it dragged the well on for about 20 expensive months. The rig became a useful landmark beacon for airline pilots and they were most upset when the well was finally finished.[16]

Rough Range No.1 found no deeper oil zones, but the more sobering news for WAPET was that a series of eight appraisal wells drilled during 1954 and 1955 were all dry holes. A tenth and last try was made in June 1957 close to the original site for the No.1 well — the one before the compromise location was chosen. It too was dry and one wonders what course the company's programme would have taken if it had been tried first.

During this period Cape Range remained on the agenda as a drilling prospect. But crestal access still posed a problem and the company decided to compromise again by sending one of the T32 rigs to drill at the end of Shot Hole Canyon. This was done in 1954 but the well was dry. Nevertheless geologists were not satisfied that this result condemned the prospect, and a concerted effort was made to find and build a suitable road to the top of the structure. As it turned out the eventual route had been mapped out several years previously. Charles Knife, Ampol's exploration overseer at the time, had already made numerous trial-and-error treks to point the way for the road that now bears his name.

> When I first arrived at Learmonth in 1949, Cape Range was featured on the four-mile series map as 'high impassable ranges'. I knew we would need to get to the apex sooner or later so I began to search out the possibilities. I obtained aerial photos and, on one or two occasions, I prevailed on airline captains to fly at relatively low altitudes over the region. As a result I concluded that there was only one ridge which ran more or less unbroken from the Exmouth Gulf side to the

> top, and that access from the Indian Ocean side really was impossible. The difficulty was in identifying the right ridge on the ground because they all looked the same.
>
> Accompanied by Jack King from the Learmonth base I tried on several occasions, but was always turned back when the ridges petered out. I knew the BMR geologists had reached a point close to the top by a circuitous route in the south, but another flight over the range convinced me to try further north. Several more treks were aborted by sheer precipices until finally I found the ridge I had seen from the air. It did go all the way, although at one point it was only 20 ft [6 m] wide with a sheer drop of 300 ft [90 m] on one side and 400 ft [120 m] on the other. We marked the way with toilet paper and later went back to build more permanent stone cairns and daub them with aluminium paint.[17]

During early 1950 Western Australian Main Roads Department engineers examined Knife's discovery and declared it impracticable. However, when necessity prompted WAPET to engage surveyors to search for a route in 1954, they could not find anything better than the one Knife had pioneered. Finally, the Perth-based firm Bell Brothers inspected the proposal and agreed to tackle the construction job using a 30-man crew and high-powered diesel trucks and earth-moving equipment. Working in scorching summer heat, the job was arduous and at times dangerous as the men hacked and filled their way up the 300 m high ascent. Contractors Alec and Rob Bell commented wryly that the temperature often reached 120° F in the shade, but their men were not working anywhere near the shade.[18] Particularly hazardous were the tight curves and the narrow saddles where sheer drops of 150 m to the canyon floors were common. Despite such conditions the work was completed three weeks ahead of schedule in the early months of 1955 — a praiseworthy effort on what was described at the time as the toughest road ever built in Australia. Even now, upgraded as a tourist route in the 1980s, the road is still a tortuous, winding nine-kilometre climb that in some places must be tackled in first gear. In 1955 it served its purpose as a route to the Cape Range No.2 well, although not without mishap and one tragedy. Jesse Haddock was still at Learmonth when these occurred.

> One time a diesel hydraulic pump failed on the low-boy truck I was driving and that meant the air brakes did not work. I had to quickly leap out and roll boulders behind the wheels to keep the vehicle from going backwards over the edge. Then I had to wait till the mechanic could get up and fix it.
>
> Another time during the drilling of the well there was a bad cyclone which flooded the lower creeks and we couldn't get the relief crew up to the rig. Even if we managed to get

> across the fords there was no traction on the road as soon as we began the ascent. In the end I got our biggest truck over the water, stacked rocks and the relief crew on the back and reversed the whole way up. It took a while, but we made it.
>
> But the worst accident was one night when a drill crew was going down the road in a station wagon after a shift. They were going too fast and ran head-on into a Mack truck coming up. The wagon driver was killed and the others badly injured. The drilling superintendent put the car — stove in as it was — in front of the camp mess as an example of the dangers.[19]

In the end it was all for naught. The Cape Range No.2 well drilled to 4 626 m before it was abandoned as a dry hole. Two more wells on the structure further south were equally unrewarding. Geologists were finally convinced that the massive structure was a surface phenomenon which did not form an anticlinal trap at the depth of the potential reservoir sands. Probably the most graphic summation of the Cape Range programme was expressed to WAPET geologist Jim Parry by one of the engineers based at Learmonth during this period, who said, 'It's 50 miles to water, 20 miles to wood... and half a mile to hell'.[20]

Although the drilling programmes in the Exmouth region were a bitter disappointment, WAPET did not have time to dwell on the lack of success. Having secured most of the sedimentary basins of Western Australia under lease, the company sent its field teams far and wide to evaluate the potential. The company had realized its limitations and relinquished the Bonaparte Basin permits in 1954. Even so, by the mid-1950s it had 20 geologists in its direct employ while a much greater number of surveyors, engineers and field hands worked for contractors associated with the exploration programmes. There were two geological field mapping parties in the Carnarvon Basin and two in the Canning, supported by three gravity survey parties and three seismic survey parties across both those regions plus the Perth Basin to the south. In addition, six well-site geologists kept abreast of the expanding drilling programme. Field camps followed the pattern set by the BMR as Murray Johnstone explained.

> Mostly we had tents — two men in each, while the chief had one to himself with a typewriter set up for report writing. Often we put up an awning between two trucks to serve as a dining area. Entertainment was via the flying doctor radio channel where we could often pick up ABC programmes like 'Take It From Here'. Many evening laughs came from listening to Jimmy Edwards and Co. in the ongoing story of the Glums.
>
> Most outback stations were supplied with milk, vegetables and the like by weekly truck at that time, so we could

order fresh food. Occasionally we'd get a sheep from the homesteads, but other fresh meat came from shooting kangaroos and knocking curious bush turkeys [bustards] on the head. The kero fridges we had were so cold many of them acted like freezers and we could keep meat for long periods. Pressure cookers were a great help in preparation of food.

But you still had to like the bush because often we were out for six months at a time. It was no good for anyone with worries at home. One bloke always left his wife pregnant just before he went away and that went on till he had a dozen kids. He quit field work after that !

There was a lot of comradeship and pride in the job too. We worked five days a week. Saturday was usually spent on camp chores. But Sunday we often went to the local station for tennis parties or picnic races — even a 'geologist's weekend' collecting fossils if there was a good locality close by.[21]

WAPET had a rule that no party should go into the field with less than two vehicles and they should carry all their water — reckoned at 4.5 L per person per day in the summer months. Geologist Robin Elliot said the company was adamant this was for drinking and cooking — not for washing.

If we found a water hole we might use it for washing, but that could be pretty rare in the dry season. One member of our team had the water ration all worked out. He'd use it first in the pressure cooker for the spuds and beef, then he's use the same water for washing up, and finally for washing his socks.[22]

Colleague Jim Parry recalled that the work days in the field were hard and long, a fact not often believed by those back at Learmonth base camp.

They used to think the field exploration parties were on some sort of bush picnic, so one time I asked if any of the base people would like to give us a hand. One bloke, nicknamed Dipstick because he was tall and thin and worked in the garage, volunteered.

We were up about 4.00 a.m. to beat the heat and just carried the basics with us because we usually stashed the heavy gear at the workplace at the end of each day. It was pretty steep, rugged country and we must have walked about 12 miles [19 km]. About midday Dipstick asked about lunch and we told him we rarely bothered with it because it meant more load to carry.

> He went a bit quiet after that and was obviously tired by the time we got back to camp. We all had a beer each before preparing dinner, but old Dipstick fell asleep. We let him alone and he slept right through till morning and missed his meal altogether.
>
> Come daylight he was so cramped and sore he reneged and wanted to go back to Learmonth. We never got any field assistance from base again, but we didn't get any more snide comments about our work load either.[23]

As part of WAPET's widening survey area, Parry and fellow geologist J.H.R. McWhae visited Barrow Island in 1954 — the first civilian group to be given permission to land there after the atomic bomb test programme in 1952 at the nearby Monte Bello Islands.

> We went across in a Navy vessel and then in a skiff for the last 4 km to the shore. It was just for a few days reconnaissance and we camped on the beach. Scouting around we checked the rock outcrops, took a few measurements and saw that it was an anticline.
>
> Historically it was interesting too because we found a cairn left by the Beagle expedition 120 years earlier. At the time the whole island was excised from the petroleum lease, but after our report WAPET applied to Canberra to have it included.[24]

The Defence Act ban was not lifted until 1962, but a second short visit by WAPET geologists took place in 1956 — the operation being serviced from Onslow. The men were issued with a dinghy and outboard to help them travel round the coastline and complete a plane table mapping survey.

> It soon became obvious that the supply line was a problem. The local vessel hired to do the job took up to 72 hours to reach the island from the mainland 80 km away. Then, during one crossing shortly after the survey began, the boat was caught in a severe storm and the skipper drowned. Perth headquarters suddenly realized the difficulties and abandoned the project till the 1960s when a landing barge was used.[25]

An equally hazardous programme was carried out on Dirk Hartog Island near Shark Bay during the mid-1950s when two rigs were landed for a multi-well stratigraphic programme (16 holes in all) while the geologists mapped the area. The difficulty was not so much with tides as with the strong prevailing wind which actually held back the water outside the Bay area and made the near shore regions very shallow. Jesse Haddock from the Learmonth base was put in charge of the operation.

> I engaged a local fisherman from Denham as pilot and we brought the company's landing craft down from Exmouth. It was a risky time what with sea mists and sand banks, but we managed it OK. Unfortunately the drilling results were not much good. All they found was water saltier than the sea.[26]

Some of WAPET's exploration forays swung closer to Perth during the mid-1950s when geophysical and geological parties examined the region between Geraldton and Gin Gin. But the main focus of attention continued to be further afield. The Carnarvon Basin programme, and Rough Range appraisal work in particular, continued to catch the headlines. Yet the company faced an even greater challenge in exploring the huge 362 600 km^2 permit covering the Canning Basin in the State's far northwest.

At the same time as the preparations were being made to drill Rough Range No.1, WAPET was setting up its Kimberleys headquarters in Derby, 1 300 air kilometres further north and a total of 2 250 km from Perth. In 1953 the town was principally used as a supply port for the cattle and sheep stations in the region. It was ideal for the exploration company's needs because it had an all-weather airstrip (essential for use during the monsoon season) and a mile-long jetty which was necessary because of the shallow water and the 11 m tides which occurred along the far northwest coast. Derby was also close to Grant Range, the area chosen as the place to drill the company's first Canning Basin well. Soon after setting up a staff house in the town, along with a goods yard to store cargo needed for the field operations, WAPET looked round for a vessel to transport its equipment to the region.

It had already bought the Federal Government's National 100 rig to use on the Grant Range No.1 location, and the most practical way to take it from Melbourne to the Kimberleys was by sea. However the ship had to be big enough to carry about 4 000 tonnes of rig and ancillary gear and have a keel flat enough to sit comfortably on the mud during unloading at low tide in Derby. Eventually the Danish vessel *Jacob Jebsen* was contracted and the massive operation began during the early months of 1954. Preparations included strengthening the Derby jetty to carry the trucks and rig sections from ship to shore, as well as upgrading the main roads out of town and constructing a new track through the bush to the drill site. A 60-man tent camp was also erected to house the construction workers brought in to actually build the rig. Bevan Cook, originally hired by WAPET as a clerk for Derby office but soon a jack-of-all trades including truck, grader and crane driver, witnessed the long preamble to the drilling of Grant Range No.1.

> Once out of their crates the rig substructure and mud tanks were all specially skid mounted and hauled into place using

the Mack trucks. But the drilling derrick itself had to be built from the ground up by bolting each piece to the next like a giant Meccano set. WAPET brought in a team of American Indians from the US specially for the job.

> Apparently certain tribes over there were well known for their stevedoring prowess. They were familiar with rigs of this type and had no fear of heights. We made up a hub on one of the Land Rovers and used it as a rough pulley to haul each section up to them.[27]

Cook said these men were all characters and he recalled that one of them came to grief in a way that caused much amusement to the others.

> It was pretty hot up there with no shade and these chaps used to shin around naked on the derrick angle iron. Well a group of MMA [McRobertson Miller Airline] hosties came out one time to visit the site. This one bloke was caught trouserless up the derrick in full view, so he figured the best thing to do was sit down. The trouble was the angle iron was red hot from the sun. While his feet were like leather, his backside was not and you could almost smell the burning flesh 100 metres away.[28]

The well itself was finally spudded during October 1954, but after all the careful preparations and a total of 14 months (including drilling time) on location, Grant Range No.1 was plugged as a dry hole. Following that experience WAPET bought a jack-knife derrick for the rig and used the fixed one as scrap. The next well, Fraser River No.1, brought logistical difficulties of a different sort. The location was in low-lying river flats prone to flooding during the wet season and, as the monsoon was fast approaching, all haste was made to transfer the 70 truck loads of rig and camp to the new site. To keep supply lines open during the wet summer months WAPET went to the expense of building an all-weather airstrip nearby. All personnel, food and materials — even drill pipe — were flown in by DC3 from Derby. Unfortunately, hopes of a discovery in this large structure were dashed when the bit encountered basement without finding the objective — Devonian-age limestone reservoir. The rig was then mothballed on site for the time being as geologists and survey teams spread out to try to delineate new targets.

During 1955 WAPET contracted Aero Services Corporation to do an airborne magnetometer survey of the Canning Desert, using a specially fitted out wartime Mosquito bomber and navigating via a spread of air photos previously taken by the RAAF as part of Harold Raggatt's national mapping programme. Two ground teams pushed into the desert just prior to the flights to make 50 accurate astronomical positional fixes as added control points. At the same time, ground parties ran gravity surveys over sections of the basin to try to outline the subsurface structure. Robin Elliot, who had previously participated in a series of mapping surveys in the coastal regions of both Carnarvon and Canning Basins, moved into the eastern Canning to run WAPET's first gravity survey in that area.

> The area covered was between Christmas Creek and the Stansmore Ranges and we set up a gravity station every eight kilometres, fixing each one using a barometer and the air photos. Back at camp the cook would record his barometer every hour so we could have a check datum for our readings.
>
> We had it down to a fine art and carried the delicate gravimeter in a special case in the vehicle. Our accuracy was good too. We actually found the Stansmore Fault using this method and later techniques did not alter its position from our calculations.[29]

The east Canning presented fewer access problems than the coastal region between Broome, Derby and Port Hedland known as Dampierland. The latter area was covered by a dense, low vegetation known as Pindan Scrub which was virtually impenetrable by conventional vehicle. Even the four-wheel drive survey jeeps and wagons had trouble making any headway. In 1956 it became obvious that vehicle breakdowns in this desolate and difficult area were costing the exploration programme too much time and money. WAPET began an experiment by sending in bulldozer crews to clear a series of rough but passable tracks to open up the region and minimize delays in gathering geophysical data. The success of this operation was immediate and led to the possibility of the gravity crews being able to continue, albeit slowly, through the wet season. WAPET decided to step up the track cutting and an array of differing makes and models of earth-moving machinery attacked the scrubland during the second half of the year.

In sweltering heat and dust, and with the threat of scrub fires and the torment of flies, the crews worked to follow the grid patterns designed to tie in with the geophysical surveys. The bulldozers could not work through the wet so the contractor needed to get far enough ahead of the survey teams to provide a bank of access trails to last at least three months. By the time the work had finished that year, more than 3 200 km of tracks had been cleared and WAPET's subsequent survey costs in Dampierland were slashed by over 50 percent.[30] But the movement of rigs from one location to the next remained the explorer's biggest headache, particularly during the wet season. Jesse Haddock was one who saw these problems at close quarters.

> Two years after drilling Fraser River we reactivated the rig. It was in poor shape after sitting idle for that time in the

> monsoonal climate and had to be refurbished, particularly the engines. That took a while and it was the beginning of the wet before we could start the move.
>
> The new well was at Samphire Marsh on Anna Plains Station not far from today's roadhouse called Sandfire between Broome and Port Hedland. Anyway it was a 400 km move and we had a line of upwards of 50 heavy trucks to shift the rig and camp. Less than mid-way through a cyclone caught us and the roads became quagmires.
>
> Some of the trucks bogged to their trays and we couldn't shift them for about two months. Others had to be coaxed, pulled and pushed through. We spent a lot of the time carting gravel to try to build a firm road base. Even forgot about Christmas Day until a bloke from town brought out some bottles of red burgundy. The wine was hot, but it went down OK with some tins of sardines and cracker biscuits.[31]

Although not successful in finding oil, WAPET gained experience in wet-weather work with every new well. Roads to locations like Meda and Frome Rocks in the region east of Derby were dressed with laterite gravel, while river crossings were filled with rock 'fords' capable of supporting heavy vehicles when under a metre of water during flood. The truck loads were strictly policed for weight when travelling on main roads (particularly after a bridge collapsed under the weight of a T32 rig's draw works outside Derby), while at the rig sites foundations were strengthened by drilling 30 m holes and filling them with concrete. Earth platforms were built to accommodate stores and rig-site buildings were raised a metre or so above the level of surrounding flood plains. When drilling through the wet season extra stocks and spares were kept at the rig in case regular supply routes were cut and, if near rivers, dinghys and rafts were kept handy. Sometimes a heavy-duty flying fox would be rigged up as extra insurance.[32]

Communication links were also vital and the crews were always equipped with radios capable of contacting Derby base via the flying doctor. As the 1950s progressed, single-side band radios were introduced to WAPET's operations and, weather conditions permitting, contact could be made directly with the town office. However these messages were not particularly 'secure', so any confidential information was posted or taken by hand. Murray Johnstone was involved in one incident which illustrated this necessity.

> In sending a normal daily drilling report from the field to Derby one time, I mentioned in passing that a waxy fluorescent type of mica had been brought up in the rock cuttings. Next thing the Adelaide Stock Exchange was on the phone to WAPET HQ in Perth wanting to know what the company was doing because its shares had shot up the boards.

> Head office had not even received my daily report from Derby at that stage and they were as mystified as the Adelaide Exchange. It turned out that a radio ham in Alice Springs had picked up my original transmission and fastened on to the word 'fluorescent' thinking this meant oil was present. From there the news spread that we'd made an oil strike.[33]

WAPET asked its field crews to be more careful in their choice or words from then on. However there was only one occasion for excitement on the Canning rigs, and it came late in the decade when Meda No.1, about 160 km east of Derby, recovered five barrels of oil. It was the first oil brought to the surface in the Kimberley programme, but it was not deemed commercial and the rig moved away. The continued lack of success was frustrating. During the middle years of the decade WAPET had joined the BMR in drilling several stratigraphic holes in the northern Kimberleys, and then two others on its own account, to try to more accurately delineate the Canning geology. This work had indicated that previous wells like Grant Range and Fraser Range had been drilled in a trough (now known as the Fitzroy Trough) where the target limestone horizons were buried too deeply for the drill to reach. Consequently, the drilling programme later in the 1950s moved up the flanks of the trough and concentrated on the ridge areas. However apart from Meda there was no oil found in those regions either.

In an effort to boost exploration incentive, the Western Australian Government Mines Minister, Bert Simpson, suggested that WAPET need only pay the State a 7.5 percent royalty on any commercial oil production instead of the statutary 10 percent. The government of the day upheld this special concession which became known as the 'Simpson Letter of 1957'. In a rider to this document, WAPET was allowed to retain 100 percent of each of its permits so long as the government was satisfied with the company's level of exploration expenditure.[34] This informal arrangement continued well into the 1960s and made it very difficult for other companies to gain entry to Western Australia. The Shell Oil Company tried unsuccessfully to win permits in the State on its own account during the second half of the 1950s and its envoys surreptitiously scouted the Carnarvon and Canning Basins for the most promising areas in readiness for a swift application if any acreage became available. When it became clear the Western Australian Government was bent on renewing WAPET's licences each year without any relinquishments, Shell changed tack to join Caltex and Ampol in the WAPET consortium itself during 1958, earning its 28.6 percent interest by paying all the group's exploration costs for the following 18 months. Effectively that brought WAPET to the threshold of the 1960s.

A review of the consortium's efforts over seven years shows that it drilled 63 wells (Cape Range No.2 being the deepest in the Southern Hemisphere at the time), built 530 km of solid roadway and carved a further 14 320 km of access tracks through the isolated West Australian inland — principally in the Kimberleys. It spent a total of £15 million on its hunt for oil during the decade, £6 million of that on the ultimately unsuccessful attempts to prove up a commercial field at Rough Range. WAPET was the first to agree that the results of those programmes had not matched expectations. Nevertheless the company's work, allied with the persistence of William Walkley, had changed the perception of Australia's oil potential and kicked post-war commercial exploration into action. The discovery at Rough Range in particular had electrified the nation and begun a scramble to put other explorers into the field.

Chapter 7
IMPETUOUS DISCIPLES

There were few explorers in the field prior to Rough Range. Most were leftovers from the pre-war period like Freney Kimberley Oil at Nerrima Dome in Western Australia, Lakes Oil in the oil shaft project in Gippsland, and some desultory efforts round Roma in Queensland. Frome-Broken Hill in the Canning and the Great Artesian Basins was a newcomer, although it was accused of being fickle in its early approach. With the exception of Maurice Mawby and Zinc Corporation, there was a feeling that the oil 'majors' within the consortium were merely using Australian exploration during the late 1940s as a staging ground while waiting for post-war re-entry into the Dutch East Indies.[1] To be fair, international companies had little reason to be heavily involved in Australia. They had cheap oil in abundance from the Middle East. In addition, their own experts had dismissed the prospects 'down under' as poor, and the attitude tended to be that Australia was a place to explore as a last resort.

Not so Papua and New Guinea. As soon as hostilities had ceased, the Australasian Petroleum Company (which included British Petroleum and Mobil plus Australian explorer Oil Search) went back to the suspended Kariava well near the Vailala River and resumed drilling. It was finally abandoned as a dry hole in 1948 but, by that time, field surveys were well under way to try to make up for time lost during the war years. Expectations were high as exploration gathered pace during the 1950s, however the dense jungle and rugged terrain were still formidable obstacles to be overcome. Geologist Keith Llewellyn was recruited to APC in 1946 and participated in a number of the post-war programmes.

> The first thing I did to get the feel of the country was to walk over the Kokoda Trail. After that I was kept pretty fit on the various surveys. Sometimes we had a central camp if we were doing one area in a lot of detail. Other times we moved on every day.
>
> We were usually off at first light which meant breakfasting in the pre-dawn. The night camp needed to be set

up before the rains which generally came in about 2.30 p.m., and we often sent some of the bois ahead to do that while we worked on till about 4.00 p.m. It was a seven-day working week and a strict routine.

Sometimes, though, it rained all day. One time I remember 117 inches [3 000 mm] fell over a three-month period. Everything was flooded, even the potholes and chasms in the limestone country where we were working. Boots got mouldy overnight and the gear was sodden the whole time.

Most of our surveys were chain and compass with the help of air photos, if they'd been taken, of the particular area we were in. Other times it was plane tabling. During this time I examined an oil seepage near Lake Kutubu and named the Iagifu anticline after an old village on the nearby Digamu River.[2]

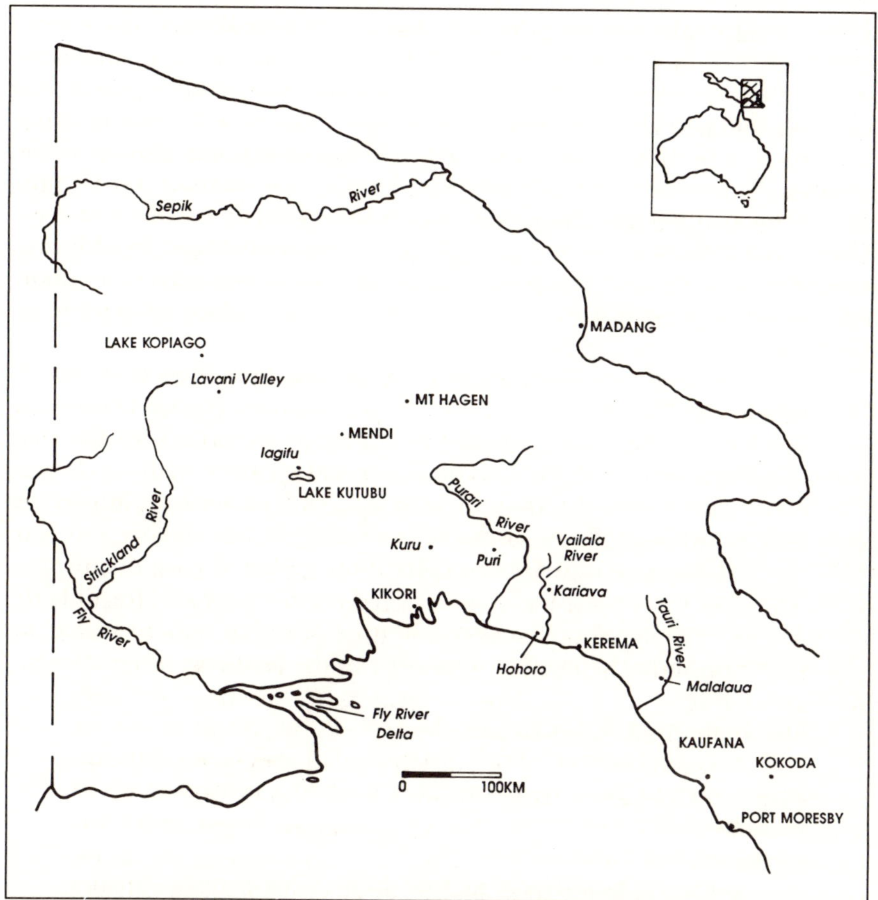

Papua New Guinea — location of key places mentioned in text.

The parties often used wartime bomb scows or motorized canoes to move up or down river. Llewellyn said that the canoes were usually fashioned on the spot and sometimes several would be strapped together.

> A special type of tree was selected and bois would take the bole end, which was the toughest part, taper it, then hollow it out. One I had built was big enough to carry 15 tons of explosives. It was huge. I sat in a director's chair in the back and the arm rests were still below the canoe gunwales.
>
> We fitted an outboard straight on the stern, but the first time we fired it up the motor wrenched the back of the canoe right off. We spent a day fishing for the motor and then drying it out. After that we fitted a transom reinforced with steel bands.[3]

One of the most celebrated survey treks of this period was in 1954 when several APC employees, including Llewellyn and fellow geologist John Zehnder, journeyed to the headwaters of the Strickland River. The patrol was of interest to the Papua Administration and hence the Australian Government in addition to APC because it traversed country not previously penetrated by whites. The survey party was flown into Lake Kutubu by Catalina flying boat as a starting point and the men then progressed on foot via Tari to Lake Kopiago, on to the source of the Strickland and finally down river again to the coast — in all a four-month trip. During a side patrol Zehnder found and walked through a 'hidden' valley now known as Lavani, but at the time more mystically referred to as 'Shangri La'.

> APC wanted to see what the high mountain potential was like as all the geological work to that point had been done in the foreland areas and in the 'foothills', if you could call them that in New Guinea. We were accompanied by a government patrol officer and a dozen or so of his police bois, plus a line of about 100 carriers.
>
> Along the way we geologists would do separate side mapping trips for a few days at a time, taking about 20 bois to help. We'd measure strikes and dips of the strata and take rock samples. The only outcrop was in the rivers, so virtually all of this geology was done in water. We reckoned we were 'dry shod' if our hats showed above water level!
>
> It was on one of these side trips that I came across Lavani — a wide, grassy valley whose people hadn't seen a white man before. I had to demonstrate what steel was and what to do with it as they only had stone axes. Even the beads we used as trade for food had to be strung for them because they had no idea what to do with them.[4]

Unfortunately the expedition was clouded by a tragic incident on the return trip down the Strickland. The party built canoes to float down river to its junction with the Fly thinking they were below the main rapids. However in a treacherous piece of water one of the craft was caught in a whirlpool and nine of the bois were drowned.

Not all APC's post-war work was concentrated on survey treks. The company also drilled a total of 18 wells between the late 1940s and 1960, most of which were based on the surface mapping done by the geological parties. The first well that Zehnder 'sat' was Hohoro on the Vailala River in 1951.

> A previous attempt to drill there had been made in 1926 and the old rig was still on site. APC wanted to retrieve it so they shipped in a gang of riggers for the job. They were tough blokes, typical of the rig builders then. Two of them arrived with a huge box of beer and proceeded to drink their way through the lot in one night. It amazed me that next morning they were no worse for wear. Another time I swear one of them actually flexed a crowbar during a fight![5]

Often a rig would be on site for 12 months or more, but up to a sixth of that time was spent in setting up the equipment itself. The cellar would be dug first, then the drill floor placed over it. After that the riggers moved in and built the derrick up piece by piece. The residences were erected about five kilometres away. Mostly they were frame and tar-paper sides with a galvanized iron roof. Residents would do their own plumbing. Usually the mess hut and store were at one end and the accommodation huts set up in lines leading away from that point.

The results of APC's drilling programme were frustrating. More often than not gas and oil indications would be found either in tests or in cores and drill cuttings. But there was nothing of commercial value. However some excitement of a different sort occurred in 1956 when Kuru No.1, drilling about 40 km inland from the town of Kikori, hit a shallow gas pocket and blew out. Keith Llewellyn was well-site geologist at the time.

> Apparently the blowout preventers were not hooked up properly and they could not close the well off. Mud shot out of the hole under about 500 psi [3 447 kPa] pressure and this was followed by scouring and channelling of gas outside the surface casing.
>
> APC contacted trouble shooter Myron Kinley and he came out from Canada to assess the damage. Gas was everywhere. We had to close two diesel engines down and an air intake line was taken out into the bush so we could stop the third one racing and get it back to governed speed. You could hear

> the well screaming from the camp three miles away. Any closer and you couldn't hear yourself think.
>
> I can remember Kinley standing with a box of gelignite on the rig floor pondering whether to blow the well and collapse the sides. In the end he decided to drill a relief well and they managed to dismantle the rig and re-erect it to one side.
>
> We installed pumps in the nearby creek and pumped 25 000 gallons [113 650 L] of water a minute down the relief hole to waterflood the reservoir and kill the wild well. As soon as the gas flow stopped, cement was run down to plug it off.[6]

The whole operation had taken five months. During that time the uncontrolled flow was estimated to be 2.8 million m^3 a day, but two more wells on the structure during 1957 and 1958 encountered only small gas shows. Both appraisals had to be abandoned due to caving problems and the inability to carry out a thorough test programme.

As was the case before the war, the 'character' of the post-1946 APC team was G.A.V. (George) Stanley. Geologist, botanist, excellent bushman, and more recently a coastwatcher in the Sepik during the Japanese occupation (for which he received the Distinguished Service Cross and the American Legion of Merit), he was well liked by his Australian colleagues. There was less admiration from British and US staff members, as John Zehnder witnessed more than once.

> He loved to take the mickey out of visiting English geologists — particularly the senior ones. He'd take them on a tour of some feature or other in the bush and get them lost. He never was himself, of course. He knew the jungle like the back of his hand and was one of the few whites who could live off the land. He was adored by his bois who called him the 'bush pig' because of his prowess.
>
> A lot of the Europeans objected to his having so much to do with the natives, but he was helpful and patient with we younger geologists, if a trifle eccentric. He used to sell empty beer bottles and he'd go collecting them round Moresby towing a trailer full of grinning, mischievous black boys as 'helpers'.
>
> He was crafty too. When he came to visit he'd often complain of a sore throat and indicate that a touch of sherry might do him good![7]

Stanley retired from APC in 1950 and for the next 12 years worked as a consultant based in Port Moresby. During this time he wrote a two-volume report for Papuan Apinaipi Petroleum Company on the history of exploration and the geology of Permit 22, a document which covered much of the Papuan Basin. It makes fascinating reading today and is still used as an essential text by modern explorers in the region. Then in 1962 he joined the Bureau of Mineral Resources. Although originally a Sydneysider, Stanley counted himself a Papuan and, like a number of whites who settled in the tropics, he did not want to leave. However he was persuaded to go to the BMR offices in Canberra to work on and write up the sedimentary geology of Papua. It was a laboured task in a climate and surroundings Stanley did not relish. Aided by friends and workmates, he went back to New Guinea as soon as he could and died there of a malignant melanoma in October 1965 at the age of sixty-three. An old friend and colleague, Peter Best, gave the eulogy in which he aptly described Stanley as 'the last of the erudite Edwardians'.

APC was not the only company which carried over pre-war connections with Papua into the 1950s. Papuan Apinaipi Petroleum was another, although it struggled to maintain its work obligations until amalgamated with the Queensland-based Associated Australian Oilfields Group of companies in 1956. Colin Laing was one of the first of the new management's geologists to work in the company's Papuan lease.

> Before going up there we received a lecture about not rocking the boat. At that time it [Papua] was the last bastion of white man supremacy. There were still some who believed that the way to handle the natives was to shout loudly in English and fire a few shots from a revolver over their heads.
>
> Luckily I had Tom Cutcliffe, a six-foot eight-inch [203 cm] tall expatriate Australian who spoke fluent Motu [one of the most widely understood languages in the Papuan Gulf region], as native labour officer. His father being a chemist in Sydney, he had a thing about medicines and our copious supplies included a one-gallon tin of cod liver oil. He used to line the carriers up every morning and dose them with a spoonful of cod liver oil each.
>
> We were working in the Tauri River area on the Gulf, mapping partly exposed limestone reefs in Miocene-age mudstone. It was possible to determine places where the line of reefs pitched under the mudstone, but the work was never followed up. After that I shifted downstream to try to confirm closure on the Malalaua Anticline and traverse up Maipora Creek where there were more outcropping reefs.[8]

As time went on without any definite plans emerging to move a rig into the area, the government increased its pressure on the company to drill a well. But despite the work of Laing and others, the only structure in Apinaipi's acreage then known to have closure was Kaufana. Access to the location was along the Aroa River from the Papuan Gulf, and a barge — the *Tarmona* — was specially built in Brisbane to ferry in the rig and equipment. But even that did not go to plan. When it arrived in the Gulf the barge was found to be grossly underpowered and took 14 days to run against the strong current. Three more days were needed to cut a track into the well site. Kaufana was finally drilled during 1958 on the west side of the river, but it was dry. Apinaipi withdrew from the country soon afterwards.[9]

During the 1950s more than 100 anticlines had been discovered in western Papua and all the wells drilled were located to test the most favourable structures. The targets varied between Miocene-age carbonates and Mesozoic-age sandstones but, although several gas finds were made along with an encouraging oil flow at Puri No.1 near the Purari River, they were not commercially viable in such rugged, isolated territory.

By this time, however, the discovery of oil at Rough Range had radically changed the exploration picture back in Australia. Where previous interest in oil had waned during the immediate post-war years, suddenly a number of enthusiastic disciples of Ampol's William Walkley sprang up and fanned out through all States. Optimistically they set up programmes and published glowing prospectuses, all declaring they were sure to be the ones to make the country's first commercial strike. Sadly, one of the staunchest backers of the pre-war oil exploration effort, stockbroker and company director Wallace H. Smith, did not live to see the results of the Rough Range well. He died a few days before they were announced. Ironically, even poignantly, just before his death he had grudgingly come round to the idea that Australia, after all, was too old to contain substantial free-flowing oil reservoirs. It was a theory he had disbelieved all his working life.[10] Smith had concentrated his 1920s and 1930s efforts on Gippsland in Victoria, Roma in Queensland and the Hunter Valley region of New South Wales because tantalizing evidence of oil and gas had been recorded at those places in the past. Others in his era had tried western Victoria and South Australia for similar reasons. Not surprisingly, the new generation of explorers began the same way.

The largest company to set up in Gippsland was Frome-Lakes Pty Ltd. Regretting its premature withdrawal from Western Australia, the Frome-Broken Hill consortium reactivated and took up leases in the onshore Gippsland region along with onshore Otway and the onshore Carpentaria Basin of Queensland. In mid-1954, urged on by the unabated enthusiasm of Maurice Mawby, the company entered a joint venture with Lakes Oil and began a drilling programme in a 3 626 km^2 area of Gippsland along the Ninety Mile Beach. At the same time smaller Australian hopefuls like Woodside (Lakes Entrance) Oil Co NL, Oil and Minerals Quest NL, Mineral Ventures NL and Pacific Uranium and Oil Syndicate NL set up in adjoining areas. Such was the anticipation after Rough Range that even Australian Paper Manufacturers Ltd held prospecting leases in the Gippsland region at that time.[11] However many of these companies preferred to sit back and wait for the results of drilling by the larger concerns. Woodside (Lakes Entrance) Oil was a notable exception and consequently its exploits were keenly watched by shareholders, the media, brokers and the general public alike. Rumours of oil strikes abounded and it was commonplace for company officials, including Rees Withers, the founder and managing director, to be phoned at the unsocial hour of 2.00 a.m. and asked to comment on a 'discovery'.

> When drilling our first wells between the town of Woodside and Welshpool we had to build an eight-foot [2.5 m] high chain wire fence around the rig to protect our operators from sightseers. We even had a sentry box inside the locked gates and employed a night watchman.
>
> The thing was we'd built a rough track along the sand dunes of the Ninety Mile Beach for access to the rig and camp, and this had opened up some excellent surf fishing sites for the locals. During fine weather there'd be dozens of cars lined up round the rig by 5.00 p.m. The fishermen would go off down to the beach and leave the kids to play round all the inviting high-voltage cables, machinery and the like.
>
> Other times there'd be pressmen and brokers or just intrigued passers-by. Naturally as soon as the fence went up there was even greater curiosity about our activities.[12]

However not all the locals were impressed with the oil business, particularly when it meant the rig and equipment moving across their land.

> After selecting one drilling site I visited the property owner and was met by the business end of a shotgun. It turned out that an oil company during the war had wanted to drill on his land and he'd lent them petrol ration tickets to help out. Apparently he'd never been paid back. On top of that the company consistently left his gates open and knocked fences down.
>
> Naturally enough he wasn't too enamoured with oil drillers. However his daughter persuaded him that we should all go inside and have a cup of tea and talk about it. Eventually I came out with permission to drill.[13]

On another occasion Withers was having a beer in a local pub with exploration manager Eric Webb and drilling manager Bill Scoble, and talking about the progress of a seismic programme being shot over the company's acreage. Suddenly they were confronted by an angry old farmer.

> 'You oil fellows?' he asked. We said yes. 'Well then, what about my ferret?' We were a bit nonplussed by that and it

> took a while to get any sense out of him. It finally transpired he'd put his ferret into what looked like a rabbit hole, but which was really a vertical shot hole drilled as part of the seismic survey.
>
> Apparently the ferret had instantly disappeared and all the farmer heard was a scrabbling at the top of the hole, followed by a long silence and then a dull thud. He waited for a long while but the ferret didn't come back, which was not surprising because those holes were about 30 m deep.
>
> Bill Scoble was on the point of collapse from laughter by this time, but the farmer was not at all amused.[14]

Although some early seismic work had been carried out over the Gippsland region at the instigation of the BMR, Woodside's first well (Woodside No.1) was selected by its consultant geologist, Professor E. Sherbon Hills of the University of Melbourne, purely on surface geology. But the exact spot was not sacrosanct.

> The Minister of Mines of the time came out to have a look round. He must have been in a playful mood because he took it into his head to shift the location marker peg 15 ft or so [5 m] to a place more to his fancy. It was all the same to us, so we left it there.[15]

Communications with the rig were equally ad hoc in the early days. There were only five or six telephone subscribers in the district at the time, including the rig. The telephone operator was a local farmer and when he went out to milk his cows the switchboard was often left unattended and the rig incommunicado. However the rig camp itself, particularly for the first three wells round Woodside township, was a real home away from home. It accommodated up to 50 people and all personnel had separate rooms with electricity and bathing facilities. The mess was equipped with a big freezer and a slow-combustion stove which apparently was used to gourmet standard by a chef and assistant cook. Menus consistently included fish, oysters, crays, poultry and beef. In addition there was a flourishing flower and vegetable garden nourished by the copious flow from an artesian water well drilled to supply the rig and camp.[16]

Woodside had also taken the unusual step of buying its rig, rather than bringing in a contractor for the work. Technical expertise was supplied by two 'old hands' in the oil game — Bill Scoble, a jolly, short, rotund, hard-drinking but likeable man who had experience in drilling all round the world including Papua and New Guinea; and Clare Lowe, a laconic tobacco-chewing Texan with a drawl to match. Managing director Rees Withers, an accountant by trade, learnt an early lesson in Texan oil 'technology'.

> I was down at the rig one day watching the stream of drilling mud and cuttings coming from the well when I noticed a brown stain in the otherwise grey flow. I was pretty excited by that and rushed up to Lowe asking if we'd found oil. He just looked at me for a bit, then drawled, 'That's tobaccy juice, son'.[17]

Unfortunately there was no oil in the company's other wells during the mid-1950s either, and the company fell on financially hard times, only surviving because Withers and chairman Geoff Donaldson put their own money into the kitty. Other companies, including the financially robust but managerially difficult Frome-Lakes, were equally unsuccessful on the drilling front and one by one they began to withdraw. After drilling a series of dry holes along the coast in the four years to 1958, Frome-Broken Hill geologists formed the opinion that the onshore Gippsland sediments were water flushed.[18] The company severed the relationship with Lakes Oil and moved across to its Otway permits in western Victoria where it spent the rest of the 1950s and the first few years of the 1960s. Its thrust was mostly onshore, but the company also mounted one of the country's first offshore campaigns, located in the Otway region. Port Campbell No.1 onshore gas discovery in 1960, close to the town of Port Campbell near Warrnambool, was its only real success. But, although it was the first significant indication of hydrocarbons in the basin, appraisals indicated it was not a commercial find. Nor was it sufficient incentive to hold the participants (Esso, BP, Mobil and Conzinc) together in a lasting relationship. After some difficulties with anti-trust laws during the early 1960s, the arrangement was dissolved and the companies went their own ways.

Back in Gippsland, meanwhile, Woodside slowly struggled through its financial crisis. By 1959 the company still had not found any oil, but it had gained enough strength to acquire more acreage in the region and interstate. That year, at exploration manager Eric Webb's instigation, Woodside also took out an exploration lease over a 13 km wide strip of water just off the Ninety Mile Beach between the town of Orbost in the east and Wilson's Promontory in the west. Webb wanted a 32 km wide strip, but the company simply could not afford to stretch its still-limited finances any further and risk committing to a work programme it could not fulfill. Nevertheless, there was a growing feeling that the premonitions of earlier Gippsland geologists might just be correct — that the real oil prize would be found offshore. As if to underline this message, the BMR took the bit between its teeth during the last months of the decade and ran a regional aeromagnetic survey out over Bass Strait. In Victoria at least, the stage was set for the amazing 'purple patch' of exploration in the 1960s.

In Queensland the first concentrated post-war exploration effort largely came through the efforts of two veterans of the 1920s Roma boom

— stockbroker Wallace Smith and mining/oil entrepreneur Clarence Byrne. They believed there was still potential in southeast Queensland despite the disappointment of Shell's withdrawal and they gathered together four old Roma explorers that had somehow survived the intervening years without being wound up. In 1951 Byrne and Smith succeeded in amalgamating Roma Blocks Oil Company NL, Kalimna Oil Company NL, Roma North Oil Company NL and Australian Oil Development NL into a company called Associated Australian Oilfields NL, and this company took up a permit of 520 km^2 around the town. AAO immediately drilled a well called Roma No.1 on the old stamping ground of Hospital Hill and established a gas flow of 42 000 m^3 a day. Encouraged by this start, but realizing the limitations of the single Roma permit, the company began to spread its wings by moving up to the Gulf of Carpentaria. Byrne was familiar with that region through his directorship of the Zinc Corporation and its association with Frome-Broken Hill. At the time there was a general interlocking of mining and oil exploration, particularly on the technical side.

> I borrowed Zinc Corp's leading geologist, Harry Evans, for some work around Roma for AAO in the early days. That was after he had been out in the Kimberleys with Frank Reeves in the late 1940s and the company didn't have a lot for him to do.
>
> Then Maurice Mawby borrowed him back again after the Rough Range well came in and sent him up to the new Frome-Broken Hill leases around Carpentaria to look for oil possibilities there. Instead of oil he found bauxite on the Gulf side of Cape York and that's how Zinc Corp, or Conzinc as it soon became, got involved with the Weipa development. Evans was lost to oil after that.[19]

Another of AAO's early geologists was W.D. (Dick) Mott, a colourful yet tragic character who was instrumental in setting up the group's technical expertise and then guiding it to a place of prominence in Australian oil and mining circles during the 1950s and 1960s with an astute recruitment programme. Educated in Queensland, he was one of Australasian Petroleum Corporation's pre-war recruits for well-site and survey work in Papua and New Guinea. Roy Richter, who later joined AAO at Mott's instigation, first met him in New Guinea.

> He was a good geologist, a pretty hefty bloke, outgoing and well liked. The same as me, he joined the airforce during the war and he was a navigator attached to Coastal Command. After being demobbed he went back to APC in Papua for a brief period before gravitating home to Queensland.
>
> He joined the Queensland Government Survey and then

> did a stint as part-time lecturer at the University, eventually becoming a consultant. He told me AAO was looking for someone with my drilling knowledge and I joined in 1953.[20]

Mott's efforts to collate as much data as possible about Queensland geology were recognized by later AAO recruits like Colin Laing.

> When employed by the Government Survey, he compiled a comprehensive report which not only listed all the exploration wells in the State to that time, but also analysed and reported shows of oil and gas from water wells. He attempted to elucidate the subsurface geology of Queensland with a series of regional cross sections.
>
> The paper [later updated and published in full as GSQ Report No.299] named the Surat, Eromanga, and Carpentaria as sub-basins of the Great Artesian Basin and predicted the continuation of the Bowen Basin Permian (sediments) under the Surat Basin. He also recognized the problem of artesian water flushing of Mesozoic-age sandstones in the Great Artesian Basin.[21]

Others brought to AAO via Mott's influence were Doug Traves and John Glover (from the BMR), Sid Stormonth (from the Tax Department), Stuart Derrington and Rod Allen. Always keen to promote the oil industry and disseminate its news, Mott began the successful monthly *Australasian Oil and Gas Journal* in 1954 and remained its editor, then technical editor, for a number of years. Tragically, however, during the early 1960s he developed alcohol problems, became disillusioned with the entry of so many foreign companies and their apparent takeover of the Australian effort, and died after an accidental fall outside his Brisbane office in April 1967.

Mott's work, via the magazine and through his more technical presentations, formed the base for renewed interest in Queensland's oil potential. But, as in most States, the influx of new companies was most noticeable after the Rough Range discovery in 1954. About 18 wells were begun at locations right across the State during the 12 months immediately following WAPET's success in Western Australia. Some were sited on geological advice, others near old water bores which had yielded petroleum indications, and some relied on the dubious advice of diviners.

One of the first American entrepreneurs to arrive on the scene was a larger-than-life, self-made man from Arkansas named Omajean A. (Gene) Goff. With a self-professed 'nose' for oil and an equally high opinion of his business acumen, Goff took a liking to the Maryborough Basin located between the dividing range and the coast several hundred miles north of Brisbane. Installing himself in a suite at Lennon's Hotel in the city, he applied for a permit over the region in the name of Lucky

Strike Drilling Company. Richard (Dick) Dumbrell, who rose through the ranks to finally become chairman and chief executive of allied company Oil Drilling and Exploration, was one of the first to come in contact with Goff.

> He was a big man, with tooled leather boots and belt, western hat and a gravel-voiced drawl — a real southern wildcatter.
>
> Having obtained the Maryborough concession, he needed to drill wells, but there were no rigs in the country in 1954 other than those being used by WAPET in the west. This only caused Gene a moment's hesitation. 'Goddam,' he'd say, 'this country is going to become an oil province. I've got a goddam nose for it. I'll start a goddam drilling company'. So he did.
>
> Lucky Strike was the name of one of his companies in America, but to drill in Australia he formed Oil Drilling and Exploration (ODE) in March 1954. In turn, ODE took a 20 percent interest in the concession.
>
> This practice was (and still is) quite common in the US, particularly with small contractors, although the percentage stake is smaller. In Australia the exploration programme was much riskier, so the interest had to be higher.[22]

Another Australian to meet Goff early in the piece was C.W. (Bill) Siller, then a young geologist with the Queensland Irrigation and Water Supply Commission. He joined Lucky Strike as a field geologist.

> Gene was always full of energy and his passionate hobby was racehorses. After a mild heart attack in 1959 he decided to return to the US and took with him a noted thoroughbred called 'Noholme', giving me the job of shipping arrangements. He raced him once for a big win and then syndicated the animal. The stud fees were something over $US1 million.
>
> Anyway, he left all the Maryborough work to me and the only order I got when he returned home with his horse was 'Look after the concessions'. I had already negotiated my pay but, when three months went by without receiving any, I cabled the States to ask when I had started work and how much I was earning.
>
> That was settled and from then on I was paid the gross amount and left to sort out the Australian tax side myself. US companies and Gene in particular did not want to get mixed up with the 'goddam Feds'.[23]

Apart from his readiness to set up a drilling effort in Queensland, Goff had also pioneered commercial seismic surveying in eastern Australia by bringing in a company called Foreign Oil Exploration to run a programme over the Maryborough permit. Located close to the central Queensland coast, work in the area might sound like a fairly attractive proposition, but Siller says otherwise.

> We all lived in town and made daily treks to the rig, but it was not always a bed of roses. I remember finishing the road building 12 months after the rig had started on the three-well programme. It had come in by train.
>
> Moving around the area you had to stick to the wheel ruts. An inch off and there was big trouble. A couple of nights of rain could measure up to 14 inches [356 mm] of water — enough to bog a duck![24]

As it turned out, the Maryborough programme was unsuccessful and Goff redirected Lucky Strike's sights to the Quilpie region of western Queensland, while ODE sought drilling work right round Australia. Not long after, Lucky Strike merged with a company called Seneca Oil Co. Goff then sold out of Seneca and left the Australian scene altogether. His dream for ODE had nearly failed when the company's four imported rigs stood idle in 1956. It was a blow after the struggle to set up operations only two years before. Some of the early problems had arisen in writing contracts and bringing in experienced drilling personnel. In the beginning, few exploration companies in the country had any experience of writing a drilling contract. Many, during the first days after Rough Range, had little enough knowledge of the oil industry itself, and weeks were spent in negotiation of the documents, particularly over liability clauses in case of drilling difficulties. In addition, the contracts awarded had all been of short duration and, as the hoped-for discoveries had not eventuated, the rigs were left stranded with no ongoing work. That in turn meant the staff had to be reduced to 12 key people, and many of those agreed to accept 50 percent salary cuts until business improved.[25]

The original importation of US drillers had brought ODE face to face with the strong labour unions in Queensland, where much of the company's early work was carried out. When permission was sought for US citizens to enter the work force, the OK was given as long as the men joined a union. The question then became: which union? After a lot of discussion it was decided that because the rigs were so powerful the most relevant body was the locomotive drivers union. Thus the American drillers had to sit for their locomotive exams. The farce came full circle when ODE acquired a set of answers so they could all pass the test and get on with the drilling contract.

Another brush with unionism came when a directive was received saying that members of the steel workers union would erect the rig derricks. Upon enquiry, ODE management was told that full scaffolding was needed and the job would take about six weeks. The company chose to ignore this and went ahead with its own erection programme, completing the job in four days. Steel workers union officials were then invited out to inspect the rig. There was no further attempt at interference. In any case there was little need because during the mid-1950s ODE exchanged its cumbersome derricks for drilling masts which could be set up in one piece.

During the slump of 1956, however, the company fought a constant battle for recognition, and management virtually had to give personal guarantees that its men could and would do the job. Work on WAPET's ongoing search in Western Australia helped out in 1957 and gradually the company's fortunes began to improve. As a fledgling Australian drilling company, ODE had to compete with the major international contractors. One of the strongest rivals at the time was the US-based Santa Fe Drilling, which was hired to drill a much-vaunted prospect called Reid's Dome in the Bowen Basin southwest of Rockhampton — a region now known as the Denison Trough. The exploration permit had been taken out by a newly floated company called Australasian Oil Exploration Ltd which sported a management team more suited to mineral exploration than oil. It was formed from the Cable Syndicate, a group of mining prospectors in Western Australia headed by mine owner Douglas Cable. The exploration manager was Jack Ridgeway, a rank outsider to the oil search whose previous experience had been as a mineralogist.

The prospectus gave a cursory mention to permits in the Fortescue Basin near Wittenoom and the Officer Basin in the far east of Western Australia, plus applications in Gippsland and in the Fly River delta region of Papua. Its main thrust was in promoting the Reid's Dome prospect, an anticline mapped by and named after former Chief Government Geologist of Queensland, John Hector Reid, during the 1930s. To this end the company appointed Reid himself to the board and hired him as consultant geologist, freely using his name in a drive to raise public funds. It was an ill-prepared programme from the start. Apart from relying solely on the 1930s original geology and mapping, a local group was hired for the drilling. It had neither experience in oil work, nor the necessary equipment, so it called in Santa Fe. The US company arrived before the camp and airstrip had been completed and the American drillers were driven back and forth daily from the nearby town of Rolleston by taxi.[26] Three wells were drilled in the area, two of which produced noncommercial shallow-zone gas. Lincoln Madden was an accountant with the company during this period.

> No one wanted the gas so AOE decided to drill deeper and they hit a series of hard carbonaceous shales that are now known as the Reid's Dome Beds. The company had originally contracted to take a certain amount of core and the Queensland State Mining Engineer, Ian Morley, held them to it.

> So the work went ahead at about 15 m a day, eventually finishing at a depth of about 1 525 m. No one really wanted the cores and they eventually ended up on the Southport dump. The second and third wells went the same way. By then the total cost for the program had climbed to £2.25 million.
>
> Mind you, when the camp was finally set up, it was as though oil had already been found and the company was there to stay. The huts were steel framed with concrete floors, flyscreens and sewage.
>
> When the programme was over without success, everything had to be sold. Some of it was auctioned on site, the rest carted away and sold later.[27]

Although field reconnaissance surveys were carried out in AOE's other permits, no drilling was attempted and the company management began to look round in the industry the directors knew best. Uranium was a boom mineral in the mid-1950s and AOE seized a chance to buy into the Mary Kathleen syndicate which had made a rich find between Cloncurry and Mt Isa. Later it on-sold the deposit to Conzinc and went into mineral sands. Other companies to enter Queensland during the 1950s were many and varied. Longreach Oil Ltd was prominent in the

Location of key places mentioned in text.

west of the State, led by 67-year-old managing director John MacD. Royle, a former water driller who had bought a small rig and turned to the oil business. He had been a distinguished soldier in the Royal Australian Engineers Regiment during the First World War and had set up listening/taping devices in tunnels under the German lines in Flanders — a forerunner of later seismic recording techniques.

Another newcomer was the Sun Oil Company of California which came to drill a series of wells in the Bowen and Great Artesian Basins. Interstate Oil Ltd announced plans for exploration of the Great Barrier Reef and Humber Barrier Reef Oil had similar ideas. A group called Oklahoma Australian Oil took up an area round Charleville and Quilpie in the far west. Other US explorers like Magellan Petroleum, Plymouth Oil, Cordillera Mining and Phillips Petroleum arrived towards the end of the decade. Perhaps the most unexpected exploration group was a syndicate formed by two Queensland Country Party members of the State Parliament to search in the Mackay district. In Opposition at the time, the two were Ernie Evans and a young newcomer named Johannes Bjelke-Petersen. Bill Siller was still active in the area and was aware of the 'project'.

> Ernie and Joh used to knock around together a bit and Evans had what he called 'an area of interest' in the Mackay region round his electorate where bitumen had drifted ashore. The material had deposited on volcanic rocks and melted in. To the uninitiated it looked like a seep.
>
> No one could persuade Evans it wasn't, so the syndicate began drilling a well called Cape Hillsborough No.1 about 32 km north of Mackay during 1957.[28]

Geologist Dick Paten, who had just graduated and was newly employed at the Queensland Geological Survey, saw the other end of the story.

> I was on the Survey's 'Prospect Desk' when a representative of the Syndicate came in to report progress. The group was planning to evaluate oil shows, but when I looked at the data I recommended it would be best to stop because they were drilling in basement. The Chief Government Geologist of the time, Alan Denmead, subsequently wrote a critical report on the programme.
>
> I didn't know Evans was involved until word came back that he'd contacted William Spooner, the Federal Minister for National Development in the Menzies Government, saying that the Queensland Geological Survey was stifling the search for oil.
>
> To my knowledge Evans never visited the Survey or trusted geologists after that. And, to make it more pointed,

> the ALP was bundled out of office and Evans became State Minister for Mines — our 'boss' — a short time later.[29]

The Minister was a real rough and tumble character. But, despite his suspicion of technology and technologists, he was generally liked by the industry because he realized there was more to the business than merely pumping oil from the ground. He was a keen advocate of exploration in Queensland, but at the same time he kept his administrative feet firmly on the pavement, carefully considering the governmental and legal implications before making decisions on oil and gas policy. A former member of the Australian Light Horse Regiment, he had been a cane cutter and more latterly a cane grower. He had also been a champion axeman and even in later life his handshake was like a vice. A big man, he was fond of a drink and his hospitality cabinet was legend — open to all no matter what the hour of the visit. Ian McPhee, who was in Australia for the three company combine of Continental Oil, Amerada and Ohio Oil, remembers accompanying two of Amerada's top management on a courtesy call to see Minister Evans during one of their trips through Queensland.

> The Amerada men were from Tulsa in the US and very conservative, proper people. Evans introduced himself as Ernie and had his cabinet open in a flash, overwhelming them with his 'see me as I am' brand of hospitality. These people rarely touched a drop of alcohol, but they couldn't refuse. I'm sure it was the first time they'd taken a morning drink.[30]

With the increase in activity after Rough Range and the consequent influx of rigs, breakdowns became much more of a problem than they had been previously. Spare parts were scarce in Australia. Even keeping up a supply of essentials like drilling bits was a difficult and expensive task because virtually all had to be imported from America. There was no 'off the shelf' service in the country. The first to set up an operation to meet this need was a long-time oil field character named Jim Bayliss. Born in Worchester, England, he had travelled the world before coming to Australia with Shell's 1940s exploration programme in southeast Queensland. His oil service company was based in Brisbane where he started by picking up the agencies for the US tool and equipment suppliers during the early 1950s. Competition of sorts came a year or two later when Oil Drilling and Exploration set up its own company called Oilfield Supply Co. with an equipment yard near Bankstown in Sydney. However much of that inventory was used by ODE's own rigs. Drilling hand and equipment specialist Maurie Calderwood knew Bayliss well during that period and then joined him in the business in 1960.

> Jim was known as 'The Sheriff' because he always wore a western-style sheriff's hat. Not a 10-gallon one. It had the

> smaller crown, more like the one the Hollywood lawmen used. He'd been in Wyoming in the 1920s and must have picked up the fashion there.
>
> He was a tool dresser by trade then and a horse driver for the old skid rigs. His favourite saying was 'When I started there were iron men and wooden derricks. Now there are iron derricks'.
>
> Anyway he came to Queensland in 1948 in charge of Shell's drilling programme. When the company pulled out he stayed and set up his agencies. He'd also design drilling tools and other pieces of equipment and make them up to sell to the industry.
>
> Jim was only a small bloke — about five feet six inches [168 cm] tall — and fairly tough. But likeable with it. He was pretty profound and scholarly too and he kept a precise daily diary.
>
> I can remember he drove an E-type Jag — the first one in Brisbane — and he always carried a rifle in the car. One evening he was driving in the country to call on one of the rigs when he pulled up suddenly because he could see the red coals of a dingo's eyes. He took a pot shot and immediately there were yells and curses from that direction. He'd sent the ashes of a drovers' camp flying.[31]

Calderwood himself had seen a lot of the drilling side during the late 1940s and 1950s. He'd started by working on the railways in Queensland and then joined Shell as a roustabout, working his way into the drilling side of the oil business. During 1954 he accompanied the Federal Government rig to Derby after it had been bought by WAPET for the Grant Range drilling programme, and went on to do a lot of work for that company. But many of his experiences were in the Surat Basin round Roma.

> Life was pretty rough. I can remember being as cold as a frog one night near Roma. I wrapped myself up in blankets and then a plastic pallet bag. When I woke there was ice covering the whole bag.
>
> There was also a time when heavy rain isolated the rig for about nine weeks. No food trucks could get in and all we had to eat were pickles. When food was dropped from the air by a Hercules plane the parcels broke up and smashed everywhere. Trouble was we had no radio to tell anyone.
>
> On the drilling side there was often a problem when the wells hit the frequent coal seams. The gases thinned the drilling mud and there was danger of blowouts.[32]

There was also fun to be had, as any visitor was likely to find out to his or her cost.

> There was one time when about 30 bushwalkers came through the area and found the rig. The crew hammed it up like mad. When the bit came out of the hole some wise joker would yell down the well to an invisible worker 'You can come up now'. Then a bushwalker asked if people ever came out of the derrick. He was told 'Only once!'.
>
> The drilling supervisor often had trouble finding blokes to work on the rig, and when anyone came to have a look at the operation, he'd always ask them to bring their friends so he could try to recruit them.[33]

Although a number of exploration companies came and went in Queensland during the 1950s, the most active and best organized operation was that of the Associated Group. Along with the BMR, the Government Surveys and WAPET, Australian Associated Oilfields was the training ground for a number of geologists who went on to prominence in other companies during the 1960s, 1970s and 1980s. The group was an amalgam of small companies that were too small to support professional staffs of their own. A company called Mines Administration Pty Ltd (Minad) was formed to provide a pool of geologists and managerial people when needed. AAO's exploration during the 1950s covered the whole of northern Australia from Carpentaria across to the Kimberleys and north across Torres Strait to Papua — the Kimberleys via the takeover of Freney Kimberley Oil, and Papua through that of Papuan Apinaipi. Stuart Derrington, who graduated in geology from the University of Queensland in 1954, was one of the first to join Minad.

> The company tackled the work on two fronts, sending out geological field teams and following up with rigs and drilling crews. The survey parties were usually made up of two geologists and a field assistant with the latter doubling as cook.
>
> The field assistants were often farm lads who had a slack period between April and November which coincided with our northern field season, so it was an ideal arrangement for them. And they saw something of Australia.
>
> We lived under canvas and used to travel with two Land Rovers and a trailer. Generally it was a six-day week, four months at a time.
>
> The work consisted of surface geological mapping and some ground gravity survey. Usually we'd study the air photos first, then try to use the existing tracks to drive in as far as possible, and finally hoof it during the actual measuring and mapping.[34]

Before entering the field the parties would inform the local policeman and the cockie or station owner of their intentions and intended whereabouts. It was done as a matter of courtesy, but also for safety, particularly in the outlying areas such as those surrounding the Gulf of Carpentaria. However the surveys did not always go to plan. Dennis Clow, an operations manager with Minad who began as a field assistant during the 1950s, recalls being caught by rising floodwaters.

> We were out by Lake Nuga Nuga in the Carnarvon Ranges area of Queensland there with two vehicles and two trailers when it began to rain. We didn't realize it, but the lake comes up pretty quickly and we bogged before we could get out. The floodwaters came up right over the vehicles and we couldn't do anything except watch them disappear.
>
> They stayed that way for four days and it was a lot of muddy work extracting them and starting them again. The local cockie came up then and offered the advice that we should use horses in our surveys because they are easier to get out of bogs. I agreed, but asked him if his breed would last four days under water.[35]

Another 'damp' experience is recalled by geologist Colin Laing, who was also in the field for Minad during the 1950s.

> We were headed for the Carpentaria Gulf country for a gravity/geological survey and our party included two consultant surveyors who were only to be with us for a part of the time. We camped by the Gilbert River, but the river rose four feet [1.2 m] without warning and filled two anabranches, leaving us on an island. The surveyors had to swim two creeks with their suitcases and gear in half 44-gallon drums to get out.
>
> There were crocodile tracks in the mud by the second creek and as we were swimming across I felt something touch my legs. It turned out to be one of the field assistants with the pole we were using to guide the drums, but it has put me off crocodiles for life.[36]

In one sense the geologists in the field parties of the 1950s could be classified as the last of the true explorers because a great deal of emphasis was placed on basic field mapping. Later work increasingly relied on seismic data to pick out structure and potential drilling locations. The early drilling also tested the geologist's skills to the utmost. Minad's well-site geologists had to double as mud engineers, radio operators, first-aid men and often aircraft schedulers as well. There was usually a battle for supremacy at the rig between the toolpusher (chief driller) and the geologist. It was a case of who could bluff the other because the geological aims, like core cutting and testing, were not necessarily the same as those of the pusher, who was being paid by the foot and the shortest time it took to drill it.[37]

Towards the end of the 1950s the lack of commercial success caused activity to slacken, even for the widespread Minad team. Gradually the Associated Group began to pull back to concentrate on its original starting point round Roma. However Minad supplemented its activities by managing drilling programmes for other companies, and there was one notably unusual operational challenge thrown up in 1959 by a contract from Humber Barrier Reef Oil to drill on Wreck Island. Measuring a mere 150 m by 700 m, the drilling location was a tiny dot on the Great Barrier Reef about 85 km northeast of Gladstone. The rig was brought in direct from New Guinea on the ship *Wewak*, but it was impossible for the large craft to come in over the encircling coral and the equipment had to be set down at nearby Wilson Island. A channel was then cut through the reef so a landing craft could reach the Wreck Island beach to unload the rig and supplies.

In trips that took up to 10 hours each way the vessel also landed two trucks, a caravan refrigerator and thousands of litres of rig fuel ferried in from Gladstone. Tents were set up in the shelter of the island's luxuriant growth of Pandanus and Pinonia trees. Humber's idea was to use the well as a stratigraphic test to gain an idea of the formations in the offshore parts of the basin and their potential for oil accumulation. Plans were made to drill to 1 830 m but, to the disappointment of all concerned (not least the drilling crew who understandably voted it the best location in Queensland), the well lost circulation in the porous and friable coral and had to be abandoned at only 610 m.

It was an inauspicious end to a frustrating decade in the State. Yet at the same time the Wreck Island well was being drilled, a Sydney-based company called Australian Oil and Gas Corporation (AOG) was entering a deal that was to usher Queensland, and Australia, into the 1960s with a blaze of renewed enthusiasm. AOG, like a number of other explorers, had spent the 1950s learning the hard way. Its baptism was centred on New South Wales. The company was begun in 1954 through the hard work of T.W.H. (Bill) Dee, a licensed surveyor, ex-director of National Mapping in Canberra and former liaison officer/office manager for the early explorer Oil Search Ltd. When Oil Search management decided to move its operations exclusively to Papua and New Guinea and pull out of its permits in Australia, Dee was not keen on the idea. A determined and sometimes stubborn man, he believed that Australia had oil and gas potential and he wanted to test the theory. He began by taking up leases in the Sydney Basin, floating AOG and appointing a like-minded former Oil Search geologist, Eric Rudd, as consultant. His idea was to try to find gas reserves within economic range of the industrial cities like Sydney, Newcastle, Wollongong, Port Kembla and Lithgow.

Rudd was born in New Zealand, but spent his boyhood and early university years in Adelaide where he graduated in 1930. After working with Oil Search for several years he went to Harvard and completed an MSc in Geology in 1935. The war years were spent with the Royal Australian Engineers and he returned home to become chief geologist for Broken Hill Proprietary in 1946. Quietly spoken and interested in training young geologists for what he saw as a future need in Australia, Rudd took up the Chair of Economic Geology back at Adelaide University in 1949 and it was there that Dee sought him out to become involved in AOG.

The spudding of the company's first well, Kurrajong Heights in the Blue Mountains, was a glittering affair. As was traditional in the early years, a number of important official guests were invited to speak and to witness the kissing of the bit for luck before the rig machinery was put in motion. The *Australasian Oil and Gas Journal*'s reporter waxed lyrical over the scene.

> Probably no drill has ever started its search for oil in more beautiful surroundings within easy reach of a great capital city. Situated on the Blue Mountains, Kurrajong is a well-known scenic and tourist resort. It may be a happy augury that the first drill in the Sydney Basin was accompanied on its career by a chorus of the beautiful song of the bellbirds.[38]

Kurrajong Heights was the first of an initial programme of six wells, but when it was abandoned as a dry hole and the costs assessed, Dee saw that the company would soon run out of money if it continued with a rotary rig. He asked a former Oil Search drilling supervisor for advice and was told to try the old cable percussion method. David (Dave) McGarry, a geologist hired to AOG from the Joint Coal Board in New South Wales by Eric Rudd and who later rose to become managing director, was with the company when the percussion rigs arrived.

> We bought two of them in from America along with a number of US and Canadian drillers. The method of drilling was to attach the drill rods to a cable and then raise and drop them in the hole so that progress through the rock was made by percussion. Every now and then we'd stop drilling and send a bailer down to clean out the cuttings.
>
> But there were problems. The main difficulty was in obtaining drilling cable. There was none in Australia and it was always a long wait importing the right ones and having to go through customs. The old lines had been spliced many times and were wearing out. There was no elasticity in them.
>
> I can remember on one well the old cable was so bad that when a new shipment of line arrived we spooled it on to the rig virtually straight off the ship. We were down about

> 3 000 ft [1 070 m] at the time and I went home for tea to the house where we were boarding. While I was gone the bit broke through a gaseous coal seam and blew the tools, sinker and everything else clean out of the hole. I came back to see the brand new line — all £12 000 worth of it — strewn around like knitting wool.
>
> We didn't have any blowout equipment on the rig and lumps of coal were flying out of the hole and landing all over the place. After a while the seam collapsed and killed itself, but we had no option than to rewind the old cable onto the drum and restart the well.[39]

On another occasion a promising well had to be abandoned after a mysterious, but apparently deliberate, act of sabotage. The *Australasian Oil and Gas Journal* of June 1958 had the story.

> By June 5, 1958 the deep well Duval No.2 [in the Sydney Basin] had reached 6 294 ft [1 920 m] and was still in shale. Fishing operations were in progress to recover broken drilling cable. On June 12 the company reported that the lost line had been recovered and drilling was proceeding...
>
> On June 19, Mr Dee reported that the well had been sabotaged on the previous day while the rig was shut down between the final night shift and the 4.00 a.m. shift. While the full extent of the damage had not been assessed, preliminary investigations revealed an obstruction in the well and it was believed objects had been dropped down the borehole.
>
> The surface machinery had been tampered with, the oil having been drained from the sump, lubricating oil having been mixed with the fuel and plugging the fuel line to the water pump... It had also been found that the telephone line had been cut and about 2 000 gallons [9 100 L] of water released into the hole from the surface. The police had been advised and were investigating the matter.

Duval No.2 at that stage was the deepest bore in New South Wales and was drilling through marine formations in a deep test for oil and gas possibilities in the region. Unfortunately the obstructions could not be removed and the well was finally abandoned the following month. Despite this setback AOG went on to drill about 40 wells in the Sydney Basin using the percussion techniques during the remainder of the 1950s and well into the 1960s. However the result was invariably the same — a lot of gas indications, but none of them commercial.

There was exploration elsewhere in New South Wales during the 1950s, but not on the scale of AOG's assault on the Sydney Basin. A company called Clarence River Basin Oil Exploration Co NL drilled one or two wells in what is now referred to as the Clarence-Moreton Basin on the northeast coast, and AOG also did some work in the area. The drilling was prompted by the original gas find on the Grafton racecourse many years earlier, however the renewed efforts were rewarded only with 'petroliferous odours'. Sun Oil of California came in to look at the Hunter Valley region near Maitland late in the decade. In addition, some preliminary surveys were carried out in the Murray River area in the far southwest of the State by the BMR in conjunction with AOG, but there was no drilling in the area. Across the border to the west, however, it was a different story.

Frome-Broken Hill was the first group in South Australia to mount a post-war exploration programme when it took out permits in the northeast of the State during 1947, hoping to find gas reserves in the Frome Embayment. With the backing of the Zinc Corporation's impressive air fleet (at that time the third largest in Australia after Australian National Airways and Trans Australian Airways) a series of magnetic and gravity surveys were flown over the region and followed up by parties on the ground. During the early 1950s three wells were drilled in the southern part of the licence area but met with no success. Frome-Broken Hill decided to relinquish its acreage in 1956, citing a lack of any surface structures as the reason. This was an amazingly inaccurate statement as was soon to be shown. The licence covered the entire South Australian portion of the Cooper Basin, and quick off the mark to apply for the areas, plus those in adjacent southwest Queensland, was an energetic, outspoken and in some ways controversial geologist named Reginald (Reg) Sprigg. He was acting for an Adelaide-based newcomer called South Australian and Northern Territory Oil Search — SANTOS for short.

Sprigg, like Eric Rudd, had been a student of Professor Sir Douglas Mawson at Adelaide University in the late 1930s and was no stranger to the (then) current view that commercial oil would not be found in South Australia. Sprigg, again like Rudd, found this dogma difficult to justify and began to ponder over the little-known sedimentary basins in the north of the State. Although his early work took him into the mineral side of geology (in particular the wartime uranium operation at Radium Hill in the Flinders Ranges), his brief as a government geologist after the war included regional mapping. Thus he had the opportunity to monitor Frome-Broken Hill's exploration programme and he witnessed that company's wells along the Birdsville Track.

As in all other parts of the country the Rough Range discovery caused great excitement in South Australia. More specifically it prompted Sprigg to leave government employment and start his own geological consulting and contracting company which he called Geosurveys of Australia Pty Ltd. The same startling news from Western Australia caused a world traveller named Robert Bristowe, then resident in Adelaide, to muse over the similarity between the prolific oil field country he had seen during his travels in Iran and the terrain in South Australia's central areas to the north of Spencer Gulf. Upon mentioning this to a friend, Adelaide lawyer John Bonython, the idea to form SANTOS was born. It was a quaintly unscientific beginning, but this was rectified when Bonython met Sprigg and Geosurveys became the fledgling company's technical consultant.

The first permits covered a broad strip through the centre of the State from the head of Spencer Gulf to the Northern Territory border, with the primary area of interest being the Tertiary-age sediments overlying much older Cambrian rocks at Wilkatana near Port Augusta. It was an area where oil shows had been found in a water bore during the 1930s. Indeed the initial SANTOS drilling programme confirmed oil was present at shallow depth in this region, known as the Pirrie-Torrens Basin, and 24 holes were drilled between 1954 and 1957. But Sprigg was more interested in the little-known northeast corner of the State. Surreptitiously he extended the company's geological and gravity surveys in that direction and, encouraged by the results, he urged Bonython to take up the licences recently vacated by Frome-Broken Hill.[40]

The full SANTOS board, however, was not so enthusiastic and it was suggested to Sprigg that he seek a 'second opinion'. This he did in the person of the noted American consultant Dr Arvil Irving Levorsen, a congenial geologist in his sixties who shunned his Christian names, preferring to use the bald initials 'A I' on official documents and the nickname 'Lev' among friends. He had vast experience of oil fields right round the world and after being shown both the existing and intended SANTOS leases, Levorsen made his celebrated comment: 'Reg, you might find a Chevvy at Wilkatana, but your Cadillacs are in the Artesian Basin'. The board mellowed in the wake of the American's comments and Sprigg needed no further encouragement. Geosurveys moved camp to Haddon Downs in the far northeast of the State and arrangements were made for a detailed survey of the area, beginning with an aerial reconnaissance of the section round Innamincka, Betoota and Birdsville. Much of the aerial work was carried out by two 'flying' geologists, Helmut (Heli) Wopfner and Rudi Brunnschweiler, during the early part of 1957. Both men were pilots and they used an old Czechoslovakian Sokol two-seater, single-engine 125-horsepower fabric and plywood machine to put the rare combination of skills to good effect.

Brunnschweiler was a Swiss paleontologist from the small deep mountain valley town of Glaurus who had migrated to Australia by flying himself and his family out from Europe in the Sokol aircraft. He had been a pilot in the Swiss airforce and had flown Messerschmitts — the Sokol was apparently modelled on the German plane. Once in Australia he joined Geosurveys where Sprigg initially put him to work on the Wilkatana drilling.

> As Geosurveys expanded into the Great Artesian Basin, Rudi, Wopfner, Bruce Wilson and I mapped much of the central

western basin. Rudi was able to do a lot from the air in the Sokol which SANTOS and Geosurveys came to own jointly. He was an excellent pilot and highly individualistic.

Much of our air surveys would not have been possible today. With 960 miles [1 545 km] of operational endurance we just took off for anywhere and returned from anywhere — often almost to the limits of the aircraft's tanks.

One trip from Ooldea to Esperence in Western Australia we were flying at the base of the great limestone cliffs at the head of the Bight when the engine cut out. Rudi lifted the plane above the cliffs and the engine refired. We made it in to Esperence. We found the cylinder head cracked, spewing oil and the plane temporarily out of service.[41]

Wopfner, an Austrian who flew with the *Luftwaffe* during the Second World War, completed his geological studies in 1950 and came to Australia with his family in 1956 to begin work with Geosurveys. After a short orientation round the SANTOS leases at Wilkatana he was moved to the north and carried out a six-month mapping project in the Oodnadatta region. It was this work that produced the first post-war evidence for the presence of major anticlinal structures in the western sector of the Great Artesian Basin, although Wopfner himself said that the government geologist Lockhart Jack had written about huge fold deformations in the region as far back as 1923.[42] Jack's descriptions had apparently lain unrecognized in the literature for 30 years and they made a nonsense of Frome-Broken Hill's professed lack of surface structure. The Wopfner/Brunnschweiler aerial survey resulted in the first structure contour map of any part of the Great Artesian Basin, though the data was gathered in an unorthodox manner, as the ex-*Luftwaffe* pilot explained.

The method by which it was obtained was fairly simple. We would fly wing-tip level with the cuestas or mesa tops and read the altimeter. Then we would fly just above grass-roots along the syncline and read the altimeter again. Repeating the process several times across each structure resulted in quite good elevation control.

There were about 100 flying hours in that map. However I think we must have been a real headache for the Department of Civil Aviation because we flew low, slow and without a wireless.[43]

For the follow-up ground surveys Sprigg hit on the novel idea of providing two double-decker buses, one of which was a workshop-cum-office and the other a dining area. These were painted bright yellow and were cumbersome to move around, but they made a good field base.

Caravans pulled by four-wheel drive vehicles were also used by the field survey parties. The work was supplemented by a regional seismic survey carried out by the BMR and some scout drilling which commenced late in 1957. It was becoming clear to Sprigg and everyone concerned that the basin did indeed have a great deal of oil potential and that thicknesses of over 305 m of sediments were likely. The results of this activity were placed before Dr Levorsen, who had been retained by SANTOS, and his immediate reaction was one of optimism. However he suggested that the company try to interest an experienced oil explorer to become a partner in the continued exploration — one which would not be daunted by some early dry holes. He suggested a US company called Delhi-Taylor Oil Corporation of Dallas.

Flying over the Artesian Basin leases a short time later with Wopfner and another geologist, Brian Fitzpatrick, the Delhi technical team could not believe their eyes. On that flight the famous expression 'It looks just like Texas' was voiced for the first time.[44] In 1958 Delhi agreed to contribute funds and expertise to earn a 50 percent interest in the region. With the aid of the South Australian Department of Mines, more detailed seismic work was carried out and finally, in 1959, it was decided to drill the first well on the Innamincka Dome — a giant structure 80 km long and 30 km wide straddling the Queensland-South Australian border. A location was picked near the centre of the feature and the big National 130 rig which had drilled WAPET's original Rough Range well was shipped around from Western Australia for the job. Up to this point the technical experts had held sway but, as Brian Fitzpatrick related, politics had a hand in the final site of the historic well.

> South Australian Premier, Sir Thomas Playford, was invited up to inspect the site. He was duly impressed, but then looked up and asked about the dark line running north-south across the structure. He was informed it was the Queensland-South Australian border fence.
>
> Playford immediately insisted that the rig be moved a couple of miles further west as he did not want the well mistakenly drilled in Queensland. No doubt he had in mind the mistake which had been discovered in a recent resurvey of the South Australian-Victorian border. Anyway the shift was made, and to hell with science![45]

The Premier's visit had another unexpected side effect as Reg Sprigg recalls.

> A number of press photographers came up with the Premier to watch him drive in the spike for the first well site and record the scene for the papers in Adelaide. After these official

> photos the journalists wanted some shots of the rig men 'at play' — a sort of 'Life in the Outback' type story.
>
> They were pointed in the direction of the new swimming tank, but every time cameras were aimed the men ducked under water. The photographers persisted and eventually they had the shots they wanted.
>
> No sooner had the results been published than the SANTOS switchboard in Adelaide was jammed with calls from irate females. They had recognized men in the photos as deserting husbands, fathers of illegitimate kids, criminals and I don't know what else. The photos had blown the whistle on them.
>
> In fact the police knew some of the men had records and they warned us about a few. But they also knew they could get into little further trouble out in the desert. From the company's point of view most of the men worked hard and well, so there was no problem. It didn't appease the ladies though.[46]

Innamincka No.1 was spudded late in 1959. In the event, it was not the hoped-for first-time success of Rough Range, but some minor gas shows indicated that the region had hydrocarbon potential. What is more, the well also proved there was a great thickness of sediments in the basin. Survey work had already indicated a number of other structures in the region and so the South Australian-based explorers entered the 1960s full of buoyant hopes for the future.

Although WAPET had most of the Western Australian basins under lease by the mid-1950s, there were enough gaps to let other would-be explorers into the field. One of those 'holes' had been occupied by the Freney Kimberley Oil Company at Nerrima Dome during, and immediately after, the war. The Freney permit was kept active when the growing Associated Australian Oilfields NL took Freney Kimberley under its wing in 1954 and began to re-evaluate the Nerrima prospect. With the oil fever generated by Rough Range rampant right round Australia, the new company (Associated Freney Oil Fields NL) believed Nerrima still to be one of the country's best oil prospects, reasoning that it had never been drilled to anyone's satisfaction. Writing in AFO's prospectus, Dick Mott came to the conclusion that a new well sited at the western end of the structure was warranted, and the company made this its first project. The managing director, Clarence Byrne, sought permission from the Federal Government to import a new rig and, when this was granted, arrangements were made to bring it by sea to Derby and then on by truck to the Nerrima location. Roy Richter, who had recently joined the Associated Group as General Fields Manager after spending the immediate post-war years working for BP in the English oil fields near Sherwood Forest in the Midlands, and then in Papua, took charge of the operation.

Harold Raggatt, geologist, Chief Government Geological Adviser and founder of the BMR.

Irene Crespin, Assistant Commonwealth Paleontologist in the 1920s and 1930s who moved to the top job in 1936 and formed a close working relationship with Harold Raggatt. Courtesy Max Reynolds.

J.M. (Jack) Rayner, first Chief Geophysicist of the BMR and its director from 1958 till 1969. Courtesy *Canberra Times.*

Norman Fisher, first Chief Geologist of the BMR and its director from 1969 till 1974. Courtesy *Canberra Times.*

A BMR party making a river crossing during a northwest Australian survey in the 1950s. Courtesy Arthur Lindner.

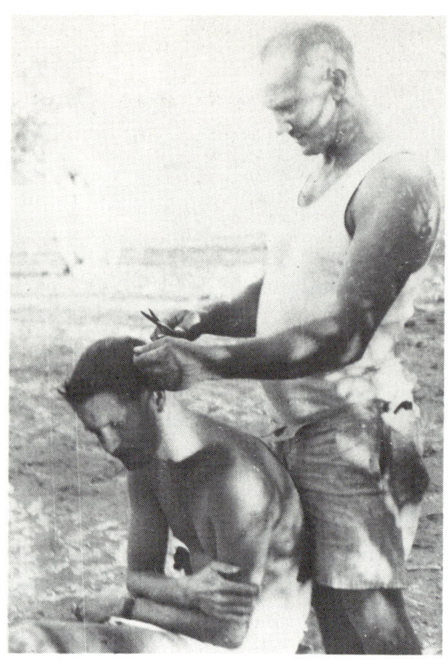

Hairdressing in the field BMR style. Barber Gordon Baker (who doubled as cook!) attends to the thatch of geologist Jack Cuthbert at a camp on Nerrima Station in the Kimberleys of Western Australia in 1948. Courtesy Arthur Lindner.

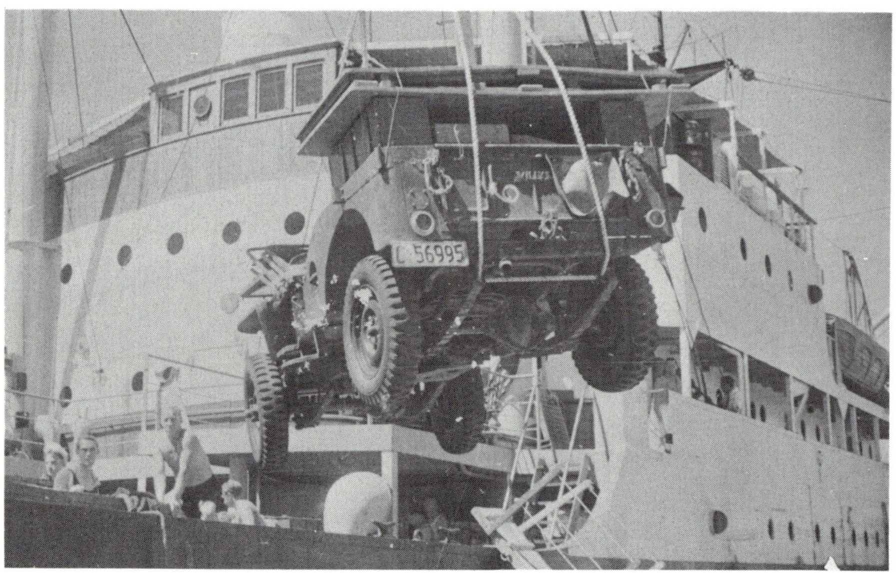

Unloading the BMR's jeeps at the beginning of its Western Australian surveys in the late 1940s. Courtesy Arthur Lindner.

Dunn's Soak, just east of Lake Dora, visited during a BMR trip across the Canning Desert in 1955. Courtesy Doug Traves.

A BMR vehicle bogged after breaking through the salt crust on the edge of Lake Dora in the Canning Basin, 1953. Courtesy Doug Traves.

BMR field camp in the Kimberleys of Western Australia during the late 1940s. Courtesy Arthur Lindner.

Maurice Mawby, head of Zinc Corporation and instigator of Frome-Broken Hill group's oil exploration in Australia. Item 2/4, Mawby Collection, University of Melbourne Archives, courtesy CRA Ltd. Photo: Peter Fox.

Harry Evans, Zinc Corporation geologist who joined consultant Frank Reeves in the late 1940s survey of the Canning Basin for Frome-Broken Hill group. Item 4/17, Mawby Collection, University of Melbourne Archives, courtesy CRA Ltd.

Frank Reeves, American geologist who led surveys in the Kimberley region of Western Australia for Frome-Broken Hill group during the late 1940s. Photo taken when in his nineties. Courtesy Prince Maccus Publishers, Virginia, USA.

Sir William Walkley, co-founder of Ampol and its first managing director. An enthusiastic architect of the early Rough Range exploration programme. Courtesy Ampol Ltd, Sydney.

Five Mayhew 1000 shot-hole drilling rigs during a seismic survey over the Rough Range area of Western Australia in 1955. Courtesy Daryl Johnstone.

Shot-hole blasting at Cape Range, Western Australia, during 1955. Courtesy Daryl Johnstone.

Geologist Jim Parry with a scrub bull shot to provide fresh meat 'tucker' during a Carnarvon Basin survey in the 1950s. Carried out with permission of the station owner, the kill was an indication of the good rapport between WAPET employees and station people. Courtesy WAPET, Perth.

The early drilling camp at Rough Range in 1954. Courtesy WAPET, Perth.

The rugged terrain tackled by WAPET's early survey teams in Cape Range during the 1950s. Courtesy Ampol Ltd, Sydney.

The Learmonth airfield buildings when Ampol took up the lease from the Federal Government in 1948. Courtesy Ampol Ltd, Sydney.

A camp cook samples his wares produced on a wood stove during the 1954 field season in Western Australia. Courtesy WAPET, Perth.

On location at Rough Range No.1 discovery in 1953. Courtesy Sami Nasr.

Oil at Rough Range No.1. Courtesy Owen Evans.

Sir William Walkley wearing his famous sombrero and surrounded by school children at Exmouth in 1953. Courtesy AIP Library, Melbourne.

The national T-32 rig convoy on its way up the Northern Highway to the Exmouth Gulf in 1954. Courtesy WAPET, Perth.

WAPET's Cape Range No.1 well in Shot Hole Canyon drilled in 1954 with a T-32 rig. Courtesy AIP Library, Melbourne.

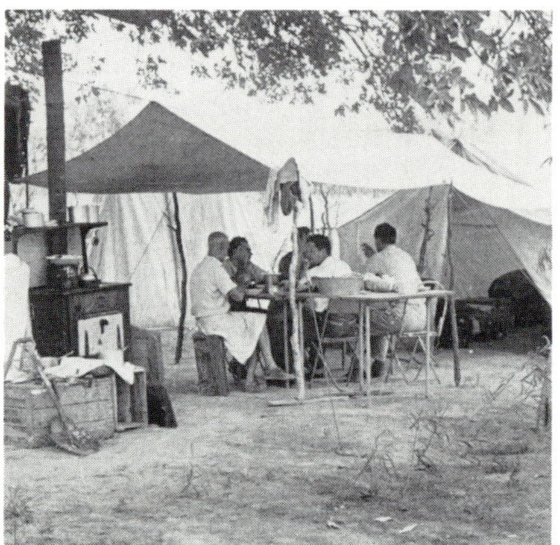

A WAPET geological camp under canvas in the Kimberleys of Western Australia during the 1954 exploration programme. Courtesy WAPET, Perth.

The vessel *Jacob Jebsen* moored at Derby jetty during low tide, illustrating the need for a flat-keeled ship to transport equipment to WAPET's Canning Basin programme in the 1950s. Courtesy Owen Evans.

W.D. (Dick) Mott pictured during the late 1930s with APC in Papua New Guinea. He later became a prominent figure in Queensland's Geological Survey and then joined the Associated Group during the 1950s. He also founded the *Australasian Oil & Gas Journal*. Courtesy Roy Richter.

The Kuru gas blowout, Papua, in 1956. Courtesy AIP Library, Melbourne.

A helicopter is carried in for repairs after a 'slight' accident during Papua survey work about 1955. Courtesy BP Australia, Melbourne.

An APC geophysicist checks a geophone station in a grass swamp during a seismic survey in Papua during the 1950s. Courtesy AIP Library, Melbourne.

APC geological survey teams in Papua often travelled in the creeks where the going was easier and cooler, and because most rock outcrops were found in the bed or along the banks. Courtesy AIP Library, Melbourne.

Canoes adapted for an APC survey party in 1957. Courtesy BP Australia, Melbourne.

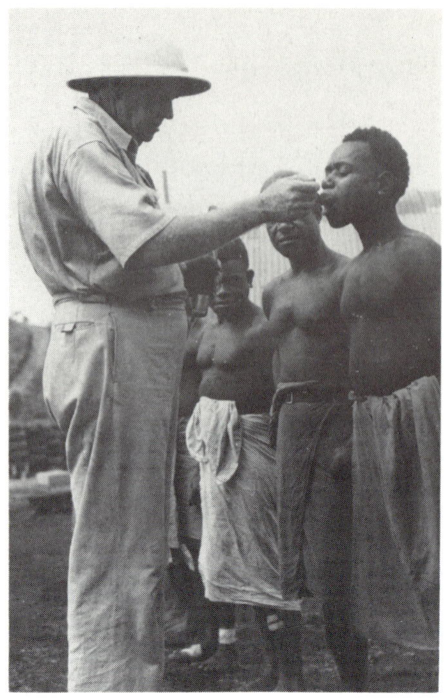

Field assistant S.C. ('Tiny') Thompson administers the weekly dose of cod liver oil to native labour in Papua. It was part of the standard ration scale and provided additional vitamins in the diet. Courtesy BP Australia, Melbourne.

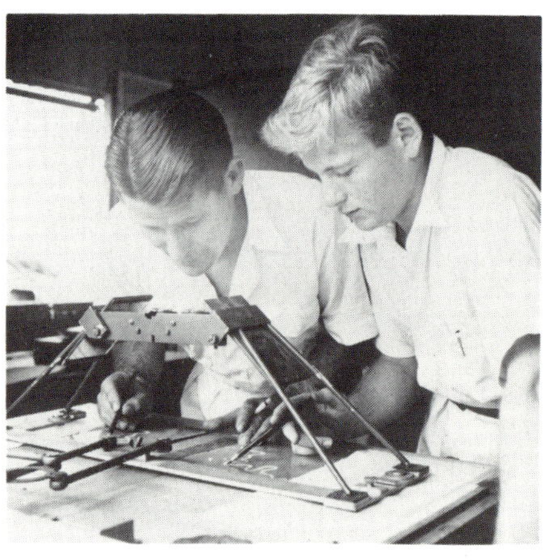

Geologists study air photos of a Papuan survey area through a stereoscope. Keith Llewellyn is on the left. Courtesy BP Australia, Melbourne.

A Beaver float plane unloading supplies for an oil survey team on the river at Komewu, Papua, in 1957. Courtesy BP Australia, Melbourne.

The barge *Gabuna* carrying rig components to the Victory Junction drill site in 1957, Papua New Guinea. Courtesy BP Australia, Melbourne.

A bomb scow carrying men and equipment along a Papuan river towards an oil survey camp during the 1950s. Courtesy AIP Library, Melbourne.

Food being retrieved after an air drop to a survey party in Papua during the 1950s. Courtesy BP Australia, Melbourne.

Native labour roll call of new recruits at APC's Port Morseby base in 1954. Courtesy BP Australia, Melbourne.

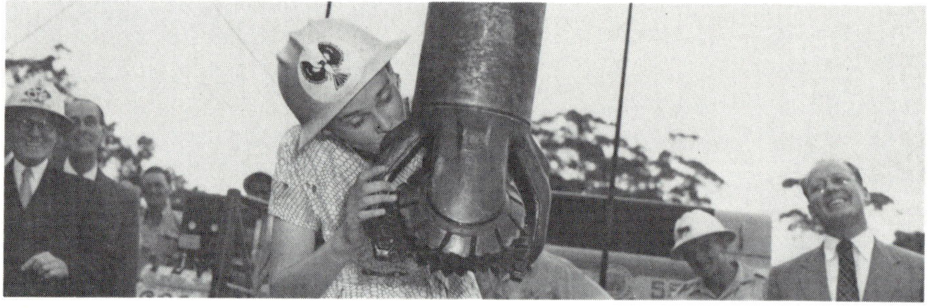
Miss Primrose Rains, daughter of the landowner, kisses the bit during the spudding-in ceremony at AOG's Kurrajong Heights well in New South Wales, February 1955. Courtesy Consolidated Press Ltd.

Papuan rig workers on the drill floor during drilling of APC's Puri No.1 oil discovery in 1958. Courtesy BP Australia, Melbourne.

Professor Eric Rudd, early exploration geologist for Oil Search Ltd, consultant geologist to Australian Oil & Gas Corporation, director of SANTOS and other companies, and Professor of Geology at Adelaide University.

Omajean A. (Gene) Goff, founder of Oil Drilling and Exploration Company and Exoil NL.

T.W.H. (Bill) Dee, first manager of Oil Search Ltd and founder of Australian Oil & Gas Corporation. Courtesy John H. Dee.

C.W. (Bill) Siller ran the Lucky Strike Maryborough Basin programme for American entrepreneur Gene Goff and later became chairman of the Exoil group of companies.

John MacD. Royle, a First World War soldier in the Royal Australian Engineers who set up listening posts under German lines, and later founded Longreach Oil Company.

Lockhart Jack, Deputy Government Geologist and Deputy Director of Mines South Australia 1912-30, laid the foundations for later Great Artesian Basin exploration with his descriptions of folding in the region. Courtesy South Australian Department of Mines and Energy.

Ernie Evans, Minister of Mines for Queensland in the 1950s-1960s who also dabbled in oil exploration when on the Opposition benches in the 1950s. Courtesy Queensland Department of Mines.

Roy Richter and family at Associated Freney Oil Field's Nerrima No.1 well site, Kimberley region of Western Australia, in 1955. Courtesy Bob McCullough.

Richard (Dick) Dumbrell, an early employee of Oil Drilling and Exploration Company who rose to chair the company through the 1960s and 1970s.

Oil explorers in the northwest of Australia dig for water to boil their billy. Courtesy AIP Library, Melbourne.

Associated Freney Oil Field's Nerrima campsite in the Kimberleys of Western Australia during 1955. Courtesy Bob McCullough.

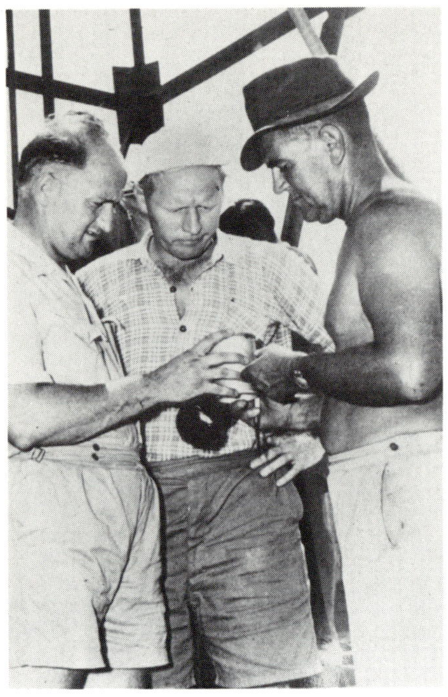

Examining core at SANTOS's Wilkatana drill site in South Australia. Geologist Reg Sprigg is on the left with colleague Rudi Brunnschweiler in the centre. Courtesy SANTOS, Adelaide.

Examining samples at SANTOS' Wilkatana drill site in South Australia. From left, Rudi Brunnschweiler (geologist), John Bonython (chairman) and Robert Bristowe (director).

American experts arrive in Adelaide to move the National 130 rig to Innamincka No.1 location. From left, Charles Easley (Delhi Australian manager), H.P. Kerr (Delhi-Taylor superintendent), H.G. Pullin (drilling superintendent), R.W. Waddell (Delta Drilling superintendent), C. Zeppa (vice-president Delta Drilling).

> It was late in the year and we were racing the coming wet to get the rig and equipment in over the Fitzroy River. Also, we wanted to make sure we had ample stores and food supplies at the rig site. An airstrip was set up nearby and it was made suitable for DC3s.
>
> When we were under way a plane came in once a week. The rig workers had their fare up from Perth docked from their first pay, but it was refunded if they stayed three months. They received a fare back to Perth on the company if they stayed for a second three months.
>
> We had four crews on rotation, each one doing an eight-hour shift. The usual stint in the field then was six months. It might seem like a long period now, but it was accepted in the 1950s and, in fact, it helped forge a team together. We had very little trouble.[47]

The camp itself was a fairly substantial one. Some of the housing used in Freney's previous occupation was still there, so it was improved and one or two new dwellings built. However finding a cool spot was a problem. Temperatures were between 43° and 46° C for weeks on end. An air-conditioning unit on its way up from Perth was caught in a cyclone at Port Hedland and all the cladding ripped away. It wasn't taken any further. Cold showers were a rarity too as the water line was laid out on top of the ground. Water was scalding hot during the day and still fairly warm during the evenings. However despite this constant heat, most agreed the camp was well appointed. It was one of the earliest attempts to introduce some home comforts into a remote area operation. Stuart Derrington, one of the Associated Group's first geologists, rated it above the Queensland facilities. He was sent out to sit the Nerrima well for six weeks — a stay that was extended to six months.

> It was a bit like a club atmosphere. Roy Richter and chief driller Bob McCulloch had their wives with them and everyone mixed in pretty well. The mess building housed a full-sized billiard table and table tennis. There was a big screen for films. We even had an ice machine.
>
> The canteen was wet, although only beer was sold, and then on a ration of two bottles a day per person. For those who liked reading we set up an external library service with about 150 books a month flown in from Perth.[48]

Much to the AFO's disappointment Nerrima was abandoned as a dry hole. The rig was then moved to the Myroodah prospect during 1956 and finally to the nearby Sisters structure the following year, but these also recorded negative results. After some further field mapping and reconnaissance work, particularly in the Kununurra and Bonaparte

Gulf region, plus a little more unsuccessful drilling, the Associated Group decided to return to its original hunting ground in the Roma area of Queensland. Other companies that flirted briefly with Western Australian exploration during the 1950s were Westralian Oil, Australasian Oil Exploration and Exoil.

Westralian began naively with two permits on the very edge of the Canning Basin that had very little prospective sedimentary section. The management, knowing nothing about oil, had a rudderless technical direction until geological consultant Leo Stach advised the company to relinquish the Canning permits and take up some areas around the Bonaparte Gulf. This done, Westralian carried out some unsuccessful exploration work for a short time before deciding to quit the oil industry and go into mineral sands in the Bunbury area south of Perth. The company reformed as Westralian Sands. Australasian Oil's main effort was in the Reid's Dome prospect in Queensland, but it sent several reconnaissance survey parties to look at permits in the Officer Basin. The field geologists were equipped with Austin Champ four-wheel drive jeeps and a three-tonne Austin four-wheel drive truck. Lincoln Madden, who was employed by the company in Queensland, had time to keep track of the brief programme in Western Australia.

> The team was supplied out of Laverton to the northeast of Kalgoorlie. That included water, although they did use rock hole water in the desert when they found it. There was no point in taking perishables into the field because of the lack of refrigeration. Everything was kept in Coolgardie-type wire cool safes. Rifles were standard equipment.
>
> The permits were close to the South Australian border and the main contact was by radio with the Warburton Mission. They did carry a big flying doctor medical kit for use in emergencies. Generally the party was out for two, even three months at a time.[49]

However none of this work was followed up by seismic survey and there was certainly no drilling. The areas were allowed to lapse when Australasian Oil, like Westralian Oil, transferred into the mineral industry before the end of the decade. Exoil's venture into Western Australia was equally brief. The company, a wholly owned subsidiary of Gene Goff's contracting venture Oil Drilling and Exploration, drilled two stratigraphic holes on the coastal edge of the Nullarbor Plain between Eucla and Norseman in 1959. These were designed to give information about the Eucla Basin, the last of the big onshore Australian sedimentary provinces to be tried in the wake of the Rough Range-inspired exploration programmes of the 1950s. But results did not encourage a follow-up programme.

At first glance it would seem to have been an anticlimactic end to a decade that had begun in such spectacular fashion. In reality though, the explorers had made great strides towards maturity. As the industry stood on the threshold of the 1960s perhaps the most noticeable change from the 'hit and miss' programmes of the immediate post-war years was in the improvement of technical skills. The most obvious benefit was in the quality of the data gathered during pre-drill surveys. But there was also a marked increase in the ability of companies to conquer the thorny and expensive problem of logistics in a country where most of the prospective areas lay in a harsh and remote environment.

Chapter 8
A Technological Awakening

Methods employed in the search for oil were rudimentary at the beginning of the 1950s. Many of the operations still employed technology that had been in use 10 and even 15 years earlier. But times were changing. Skills developed during the urgency of war were beginning to find their way into peacetime pursuits. The use of electrical instrumentation and hydraulic power drives, improvements in communications, the growing versatility of aircraft, advances in photography and photogrammetry (which enables reliable measurements to be obtained from photos), and the advent of four-wheel drive vehicles were all put to use by the post-war band of oil explorers in Australia and New Guinea. Yet the greatest hurdle to overcome was the lack of knowledge of the country's sedimentary geology. Because the most conspicuous outcrops were of ancient Precambrian rocks or metamorphosed material, there was a lingering influential body of opinion that Australia was too old to contain oil. No one realized the full extent of the poorly exposed sedimentary basins hidden under the sands of the inland deserts. Nor was there any real understanding at the beginning of the decade of the great thicknesses of sediments involved.[1]

Frome-Broken Hill consultant Frank Reeves was the first post-war geologist to suggest a ranking for the (then) known Australian basins in terms of potential for oil discovery. Using three categories, he thought the Carnarvon, Canning, Bonaparte and Perth Basins had moderate prospects, while the Gippsland, northern Surat, Murray, Great Artesian and Sydney Basins had slight prospects. The Eucla, Bowen, Clarence-Moreton, Georgina and Ord Basins he dismissed as having no oil prospects at all.[2] Overall he did not think Australia had much to become excited about. Reeves based this assessment, published in 1951, on his own field work and on reports from other workers in Australia to that time. By then the BMR parties were also in the field right across the country. Their conclusions were published in a series of reports during the latter half of the decade after collating the data from extensive geological and geophysical surveys.

Putting the basins in a rough order of perceived oil potential, they too considered the Carnarvon and Canning Basins to be at the top of the list. The next rankings were the east Queensland coastal basins, followed by the Georgina, central Great Artesian, Perth, Carpentaria, Gippsland, Bonaparte and Murray Basins, tailing off to the Torrens Basin in South Australia at the end of the line.[3] The BMR reports, along with similar conclusions published by two WAPET geologists, Phillip Playford and Murray Johnstone, in 1959,[4] had the benefit of the Rough Range discovery along with well data from the subsequent rash of drilling across the country. They also had the benefit of advances in the use of microfossils for rock identification, age grouping and the understanding of depositional environment. A major breakthrough in this area occurred in the mid-1950s when palynology (the study of fossil plant spores and pollens) was recognized by the oil industry as a way of dating rocks which have little or no marine content.[5]

Apart from the BMR's paleontological team in Canberra headed by Irene Crespin, the State Geological Surveys and often the universities carried out a valuable service by analysing and identifying the fossil samples sent in from the field. Private oil companies as well as the government teams availed themselves of this expertise. Dorothy Hill,

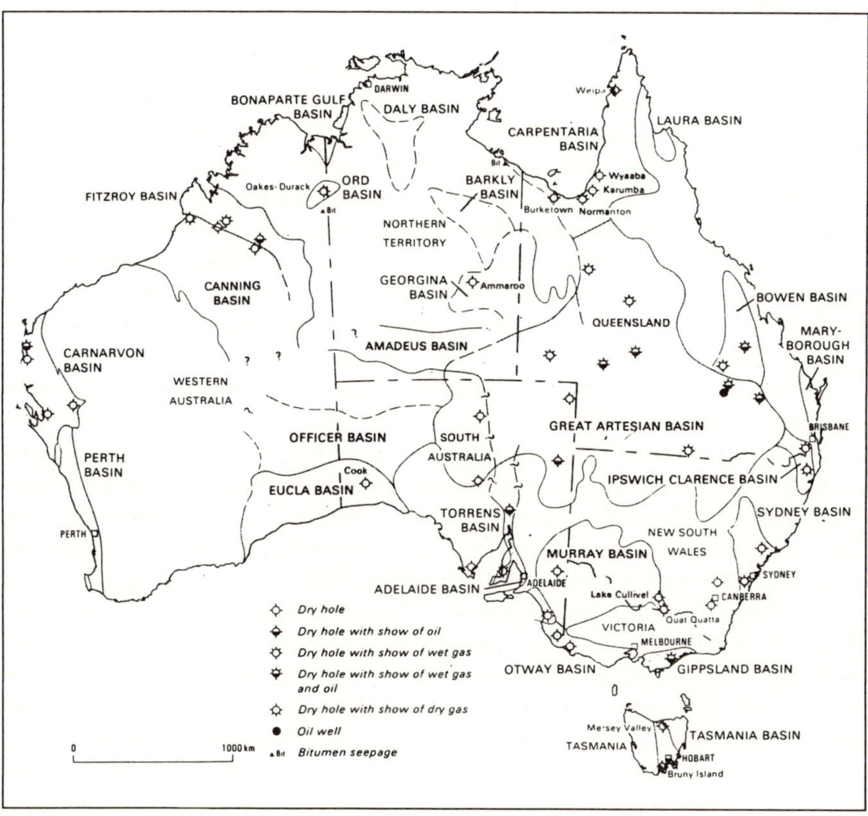

The sedimentary basins of Australia and reported petroleum occurrences (after BMR 1960)

a lecturer and researcher at the University of Queensland who was later appointed Professor of Geology, became involved during the mid-1950s. Her specialty was fossil corals, but she had to spread her knowledge right across the paleontological spectrum to analyse the specimens sent to her.

> I needed permission from the head of the department to do this work, because it was outside the university curriculum. But that was never refused. The quality of the samples sent in to me varied a lot and some of the time the results were fairly approximate. In really difficult cases there had to be a bit of spot guessing. But I would normally furnish a two- or three-page report and sometimes be sure enough to name a rock unit characterized by a particular fossil species.
>
> I was allowed to keep the fossils afterwards too, so my collection became quite large and comprehensive after a while.
>
> There were a lot of foreign geologists about then, particularly Americans, and the competition was pretty fierce. They were a friendly bunch, but they all played their cards close to their chests. All I'd get was the bare sample or piece of drill core to work on. I always wanted them to be a bit more indiscreet with their knowledge so I could learn something of the basin structures. They never were though.[6]

While geological mapping and the establishment of stratigraphic columns were vital initial steps in understanding and delineating the sedimentary basins, a large portion of the gradual build-up of subsurface structural data was contributed by geophysical surveys. At the beginning of the 1950s most of these were run by the BMR but, as the decade wore on, private companies contracted survey teams for more specific or detailed work on their own prospects. They began with gravity and magnetometer traverses to give a broad picture of the depth of basement rocks and hence an idea of the thickness of sediments in various parts of a basin. Then came the seismic surveys, with the first contract crew (Seismograph Services of London — SSL) arriving in the country during 1951 to help evaluate WAPET's Exmouth Gulf permit before moving north to the Fitzroy Basin during 1953. Three more crews, from the Texan firm of Geophysical Services International (GSI), were hired by WAPET during 1954 to examine a wider area of the Carnarvon Basin in the wake of the Rough Range discovery. They also ran a preliminary evaluation of the Perth Basin further south, and continued the Canning Basin programme in the north. Although seismic techniques had been in use overseas for a number of years, beginning in the USA during the 1930s and developing swiftly in Germany and the UK during the Second World War, there was little knowledge of them in Australia. No universities taught the subject, so local 'geophysicists' had to learn on the job. Queensland geologist Daryl Johnstone was one of the first to join WAPET's Carnarvon work.

> There were only two or three text books on the subject available in the 1950s and they weren't particularly good. University courses didn't incorporate geophysics, and more particularly seismic technology, until the 1960s.
>
> I found the best way to learn was to talk to the contract people out on the crews and in the office. I was lucky to have done a fair bit of university maths and physics with my geology, and it gave me a good start. Companies like GSI did run some basic courses in various aspects like surveying and instrumentation but, other than that, it was a case of learning on the job.
>
> The energy source used for all the early surveys was dynamite. In the Rough Range area we worked with 36-hole patterns for one shot. That meant six lines of six holes, each drilled to a depth of about nine metres. They were then all charged with 10 to 20 pounds [5–10 kg] of explosive, tamped down and fired at once. The recording would be made via geophones laid out on the ground and attached to the instrument truck by electric cable.
>
> It was tedious work, particularly along the West Australian coast where porous surface limestone tended to dissipate the 'shot' energy horizontally instead of allowing it to travel down through the subsurface to the older rock layers below. Often we had seven shot hole drill rigs working together, and even then we sometimes only managed to fire two patterns in a day.[7]

Dynamite surveys had the drawback of causing potential subsidiary environmental problems like caving and cratering in the limestone. The system also meant there was a lot of heavy vehicle traffic along each of the seismic lines because of the need for transporting the drilling rigs, water trucks and recording van to each shot point. However, in the latter part of the 1950s, the shooting technique advanced to what was called the 'dynaseis' method, whereby a bell-shaped apparatus directed a dynamite charge directly onto a heavy plate laid on the ground. The result was a single 'thump' vibration transmitted into the earth. This method was the forerunner of the modern mechanical vibroseis technique. Data acquisition and retention was an equally important side of seismic work.

> When SSL arrived in 1951, the seismic recordings were 'wiggle traces' on paper rolls. The first magnetic recordings came with one of the GSI crews a couple of years later. That team was

> equipped with magnetic 'tapes' which were large floppy discs about the size of a long-playing gramophone record. It took a while to get the system to work but, once that was sorted out, they were a big improvement. Another technique was to use what was called a 'techno tape' — a tape about 24 inches [61 cm] long and six or nine inches [15–23 cm] wide wound on a drum. Needless to say there were no computers then, but you could get a paper print if you wanted.[8]

These took the form of dyelines, exposed in the sunlight and developed with ammonia. A rough sun screen was used for this process and the geophysicists became adept at judging the right exposure period and using their outstretched fingers to shade the paper. The advantage of this process was that results could be seen quickly and any malfunctions or misfires detected. Also, any adjustments that were required to the shot or geophone pattern in order to improve the data record could be done immediately. In WAPET's case the daily record sheets were also sent on the plane to Perth so that progress could be monitored at head office. The use of magnetic recording was arguably the most important seismic advance of the 1950s because it allowed the geophysicist to replay the tapes through the amplifying system as often as desired. Unlike the 'one-shot' wiggle paper trace recording, it enabled multiple experiments to be carried out with various filter types and frequency ranges to enhance the record. Even so, the best results were still obtained by carefully matching the shooting technique and the spacing of geophones with the characteristics of the soil or terrain being traversed.[9] The clarity of results obtained with seismic techniques during the 1950s bears no resemblance to the high quality digital data records produced with modern computer manipulation. Nevertheless, the advances made over that 10-year period were substantial, and were quickly adopted by survey teams throughout Australia.

In 1959 the Frome-Broken Hill group even felt confident enough about progress to contract GSI for a shallow-water seismic survey in the Otway Basin off Warrnambool on the coast of western Victoria. It was an unsophisticated programme using gas guns as the energy source, but it marked the first attempt at offshore seismic work in Australia.[10] Around the same time a BMR team conducting short-line seismic traverses east of the town of Surat in southeast Queensland discovered a structure called the Cabawin nose. Simultaneously, the Associated Group of companies shot some lines in the Roma and Denison Trough regions further to the north. All three programmes proved to be the industry's springboard to the 1960s.

Moving from the preliminary surveys to the actual probes for oil, Australia's drilling programmes during the 1950s were among the most sophisticated in the world. Most of the equipment was brand-new and the imported expertise — in the person of US, Canadian, European and some returning overseas-trained Australian superintendents, toolpushers and drillers — was high in quality and long in experience. The bulk of the work was carried out using rotary techniques, with hydraulic controls replacing the steam drive of earlier times. A notable exception was the Australian Oil and Gas Corporation's percussion rig programme in the Sydney Basin of New South Wales, where the high costs of rotary drilling in a multi-well programme were prohibitive for the fledgling company. Twelve rotary rigs worked in the country at various times during the decade with the two most common manufacturers' names being the US companies National Supply Company and Failing.

The National units were designated by a figure related to a tenth of their horsepower rating such that the T-20 rig was powered by a 200 hp plant (the 'T' signifying that the unit was trailer mounted), while the big National 130 which drilled Rough Range No.1 was powered by a 1300 hp plant. The rig capacity was determined by the amount of drilling rod that it could lift out of the ground, so the horsepower was related to the length and weight of the drillstring rather than the ability to turn it in the hole. The Failing rigs, on the other hand, were described according to their depth ratings, so that the Failing 1500 was capable of drilling to a minimum of 1 500 ft [460 m] and the larger Failing 2500 unit was rated to at least 2 500 ft [760 m]. In fact, rigs of both brands could drill to much greater depths than the fairly conservative ratings they were given by the manufacturers. The Cape Range No.1 well, for instance, was taken to a depth of just over 8 000 ft [2 440 m] with a T-32 rig which was rated at between 3 000 ft and 5 000 ft capability [915 m–1 525 m].[11]

During the first tentative programmes of the early 1950s there was scant knowledge of the depths of potential oil-bearing reservoirs, so the bigger rigs were brought in first. Later, when indications were found at relatively shallow depths, some smaller rigs were imported to use in follow-up programmes. Mindful of the greater costs in rotary drilling, the explorers began to tailor their programmes to their specific needs. WAPET, for instance, employed the small, mobile Failing 1500 rig to drill its 16-well stratigraphic programme on Dirk Hartog Island during the mid-1950s, but reserved the big National 100 for deep tests of over 350 m in the Kimberley region.[12] High operating costs also dictated that rigs be worked 24 hours a day and, as much as possible, precautions were taken against loss of operational time. These included the building or upgrading of roads and airstrips to allow rapid transportation of equipment and supplies.

The new sense of urgency in drilling, however, was not only to do with financial savings. The rotary drilling process involved the use of chemically treated mud, and long periods of 'open hole' conditions (where this material was in direct, static contact with the rock formations) were avoided as much as possible. It was quickly realized that adverse reactions could occur, causing eventual caving of the well sides or damage to the pore spaces in the reservoir rock. The latter event could prevent any oil or gas being produced or detected.[13] As a rider to these potentially ruinous circumstances, explorers became aware that the techniques of downhole observation and management also had to be improved. In looking back over the years, geologist Murray Johnstone believes Australia has always been at the forefront of this technology, which involves logging a well using various electrical, radioactive and temperature-sensing devices to determine rock properties like porosity and permeability.

> In 1953 the first logging unit available in the country was a two-conductor cable (supplied by the US firm Halliburton) connected to an oscillating pen and paper trace on the surface, but which only had the capacity to run two logging operations at once... In 1954, the French company Schlumberger imported a six-conductor cable which enabled a complete electrical log to be recorded in one run, and the microlog, temperature log and neutron log were added to the repertoire. The sonic log, dipmeter and induction log were introduced in the late 1950s, and the density log came in during 1965.[14]

The latter unit had a less than auspicious entry into the country as Johnstone recalls.

> The first density log in Australia was lowered to the bottom of the Gin Gin No.1 well in the Perth Basin of Western Australia not far north of Perth. It is still on the bottom of Gin Gin No.1 after being lost down the well in that operation and is now cemented in. It took some time for another logging tool to be made and imported as they were few and far between in those days.[15]

During the 1950s it was usually the well-site geologist who took on the additional tasks of logging and reservoir engineering. Mud logging was one of the most time-consuming jobs. This involved the collection and examination of rock cuttings brought up in the mud flow from the bottom of the well and tests for the presence of hydrocarbons using a mud gas detector. This instrument consisted of a Wheatstone bridge electrical circuit where one of the arms was a glowing platinum wire. Any combustible gases were ignited by the hot filament, thus upsetting the electrical equilibrium and registering on a galvanometer. The instrument had a limited analytical range, but it was in use for much of the 1950s.

Although still fairly basic by standards set in later decades, mud logging began to modernize around 1958 when the first specialist companies from the US arrived in Australia. Even so, the mud logger's unit began as little more than a skid-mounted caravan transported to location on the tray of a truck.[16] Cementing of the steel casing used to line the wells also involved a lot of improvisation in the early 1950s. Everyone on site helped out as the slurry was mixed at the wellhead from bags of raw cement. The petrol-driven rig pumps were used to force the cement down the well and up the outside of the casing to bond against the bare rock, but they often suffered from vapour lock in the high northern Australian temperatures. No additives were used and so the time spent waiting for cement to dry was measured in days. A successful operation called for a little luck as well as good judgment by the engineer in charge. A lot of things could go wrong, although few mistakes were as bad as the 'ultimate cement job' in the Canning Basin recorded by Barry West of the BMR in Canberra.

> The rig pumps were being used to cement an intermediate string of casing at about 335 m. The engineer in charge miscalculated the number of pump strokes required to bring the cement up into the annulus around the casing and he kept the pumps running. Most of it came back into the mud tanks and was pumped down the hole. After duly waiting for setting time, it was found that everything was seized up.
>
> The kelly hose was full of rock-hard cement; the pumps were full of cement; all the mudlines were full of cement; and when the upper joint of casing was unscrewed and raised, there stood a magnificent column of cement rising out of the ground to about two metres above the rig floor.
>
> It was several days before all of the cement was chipped out of the system and drilling was able to be resumed.[17]

When few of these ancillary downhole services were available in the mid-1950s, the Australian drilling contractor Oil Drilling and Exploration set up its own cementing and well-testing equipment. The cementing unit was mounted on a Mack B61 truck and included a big plunger pump driven by the truck engine. It also had a separate mixing pump driven by its own four-cylinder engine, so that the whole unit was independent of the rig itself.

By the end of the 1950s technological advances were beginning to noticeably improve the quality of both the pre-drill survey data and the drilling of wells themselves. At the same time, technology was also making inroads into what some explorers saw as the thorniest problem of all — that of overcoming the remote and harsh conditions that seemed to predominate in the Australian and New Guinea programmes. High on the list of factors affecting survey costs, and certainly drilling costs, was the accessibility of the search area and, naturally enough, such costs increased as the explorers penetrated further and further inland from the coasts. Mountainous jungle operations could be just as costly to set up and maintain as those in the barren deserts. Probably the single most important technical advance in the 1950s oil search was the introduction of the helicopter. Nowhere was its presence more effective than in the jungles and swamps of Papua New Guinea.

By 1954 the Australasian Petroleum Company had three of these machines supporting its renewed survey effort in the Papuan Basin, particularly along the Fly River and up in the highlands round Lake Kutubu. These were the early perspex bubble-fronted, single-engine Bell type which could carry two people including the pilot. Capable of lifting a cargo weight of around 180 kg and travelling at speeds of up to 100 km an hour with a range of about 320 km, they soon became an indispensable part of the exploration programmes, as the Commonwealth Oil Refineries house magazine *The Accelerator* indicated.

> On a recent sortie... two shuttling helicopters moved the complete seismic equipment, tentage, ancillary gear, shooters and observers, on to a line 16 km away in six hours.
>
> On another occasion the same machines did in just over two hours a task which would have taken 100 native carriers four days to do.[18]

During operations in the headwaters of the Fly River, all materials were carried the first 320 km upstream from the Papuan Gulf by steamer and smaller craft, including war-surplus landing craft and bomb scows, to a base camp which had been cut out of the jungle. From there the helicopters flew equipment and supplies in to the small advance parties conducting the seismic surveys. In an average month one machine would ferry between 90 and 100 tonnes of food, supplies and equipment between base camp and the bush survey teams. One of the essentials was dynamite, of which the parties used about 15 tonnes a month. Along with other necessities it was strapped onto the helicopter frame just above the landing skids or pontoons for the flight. As a safety measure small emergency landing areas were cut in the jungle every three kilometres along the line of flight. When fitted with pontoons the helicopters could land on rivers, swamps, grass or, more often in highland work, on wooden platforms about three metres square constructed in jungle clearings.

Although the helicopters nearly doubled the actual cost of a seismic survey operation on a daily rate basis, they paid for themselves by reducing the number of man hours taken on the survey itself. APC calculated that about 50 km of seismic line a month could be accomplished with helicopter support in jungle terrain as against a mere 14 km moving all equipment and supplies along the ground. Taking into account the high cost of operating the machines, the overall break-even cost was reckoned to be at the 40 km point. However there was still no alternative or easy method for actually cutting the seismic lines and, even with the aid of air transport, the going was tough. Through some of the bamboo thickets, 200 m or 300 m a day was considered good progress

— and a complete traverse could be up to 240 km long. Nevertheless the air support was a major breakthough and APC added to the helicopter ferry service by moving its personnel in and out of the base camps by regular Qantas-charter Catalina flying-boat runs from Port Moresby.

With the rapid advances in helicopter technology it was not long before APC made another breakthough in its battle with Papua New Guinea logistics — the advent of the heli-rig. In 1957 the company brought in a fleet of Sikorsky S-58 machines, then the largest helicopters in commercial use in the world and the first to be commissioned for work outside the United States. The *Australasian Oil and Gas Journal* of May 1957 recorded the inaugural use of these work horses of the air at the Komewu drilling location in western Papua.

> The Komewu site is located... about 130 km from the coast and some 13.5 km beyond the limit of navigation of the Aworra River with intervening jungle cover and low lying swampy terrain. The building of a road across this 13.5 km stretch of country would have proved most difficult and would have taken many months to complete. But, in the course of five weeks, Australasian Petroleum Company has landed on site a complete drilling outfit capable of reaching 3 000 m, earth moving equipment, building materials and sufficient supplies to commence drilling.
>
> These machines [Sikorsky S-58s] are capable of lifting loads of approximately 1 800 kg, either internally or externally, with sufficient fuel for 160 km flying, and have seating capacity for 13 passengers. For the shorter haul from the Aworra River they are equipped to lift 2 270 kg and, in fact, loads of 2 088 kg have actually been carried into the Komewu site.[19]

Flying in the cargoes was a continuous process because several trips could be made between the river and the rig site without the helicopter landing at either. All equipment and supplies were assembled in cargo slings or nets at the landing depot and a loading marshal directed the helicopter into position for the hook-up. Release at the drilling site was done in the same fashion and the journal pointed out that one machine flew 33 tonnes of cargo into Komewu in 10.25 hours flying time. The rig itself, a National 50, had been redesigned in such a way that all components could be broken down to individual loads of just under two tonnes. With the successful operation at Komewu under its belt, APC set up a second heli-rig and kept three Sikorskys to service them both during drilling programmes in the last years of the 1950s.

In Australia too, air support was a vital component of the oil exploration programmes. Aerial surveys had been utilized since the pre-war days and during the 1950s revival the use of aircraft was broadened to include the swift movement of personnel and equipment to outlying areas. In the case of WAPET's far-flung Canning Basin work, for instance, where the regional base at Derby was 2 730 km from Perth by road or ship, practically all personnel for the operations were moved by regular commercial DC3 and Fokker Friendship flights. Chartered aircraft were also used for large numbers or for destinations outside the regular routes. Airlifting of freight was often a viable option during the northern drilling operations, particularly in the case of emergency spare parts. Oil companies soon found that the cost of airfreight was small compared to the lost-time factor when possibly a whole drilling crew could be held up waiting for replacement items. One of the most unusual cargoes to be carried by air during the 1950s consisted of lengths of heavy well casing required urgently at one of the Associated Freney group's Kimberley drilling sites. Road transport was impossible because of the 'wet', so a DC3 aircraft was chartered for the purpose — the 9 m lengths of pipe only just fitting along the plane aisle.

Helicopters also made their appearance, although they were regarded as a less imperative part of operations in Australia than they were in New Guinea. However in one survey for WAPET in the vicinity of Lake Betty on the edge of the Canning Desert, surface transport was virtually impossible and certainly not practical. Use of a helicopter allowed a geologist to examine more than 100 points of interest over an area of 26 000 km^2 in a few days — a task that would have taken months via surface transport. Nevertheless road vehicles were still a vital part of the explorers' mobile equipment and the variety in use was extensive.

By the late 1950s WAPET's fleet numbered 150 vehicles and cost about £120 000 a year for maintenance, fuel and running costs. It comprised 29 Land Rovers, 6 Mack trucks powered by special 200-horsepower diesel engines, roadmaking equipment, cranes for unloading barges on beaches, bulldozers for preparing drill sites, mobile drills for seismic field work, recording vans and all shapes and sizes of utilities, and smaller trucks fitted out for survey parties and work in desert regions. The distances travelled were staggering. During the finale to the drilling programme on Dirk Hartog Island, WAPET contracted a National T-20 rig owned by Oil Drilling and Exploration to drill one deep hole as a follow-up to the shallow stratigraphic work. At the time the unit was in Queensland so, in what was believed to be the longest rig move in the world, it was taken 14 500 km across the country for delivery to the island location.[20]

In many cases time was at a premium. ODE had an instance where a rig was due to spud a well near Grafton, New South Wales, three days after it finished a job at Yallourn in southern Victoria about 1 600 km away. The transport crews managed to have enough of the unit in position to enable the oil company to stage its official spud-in ceremony by noon on the third day. Apart from the four-wheel drive jeep, the most common war-developed vehicle adopted by the oil explorers was the landing barge. It proved to be invaluable in Papua New Guinea programmes, but really came into its own when WAPET needed to position a rig on the isolated coast of Exmouth Gulf and again later during the same company's Dirk Hartog Island project. In the latter exercise the landing barge travelled more than 1 300 km carrying the rig and equipment from the mainland, plus the follow-up supply runs for the 16-well stratigraphic programme. WAPET's own house newspaper, *WAPET News Digest,* described the vessels as 'bulldozers of the sea'.

> They don't slice through the waves, wind and the currents with the shapely grace of traditional vessels. They bash their way through the water by sheer horsepower and beef, with a tremendous uproar, frightening the fish for half a mile and riding like a tram-car with square wheels.[21]

Yet without these overpowered floating iron boxes the drilling on Dirk Hartog Island might have been too costly. Nothing else could have ferried the rig, motors, trucks and ancillary equipment across Shark Bay from Denham to the island beaches. High tides and a strong prevailing wind which banked up the water outside the bay entrance made the island coastal fringe extremely shallow and difficult to negotiate.

> On one occasion the barge bashed its way across the bay in the usual three and a half hours, and then took 10 hours to make the last half mile ashore. They [the vessels] are not amphibious, but they are pretty close to it.
>
> When you go aground you don't sit there and signal for help. You set the huge diesels roaring, driving ahead and then astern, and the screws churn great lumps out of the bottom of the sea. You juggle the wheel and the throttle levers until you're stern first to the beach, and then gouge your own channel in.
>
> They're savagely utilitarian, shaped like coffins for elephants, and their thirst for diesel fuel is staggering. But they're cheaper than defeat in war, and cheaper than harbour works in places like Dirk Hartog.[22]

Partly due to these technical advances, the Australasian oil exploration industry on the threshold of the 1960s had a vastly different appearance to the one which had struggled to regain momentum in the years immediately following the Second World War. There was certainly a much greater emphasis on running geological and geophysical surveys before choosing a site for drilling a well. Previously there had been little preliminary technical work done and the resultant bores were usually 'off structure' with little hope of finding oil or gas.

At the end of 1959, eighty companies or individuals held some form of oil permit, and the total area concerned covered nearly 4 million km^2 of sedimentary basins. Ten years earlier the number of companies involved did not reach double figures. But the activity had not increased evenly during the decade. Undoubtedly the Rough Range discovery had a significant bearing on the initial explosion of companies into the field during 1954. But, when that ultimately turned out to be noncommercial and no other finds were made, interest began to wane after a couple of years. More concerned than ever at Australia's lack of indigenous oil reserves, the Federal Government was prompted to step in once more with a new scheme to encourage companies to continue exploration — and the 1957 Petroleum Search Subsidy Act was born.

Chapter 9

INDEPENDENCE OR BUST

Despite the flurry of exploration following Rough Range, the vision of rapid 'oil independence' for Australia did not materialize, and activity began to fall back towards pre-1954 levels. Although companies like WAPET, Oil Search, Australian Oil and Gas Corporation, Woodside and the Associated Group were still in the field, total exploration expenditure dropped from a high of around £9.5 million in 1956 to £7 million the following year, and lower still, to £5.5 million in 1958. There were even fears that companies might withdraw completely by the end of the decade. The old problem of having nothing to offset the rising costs of imported petroleum products was once again causing concern in Federal Government circles. It was the same dilemma that had been faced by the Nationalist Government of Prime Minister Billy Hughes back in the 1920s, except that by the latter half of the 1950s Australia was in the throes of a post-war revolution in terms of petroleum usage. The country spent a staggering £90 million on imported oil and products during 1956–57, a figure that represented 12.3 percent of the nation's total import bill. And it was rising.[1]

The Menzies Government in Canberra realized that apart from reducing Australia's dependence on overseas sources of oil, the discovery of a commercial indigenous supply would also release foreign exchange for the purchase of other goods from overseas. This was deemed an important factor in a country where the goal of rapid development and ultimate improvement in living standards meant that the demand for overseas goods and equipment far exceeded the ability to pay for them. There was also the consideration that a major oil discovery in Australia would be of great significance in terms of national defence strategy. Although building up a refining capability was an important part of such a plan, it was weakened if the crude oil feedstock had to be brought in from elsewhere. Therefore, for both economic and defence reasons, the search for oil assumed a high priority in the Federal Government's list of policies. The vital question, becoming more urgent by the month, was how to restimulate exploration activity.

The BMR was still making a notable contribution to understanding

Australia's geological setting, as were the various State Mines Departments and Geological Surveys. But it was not enough. A new stimulus had to be devised.

The first inkling of a solution came during the visit to Australia by American consultant Dr A.I. Levorsen during 1957. Brought out by SANTOS Ltd to give an opinion on that company's leases in South Australia, he met a number of government officials during the course of his visit, including the head of the Department of National Development, Dr Harold Raggatt. Reg Sprigg, geological adviser to SANTOS, had been largely responsible for the American's trip.

> Raggatt took advantage of Levorsen's presence and his acknowledged leadership in the world industry. In particular he sought Levorsen's opinion on ways and means to stimulate the country's flagging exploration effort. Raggatt was conscious of the consultant's recent visit to north Africa to view French discoveries in basins purportedly having geology comparable to much of inland Australia.
>
> Levorsen had already told SANTOS that, in view of the almost complete lack of knowledge of sediments below the Cretaceous-age aquifer in the Great Artesian Basin, the company should drill a series of stratigraphic wells in the areas of presumed deeper sedimentary section. He expressed the same view to Raggatt, suggesting that the Federal Government instigate a subsidy scheme to encourage private enterprise to tackle the task.[2]

Raggatt was perfectly aware of the problems associated with prewar subsidy schemes (the most recent being the Petroleum Oil Search Act 1936), where lack of geological structural control made a nonsense of some of the proposed drilling locations. However, after mulling the idea over, he decided to broach the subject with Minister for National Development Senator William Spooner and Prime Minister Robert Menzies. The idea gained acceptance readily, and the importance of oil in government thinking at the time was illustrated very clearly by Senator Spooner during the second reading of the resultant Petroleum Search Subsidy Bill 1957.

> This is a Bill which aims to encourage the search for petroleum in Australia by subsidizing stratigraphic drilling. There is little need to establish the importance of finding oil in Australia. Even now, with our oil refineries in full operation, we are spending about £120 million per annum on the importation of petroleum and petroleum products.
>
> Our consumption of petroleum products increases each year. In the five years ending 31st December, 1957, the average

> rate of increase will have been about 10 percent per annum. The national benefits — both from the point of view of defence and general development — that would follow attainment of self-sufficiency in petroleum supplies are almost incalculable.[3]

The Bill was passed on 12 December 1957 and applications to participate in the new scheme were invited. The subsidy itself was limited to stratigraphic drilling with the idea that the government would provide 50 percent of the money for operations likely to provide new stratigraphic information. Former BMR field party leader, Alan Condon, was asked to head the administration group within the Department of National Development, while Harry Taylor-Rogers, from the BMR's Petroleum Technology Division, had the job of visiting all the subsidy recipients' drilling operations.

> The government did set down a number of conditions. For a start, the subsidy was only paid if the government was able to have full use of all the well information. This was done in the form of monthly reports submitted from each operation and it meant that the BMR was able to increase its store of geological knowledge. After six months the data was put on 'open file', which meant that it was accessible to all interested parties.
>
> Another condition was that a full drilling programme and prognosis had to be submitted to the department prior to any work being done. This was analysed and vetted, and if it was not considered up to standard, the application was rejected, or the amount of subsidy was reduced.
>
> Any changes to the original programme — for instance, plugging back or sidetracking a well — had to receive ministerial approval before they could go ahead. Admittedly this did make life difficult for the oil operators who worked round the clock, because the public service was only available between 9.00 a.m. and 5.00 p.m. Sometimes the Minister could not be contacted till after a weekend.
>
> The system of using graticular blocks [based on an area bounded by five minutes of latitude by five minutes of longitude] was introduced as the basis for drawing up permit boundaries in 1958 and it formed part of the regulations surrounding the Subsidy Act. Previously, all the different States' titles had been set up under individual Mining Acts with the exception of Queensland where petroleum had been produced during the 1920s.[4]

Despite its administrative difficulties at times, Taylor-Rogers believed the subsidy scheme was the best thing the Menzies Government ever did for the Australian oil industry. It certainly appeared to be the right approach for the time because operating companies quickly put forward more drilling proposals than the limited funds available to the scheme (£500 000 a year) could cover. However not all applications fulfilled the requirements and the government was able to honour its commitment to the remainder. During 1957–58 five wells (including the Puri No.1 oil discovery in Papua New Guinea) received subsidies totalling £263 634; in 1958–59 seven wells received £463 385.

By that time, however,the old problem of poor preliminary survey control prior to drilling was causing concern once again and the scheme was broadened to include a 50 percent subsidy for approved seismic, magnetic and gravity work. It also provided for a subsidy equal to two-thirds of the well costs for any 'off structure' drilling and a 100 percent payment for electric logging of water bores. An additional incentive was offered to oil explorers when the government allowed a full remission of tax on any investment by shareholders in an exploration company. The amount available in the subsidy kitty was increased to £1.5 million per year. Although in 1959–60 the total subsidy payment was only £359 893 (for projects including 11 wells, 16 seismic surveys, 8 gravity surveys and 2 aeromagnetic surveys), it rose during 1960–61 to reach more than £1 million for the first time, as companies took advantage of the broader parameters.[5]

The revised scheme received a positive response from the industry, but it was not without controversy. While raising few objections to details like the requirement for a full suite of downhole logs for each well, operators did baulk at the BMR's call for coring at 90 m intervals. They argued that this was unnecessary and that it raised well costs unduly. Nevertheless the regulation persisted until more subsurface information became available and log interpretations became more reliable. Even then cores were still required through known reservoir zones and a bottom-hole core was always taken.[6] Unfortunately not all explorers fully understood the subsidy rules as geologist John White discovered soon after he had joined the BMR.

> I was called out to inspect a well which had been drilling for a while in New South Wales. When I arrived I found that it was on the property of two old spinster sisters who were funding the work from the proceeds of their orange grove. Someone had apparently told them they might now obtain government assistance through the subsidy scheme.
>
> After having a look I told them it was not possible to have a grant because regular cores had to be taken during drilling. Well they brightened up at this news, saying they had plenty of cores round the back of the shed.
>
> They took me to see... The cores were there all right — all piled up in a heap, unlabelled and completely useless.[7]

BMR Director, J.M. (Jack) Raynor, who had taken over the reins when P.B. Nye retired in 1958, conceded that the attractions of the scheme did vary from company to company.

> Probably the most general attraction is the release of information from subsidized operations six months after completion of the operation. This rapid dissemination of information leads to a better understanding of Australian geology and to more efficient exploration. I might mention that I have heard the subsidy scheme referred to as a governmental mechanism to purchase information.
>
> At the inception of the scheme it was my view that it was primarily intended to help the weaker brethren. It was at first something of a surprise to me to find that no company from international major to indigenous minor was too proud to apply for subsidy. However the financial contribution is no doubt particularly important to those companies which finance their exploration from capital. Not only does it help them directly to finance their exploration, but it also helps them to raise capital through demonstrating the government's confidence in the ultimate success of the search.
>
> Another factor which should not be overlooked is the effect that subsidy has had in influencing foreign companies to explore in Australia. Representatives of some major companies have told me that the fact that the scheme would enable their exploration dollar to go further has figured quite prominently in the decision of their companies to come to Australia.[8]

The subsidy scheme, particularly in its broadened 1959 form, did all that was required of it in reviving Australia's exploration activity, while the influx of foreign companies and foreign expertise was very much to the Menzies Government's liking. At the same time, however, there was growing disquiet among the small Australian explorers who felt that they were being swamped by the invasion of overseas majors. Worse, their rights and needs might be totally ignored by government. In 1959 a small group of local companies began a move to assert their own brand of independence — a move that was eventually to grow into the powerful oil industry lobby group called the Australian Petroleum Exploration Association, better known simply as APEA. The Association's beginnings were no more than conversations between the few serious Australian independent explorers about the state of the industry and the difficulties they all faced in carrying out effective search programmes. Although the subsidy act provided funds, the money was not paid until after the wells or surveys were completed. That meant the small independents still had to finance the work in the first place, and it was not an easy task when they relied heavily on the stockmarket for capital.

The same lack of monetary resources made the huge permit areas allocated during the brief post-Rough Range boom virtually inaccessible to the locals, and there was concern that the major companies were just 'sitting' on the acreage. A feeling of 'one rule for them and another for us' began to emerge, so that the initial impetus for Australian companies to band together was born out of a growing frustration at the perceived favouritism granted to foreigners. It was also felt that the government had been persuaded by the majors that there was little chance of discovery outside some areas of Western Australia and Papua New Guinea. Alan Prince, a director of Papuan Apinaipi experienced in pre- and post-war government attitudes to Australasian exploration, noted that the locals had a surprisingly widespread following.

> The formation of APEA was not just a revolt by exasperated local explorers. There were also some academics and scientists within the BMR who felt that the large companies were deliberately manipulating the oil search to suit themselves. And they also thought that it was a geological nonsense to say there were only two places in Australasia to look for oil.
>
> While they could not readily go against the government line, these people did lend tacit support to what was a commercial, financial and, to a large extent, professional rejection of an illogical official view.[9]

Rees Withers, managing director of Woodside (Lakes Entrance) Oil, was one of the prime movers behind the push for the establishment of an independent voice.

> There were only six or seven Australian companies in the field coming out of the mid-1950s, and it took a few years before we all started to think seriously of presenting a united front to air some of our problems. Apart from the difficulties of participating in large permits, there were also constant hassles with import licences for essentials like drill casing, chemicals and the rigs themselves.
>
> We felt the government was giving us little encouragement. Several of us, at different times, talked over the issues. I kept in touch with John Fuller, who had been on the Woodside board and then moved to start Planet Oil, and finally it was he who suggested a proper meeting of all the local Australian groups to thrash out a course of action, instead of just talking about it.[10]

The historic first meeting was held in the Planet Oil offices in Sydney on Tuesday 20 October 1959. The story goes that Reg Sprigg from Geosurveys in Adelaide was late and the only remaining seat was at the head of the table. The group had unanimously elected him meeting chairman in his absence. When he sat down, one of the first motions on the agenda was to form APEA. It was passed with little debate — even the name had been thought out beforehand — and, when the gathering broke up, legal details had been initiated through the firm Minter Simpson Ltd. Four weeks later a more formal meeting was called to sign the Memorandum of Association and Articles and to elect office bearers. That first APEA Council consisted of seven members, all of whom were already well-known local explorers and characters in their own rights.

Not surprisingly, Sprigg became the Association's first chairman. The choice was not an ill-considered one. Sprigg was the obvious head because he was a leader in the industry, had a sound technical knowledge, formidable powers of oratory, and no shortage of vitality. In accepting the position, he made plain his desire that the exploration industry produce its own scientific and technical publication. Having had some experience in the Great Artesian Basin, he was deeply concerned that far too little was known about the country's sedimentary geology and that no progress could be expected if the big foreign companies had their way. In his passion for technical explanation, the chairman sometimes set himself up as the butt of humour, as fellow councillor Bill Siller recalled.

> Even then Reg was a little sparse of hair on top. So the spectacle of him explaining the concept of a bald-headed anticline to show why some wells on top of a structure were dry, while down flank they found the reservoir formation, was viewed with some amusement even though he was geologically correct.[11]

Sprigg was supported on the executive by Rees Withers, the determined founder of Woodside; Eric Avery, an imposing man who was chairman of the Associated Group and a former managing director of Shell in Australia; John Fuller, a dynamic ex-Wallabies rugby player turned oil entrepreneur who had formed Planet Oil; C.W. (Bill) Siller, the Queensland geologist who had joined Gene Goff's Lucky Strike company and then gone on to head the new ODE-backed explorer called Exoil; Alan Prince, an experienced manager and a director of Papuan Apinaipi; and finally, Lou Smart, a wheeling-dealing oil man in the true American mould who had formed his own drilling and exploration company called L.H. Smart Oil Company.

The early membership could be described as an enthusiastic band of political amateurs, who nonetheless unashamedly set themselves up as a lobby group to tackle the government in Canberra on oil exploration issues. The early meetings were rather ad hoc affairs and they were held 'when issues needed to be discussed' rather than being scheduled on a regular basis. Generally they were in Sydney, although venues in Melbourne and Brisbane were occasionally used to ease the strain of air fares on some members' budgets. One of the prime reasons for APEA's formation was to present the Australian exploration industry to the general public, so public relations and press coverage were important topics right from the start. Individual members (particularly Fuller and Sprigg) were regular contributors to newspapers and magazines and, in the early days, APEA would speak at various clubs and organized public meetings.

Government reaction to what some parliamentarians called the 'group of self-interested crusaders' was dismissive at first. But after a short time this changed to a grudging, if not outright, respect. There was certainly support for APEA's geological assessment of the country from within the BMR, although this was rarely expressed publicly. Senator William Spooner, the minister responsible for the oil portfolio, was often heard to label the organization a 'bloody nuisance', but even he gave it due attention. Quite possibly this was because of his good relationship with the head of his department, Harold Raggatt. Whatever the reason, Spooner was always accessible to the the local explorers when they came to Canberra and has been described in retrospect as the best minister the oil industry ever had. Norman Fisher at the BMR was associated with Spooner throughout this period and on into the 1960s.

> An accountant by trade, he was a rough and tough customer politically and very strong in a debate. He did not know much about minerals or oil, but he would listen and take advice. Often he would play the devil's advocate. Yet if he was convinced by an argument he would fight hard for it in Cabinet. He seldom lost in his submissions.
>
> Physically he was fairly portly and heavy, but this did not stop him showing an interest in the detail of his portfolio. He was always available for travel and liked to see the mines, rigs and prospects for himself.[12]

Above all, Spooner was fair in his dealings with the oil explorers, as Bill Siller found on one occasion when he and Alan Prince met the Minister in Sydney.

> He enjoyed the personal contact with the industry but, when arguing a point, personalities never entered or influenced the debate. During the Sydney conversation Alan and I were trying to convince the Minister of a point, and he was adamantly defending the opposite because his department officials had briefed him on that counter view.
>
> Spooner: 'My people tell me...'
>
> Prince interrupting: 'Minister, we are your people'.

> Spooner was thoughtful for a moment. Then, chuckling, he said, 'You're right'. And he began to take our comments on board.[13]

At that time the majors were mostly clustered round WAPET and the Frome-Broken Hill group and, in the beginning, the foundation members of APEA did not want to have any of these foreign companies involved. The same attitude applied to the Australian Institute of Petroleum, largely a producers' and refiners' organization. For their part, the overseas explorers tended to ignore the fledgling organization as irrelevant anyway. However this segregational thinking did not extend past the corporate level and, on a personal basis, relations with the foreign explorers were very cordial. Probably the best example was the formation in Brisbane of the Queensland Petroleum Exploration Association, more familiarly known as QUPEX, whose first meeting was held in the city's National Hotel on 3 December 1959. The most apt description of this loose-knit group was published by the Association's steering committee for the 25th anniversary celebrations in 1984.

> [It is] a gathering with no constitution, no memorandum, no articles, no written rules or regulations... which continues to baffle the experts...
>
> [In 1959] the Americans had discovered that large tracts of sedimentary basins were available for exploration and Herb Harris and Roy Hopkins of Magellan Petroleum, Joe Tanner of Phillips, Weldon Hightower of Austral Geo Prospectors, to name a few, had made their headquarters in Brisbane. They missed their friendly petroleum clubs and Herb Harris especially found scouting [the practice of obtaining information about other operations] difficult, having to visit companies weekly to ask what they were doing instead of finding out over a friendly drink.
>
> He circulated a letter... proposing the formation of a petroleum organization 'able to set up committees for scout checks, employment possibilities, standardization of drilling supplies, bulk ordering of oil field supplies, relations with government agencies, exchange of information etc.'
>
> A band of five, consisting of Tanner and Harris from America and the locals Bill Siller, Dick Mott and Doug Traves formed into a Steering Committee to get QUPEX on the road.[14]

Joe Tanner, manager of Phillips Petroleum in Australia, became the first president, with Doug Traves of the Associated Group as vice-president. It was understood that the Chair would alternate yearly between Americans and Australians. Attendance at the first and subsequent meetings contained a mix of foreign and local explorers, Queensland Mines

Department officials, as well as visitors from interstate, while both ABC radio and ABC television gave coverage in their main evening broadcasts. Although formed just ahead of APEA, QUPEX was not meant to be a rival organization. Bill Siller was closely associated with the early years of both.

> APEA was basically a lobby group with membership on a company basis only. QUPEX, on the other hand, was a loose association of individuals... It is a living example... that people of goodwill do not need a constitution, Articles of Association, or whatever, unless the laws of the land impinge on their basic freedoms.
>
> It [QUPEX] played a great part in creating personal friendships between the locals and overseas explorers, the vast majority of whom were American... there were times when I was sitting on a barbed wire fence. One day I remember being in the old Lennon's Hotel and being savaged by a prominent local as pro-American and accused by an Oklahoman of being anti-American... but it was a problem which was never allowed to get out of hand.
>
> As a non-specialist organization, QUPEX is in a position to float new ideas, be they technical or commercial, without any of the hang-ups which one-direction professional organizations must have by their very nature.[15]

Although superficially different, organizations like APEA and QUPEX did have a common thread — and that was to find indigenous oil. The aim — virtually a dream in those early years — was to build up a successful national exploration and production industry, independent from the rest of the world. The intriguing aspect was that this same goal was envisaged in government circles, only the paths towards it varied and, despite the explorers' fair hearings in Canberra, there seemed little likelihood that they would ever completely coincide. Certainly, as the 1950s came to an end, no one could have predicted the spectacular fulfilment for all parties that lay waiting in the decade ahead.

PART THREE

THE 1960s
DREAMS COME TRUE

Chapter 10

THE ONSHORE COMEBACK

The dying days of 1959 provided the decade ahead with a running start when Frome-Broken Hill's Port Campbell No.1 well on the west Victorian coast came in with a significant gas flow. Drilled during December, the find was the highlight of the year. At least one writer went further and described it as the first significant emission since the Roma No.1 gas discovery in Queensland during 1952 drilled by the (then) newly formed Associated Australian Oilfields group.

> Government analysts have reported that the gas at Port Campbell is in the ethane-pentane range and is of relatively high quality. That is, it has a relatively high proportion of higher hydrocarbons.
>
> Thus it differs from the gas discovered in several bores drilled in recent years in the Sydney Basin where the substantial volumes encountered, sometimes at good pressure, have been of methane originating in the black coal deposits of the basin.
>
> In the opinion of Victorian Government and other geologists, the gas flow at Port Campbell could mean that oil is also present in the region.[1]

Following exhaustive tests (which included three more wells) Frome-Broken Hill concluded that, although promising, the gas accumulation was too small to be commercial even in such a relatively populated part of Australia. However the find restimulated flagging interest in the nation's exploration effort, particularly along adjoining coastal areas of Victoria and South Australia containing the same Cretaceous-age marine sediments. One of the first to see an opportunity was Leo Stach, a geologist with APC in Papua before the war who then joined General Macarthur's occupation forces in Japan and moved on to the United Nations scientific staff during the 1950s. Back in Australia in time for the Port Campbell discovery, Stach thought the Nepean Peninsula closer to Melbourne might have similar potential. Living in that city he was able to act swiftly.

I went to the Mines Department to check on the tenement situation and found that Frome-Broken Hill had all of western Victoria to Cape Schank under lease with the exception of a small permit round Torquay and Anglesea. It had apparently been held by a Perth syndicate, but had lapsed only six months previously.

I asked what was needed to apply for the area and was told to mark the four corners and register the claim. It was late on a Friday evening by this time, so first thing next morning I jumped in my old microbus and went to the hardware store to buy stakes, nails and galvanized iron cut to the right size for markers. Then, armed with a map, I sped to Geelong area and staked my claim. On the following Monday I lodged the application and was duly awarded the lease.

Apparently I was just ahead of the game because not long afterwards a chap came to my door saying he had a lignite mine in the Anglesea area and would I transfer the lease across to him. After a bit of 'horsetrading' we worked out a satisfactory financial arrangement. The other party turned out to be Jim Roche of Oil Developments, and not long afterwards the Anglesea permit became one of the features of the Oil Developments NL float.[2]

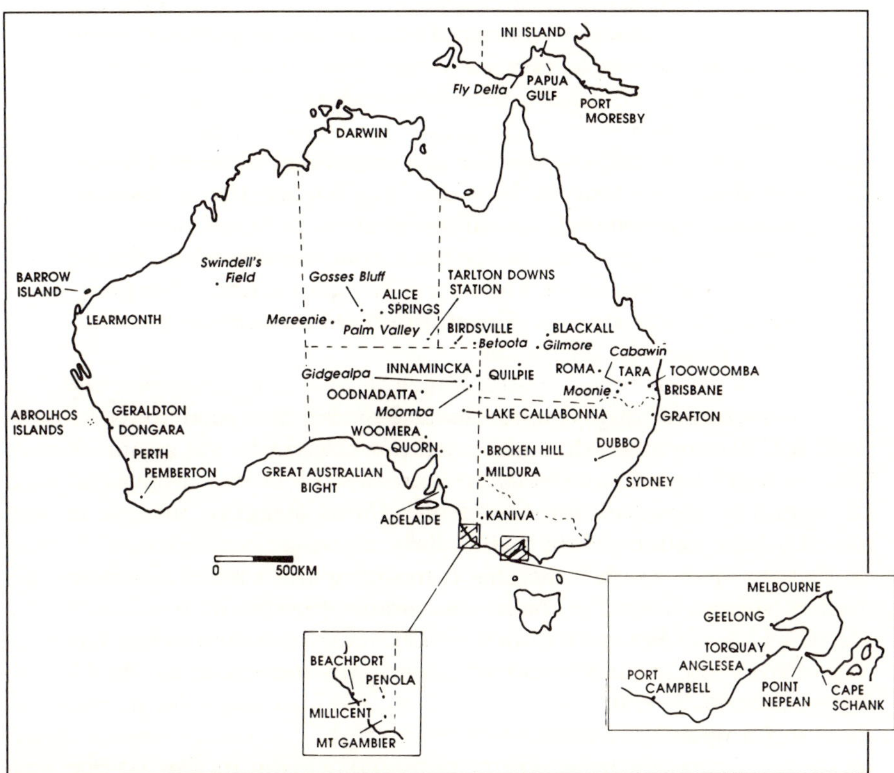

Location of key places mentioned in text.

Stach kept up an association with that company through his own consulting work for the firm Reg Hare and Associates which, in turn, had been appointed its technical manager. Oil Developments had its eye on several prospects along the coast in the wake of the Port Campbell discovery. Two others (Penola and Beachport) were in South Australia and were drilled before the Victorian Anglesea well. Unfortunately, enthusiasm within the fledgling company was no substitute for knowledge in the oil patch, and the programme suffered through inexperience. Jim Cundill, a geologist from Western Australia who had returned from several years oil work in Canada and later went on to form the successful Cundill Meyers consultancy firm, was employed to 'sit' the South Australian wells.

> All of the Oil Development management were mining people, so the extent of knowledge about oil drilling was not great. Apart from the toolpusher and driller, few of the drilling crew knew much about it either. Consequently, even things like mixing the drilling mud left a lot to be desired. At Penola in particular there were lumps the size of pumpkins floating round in the mud tanks. We stuck in the hole a number of times before drilling through the Tertiary section into the Cretaceous sediments. The well was dry, although it did show the Cretaceous rocks were present.[3]

However the next well, Beachport No.1, was a disaster, drilled with an offhandedness that would be totally unacceptable today. Located near the crayfishing town of Millicent, it was financed by a group called the South East Oil Syndicate, a joint venture of Oil Developments, Beach Petroleum (newly started by South Australian geologist Reg Sprigg) and the Kaniva Syndicate (a water-divining group of farmers from the Wimmera region of Victoria). Jim Cundill recalled the episode.

> The well was begun with a Mines Department percussion rig to set surface casing, partly to save money and partly because the Penola well had knocked the rotary rig round so much it needed repairs. A concrete apron was put down over the sand on the site, but the sand riffled underneath it, collapsing both the cellar and the well collar.
>
> That was just the beginning of what became a comedy of errors. When the rotary rig came in for the main hole we had 11 fishing jobs during the course of drilling. The first two or three were caused by dropping spanners and such down the well. But then a whole set of logging tools went down, followed by the fishing tools themselves.
>
> I can remember watching the crew lowering the fishing tools down. I turned away for a moment to attend to something

> else and next time I looked the safety joint was missing and the whole lot had gone down. There was so much iron in the hole you could have started a metal mine.
>
> At another point the bit became stuck in flint nodules at the bottom of the hole, so it was decided to try to break free using acid. Sixty gallons [273 L] of hydrochloric acid was purchased, packed in stoneware crocks, from a local Millicent merchant and pumped down in stages. It didn't free the pipe, but the carbon dioxide pressure built up and blew the drilling mud out of the hole as high as the monkey board where the derrick man was. Lucky the acid reaching him was spent.

The Beachport well also demonstrated how little the management knew about oil-field work. Cundill sent a telegram to Oil Developments in Melbourne after one fishing job saying that he had 'successfully backed off', meaning the bit had been freed. He received a terse reply asking who had given him permission to move the rig! However another incident, which could have tragically compounded the troubles, is paradoxically remembered as one of the lighter moments. Cundill referred to it as the 'twelfth fishing job'.

> The driller had his wife and child with him at the time and one morning she drove down to see him about something that was urgent. In a hurry, she left the baby in the car, but hung onto the feed bottle for some reason, and made a beeline for her husband on the rig floor.
>
> The Beachport well had mud pits dug into the ground so that the surface was level with the concrete apron round the rig. Anyway the lady raced across what she thought was concrete, but was in fact thick mud on the surface of the pit. In a flash all that we could see of her was an upstretched arm holding the feed bottle aloft like the Sword of Excalibur. It was just as well too because at least we could see where she was.
>
> The mud had an astringent quality that gave it the tendency to set quickly, so when we got her out and stripped off her clothes — there was no time for privacy — the handiest thing to clean the stuff off with was the high pressure rig hose. After that ordeal I fitted her out with a pair of trousers and a shirt. It was only later that the ribald comments were heard.

It is difficult to ascertain now whether the Beachport well should be singled out as an example of shoddy workmanship. The drilling practices of the time were generally good in areas that had a record of accident or failure, but there was a prevailing 'hell, she'll be right' attitude where things did not go wrong very often. The drilling regulations themselves were based on 'accepted oil field practice', and were a straightforward adoption of work practices in the US and Canada which, shortsightedly, did not really make any adjustments for local conditions. At times, too, miscalculations were made by the geologists — Beach Petroleum's Grange No.1 well in the suburbs near Adelaide during 1962 being a case in point. Jim Cundill was again in attendance.

> The well was programmed to 7 500 ft [2 300 m] and, as was customary, I went to log a water bore nearby to get an idea of the stratigraphy. In it I saw cuttings of metamorphic rocks and this worried me a bit. But Reg Sprigg at Beach was quite sure the geology at Grange was OK and the metamorphics were explained by a fault through the area of the bore.
>
> About two days after the opening ceremony there was a clunking sound downhole and analysis of the cuttings showed a preponderance of quartzite. We had hit metamorphics at just under 2 000 ft [600 m]. True to character, Sprigg was the first to believe me and the last to worry unduly. Ever the optimist, he was already thinking about the next project.

At the beginning of the 1960s there were still very few Australian crews experienced in oil work and most of the toolpushers and head drillers were American. These 'foreigners' knew their jobs, but they stamped their own brand of lifestyle on unsuspecting local communities. A prime example of the breed was a driller nicknamed 'Tip', short for Tipper, who worked with Cundill on Oil Development's 1962 Mt Salt well near Mt Gambier in South Australia.

> Tip was a bachelor who defied convention in every possible way. He smashed the company car on his way to the rig on the first day. The local 'beak' asked him if he had anything to say. Yes, Tip had. 'Your roads are too narrow. Your cars are too small, and everyone drives on the wrong side of the goddam road.'
>
> He was quaintly naive too. During drilling at Mt Salt I told him about the extinct volcanoes in the region. One of them, Mt Burr, had thick scrub in its crater which caught alight in a local bushfire just at that time, sending up lots of flame and sparks. Seeing this from close quarters one night and thinking the dormancy was over, old Tip rushed round calling everyone out of bed to flee the scene.
>
> But perhaps his best performance was in the local boarding house. He was an early riser and always had a shower before breakfast. Being so early he trotted down the passage

> from his bedroom to the communal bathroom stark naked. I told him he ought to wear something just in case the landlady came by. Old Tip just looked at me, clamped a hard hat on his head, chomped a cigar out of the side of his mouth and waltzed off down the passage as before.

Neither Beach Petroleum nor Oil Developments, nor even Frome-Broken Hill itself, was able to follow up the Port Campbell discovery successfully on either side of the Victoria–South Australia border in what became known as the Otway Basin. Even the much-vaunted Anglesea No.1, finally drilled during 1962, was dry. By that time, however, Australia had made its first commercial discoveries, and the spotlight had shifted to southeast Queensland.

First blood there was claimed by the Associated Australian Oilfields group, which, in the late 1950s, had pulled back from its programmes across northern Australia and Papua New Guinea to concentrate on its original starting point round Roma. Knowledge of the region was better than any other in the country by that time and AAO had added to it with Australia's first commercial seismic survey in 1959. The company was able to drill 1 200 m wells in 10–14 days. Timbury Hills No.2 in March 1960, located only five kilometres from the town, was the breakthrough well, coming in as a gas discovery with a test flow of 35 000 m^3 a day. It was followed three months later by the even larger flow of 182 000 m^3 a day from Pickanjinnie No.1, 32 km from Roma.

In August that year a decision was made to use the two finds as feedstock fuel in the Roma power station for an initial one-year trial period. The short pipeline (the first of its type in the country) and various elements of the project were engineered and fabricated by the rig crews of the area without the aid of specialist pipeline contractors.[4] But although gas, even on this small commercial scale, was welcome, the big hope was oil. AAO geologists were optimistic about their chances in the programme ahead because the gas had high concentrations of condensate, while oil itself had been seen bleeding from cores taken from just below the reservoir section. Indeed the first commercial oil discovery in Australia was made nearby, but the coveted prize was won by another consortium.

When T.W.H. (Bill) Dee floated Australian Oil and Gas Corporation during the mid-1950s boom, he concentrated the company's initial exploration in New South Wales. However he had also taken a lease out on 155 400 km^2 of the Surat Basin in southern Queensland centred on the small town of Tara to the south of Roma. From the beginning Dee realized that exploration would be expensive, and geological consultant Professor Eric Rudd was empowered to seek a farm-in partner during a trip to the US.

> I found the Americans sceptical during that period of the late 1950s. They were used to dealing in much smaller blocks of

> land and thought that such a large permit couldn't have any value. There were two Venezuelan colonels hawking blocks of only a few acres in South America at the same time for about $US1 million an acre, and it is ironic that the US companies took far more notice of them than me.[5]

However two Californian corporations — Union Oil Company (which had exploration, production and refining interests spread through nine countries including the US) and Kern County Land Company (which had extensive ranching and oil interests in the US and Canada) — did show an interest after being 'introduced' to the acreage by a go-between named E.A. (Earl) Noble. An agreement with AOG to explore the Surat permit was signed in October 1959. The deal was that for two 3 660 m wells Union-Kern could earn an 80 percent interest. Noble would receive an override interest on any commercial discovery. By that time some preliminary surveys had been run and three shallow stratigraphic holes drilled near Tara. In addition, a BMR seismic survey had found a structure in the vicinity which was dubbed 'Cabawin'. Taking up the operatorship, Union Oil conducted more detailed surveys before deciding to make Cabawin the first target. Australian contractor Oil Drilling and Exploration won the drilling contract and spudded the well in early October 1960. Two months later the operation was interrupted in spectacular, if not welcome, fashion when gas encountered at 3 050 m blew out of control. For a time it appeared that the well would be lost. Dave McGarry, a geologist with AOG, was in the midst of the tension which followed.

> When the blowout began the rig hydrill slammed shut and cut off a lot of the flow, but it was really only recommended to be in place for 36 hours. By the time we had things under control it had held for 12 days.
>
> The gas cloud was at knee level round the rig and it just sat there like an early morning mist. We had to post men at the perimeters to stop people coming into the danger area. Sightseers were actually camping nearby and they treated the men like heroes.
>
> The trouble was we had little barytes on site to weight up the mud and kill the flow. About 20 tonnes was bought from the Richter Drilling people at Roma, but that was not nearly enough.
>
> A tractor was hooked up to the rig to pull it clear in case of fire.[6]

Richard (Dick) Dumbrell at ODE was also caught up in the feverish fight to save the well.

> It was the only rig for miles and mud stocks were soon exhausted. We called up Quorn in South Australia, where the barytes was mined and milled, to order urgent supplies. In the end we ran their stocks dry too and we were carrying the stuff away as soon as they'd mined and processed it.
>
> The difficulty was getting the barytes trucked up to Queensland and there were 30 semitrailers strung out along the outback roads between Quorn and Tara at one time. The fact that it was Christmas compounded the problem and we had trouble convincing the drivers of the urgency. We were on the phone for a week ringing the police all the way down the track to roust drivers out of the pubs and truck stops. Some had also decided to go home for the festive period.
>
> It was pretty desperate there for a while and we even threw local earth down the well to try to stop the flow. A second worry concerned the integrity of the casing, which was secondhand tubing from AOG's percussion hole work in New South Wales. There could easily have been leaks into shallower formations which would have made drilling in the area more dangerous.[7]

Eventually enough barytes had been pumped down the well to stem the flow and drilling resumed. In May 1961 the well tested 14 000 m^3 of gas. More importantly, in the eyes of the Union group, there was also a 60 barrels a day flow of light oil, while a sand at the base of the Jurassic section higher in the well exhibited excellent reservoir characteristics and oil staining. Reaction in the local farming community to this result was one of excitement as more than 50 percent of the residents in the district had shares in AOG. In Sydney the reaction was mixed. Outside the stock exchange building someone had daubed the message 'Yankee Oil Swindle' in large letters spread over 4.5 m. However inside, AOG stock went wild with 21 000 shares changing hands within a quarter of an hour — the overnight price rocketing from 20 shillings to 28 shillings before profit takers eased it back to 23 shillings. During the mid-morning session the exchange chairman had to call a five-minute recess to enable brokers to catch up on the book work.[8]

Naturally enough everyone's attention was focused on the next well, Cabawin East No.1, some six kilometres away. But it was disappointingly dry — in hindsight, mistakenly drilled on the other side of a fault where the reservoir sands had silted out. Although disappointed, Union's Australian manager, Doyle Graves, was keen to move on and try the Jurassic sands which had shown such good reservoir characteristics when seen above the hydrocarbon accumulations in the Cabawin No.1 well. Consequently, he began to look for a structure where these sands were present in an anticlinal trap position. He found it 32 km further south where the seismic profile gave a reasonably distinct outline of a feature which the group called Moonie.[9]

Moonie No.1 spudded in during November 1961 and a month later it flowed a mixture of oil and water under test at around 1 765 barrels a day. A sustained test programme over the Christmas–New Year period indicated an average flow of 1 000 barrels a day of good quality light crude. Christmas that year was a marked contrast to the one before when the company was desperately fighting to stem the Cabawin blowout. Once again the stock market went wild. But, mindful of the Rough Range disappointment eight years earlier, Graves anxiously called on speculators to calm down, saying that a lot of appraisal work was needed before the find could be called a major strike — or indeed a commercial discovery at all. Few took any notice. Small bottles of Moonie crude were sold to raise money for charities. They were produced in hotel bars throughout the outback and proudly displayed on the mantlepieces of private homes. Moonie ash trays, Moonie pen knives, Moonie pens and pencils, Moonie tea sets and Moonie tea towels suddenly appeared on the shelves in novelty shops.

Union was inundated with requests for the use of the Moonie name in all manner of bizarre promotional schemes — even the latest teenage craze in twist-and-stomp dance records. Bus tour companies advertised excursions 'to the oil fields' and self-styled experts mushroomed all over the country, bursting into print with profound analyses of the significance of the discovery. Rumours abounded and at least one stockbroking firm monitored messages sent by field radio to Union's operations office in Toowoomba. Geologists at the rig resorted to transmitting in Spanish, but it took the brokers just 24 hours to find an interpreter.[10] On the other side of the coin Union and Kern became the butt of anti-American slogans and the two companies were accused of swindle. During the appraisal drilling programme which followed the discovery, a number of wells happened to be completed at weekends, but the hysteria was such that this was interpreted as a deliberate ploy to give the American operators an unfair advantage in share market dealings.

The press, while magnifying the find, also showed its ignorance of oil field practice by accusing Union of running tests through differing choke sizes to deliberately obscure the true flow rates from the public. There was also criticism of the lack of publicized reserve estimates and insufficient data in general. However Doyle Graves refused to be stampeded. The 42-year-old American geologist could just as easily have been a diplomat because his handling of the euphoria remained calm throughout the 18-month appraisal/development programme. Patiently he explained the reasons for each step and fended off criticism that Union and Kern were trying to monopolize the Australian oil industry. In this he found an ally in Federal Minister for National Development, Senator William Spooner.

> The [Moonie] strike is a moment of history. All is going well at the moment, but I just want to sound a note of caution

> that it is not only a case of finding oil, but of finding out how wide and how large the area is before an estimate can be given of the size of the deposit.[11]

The cautionary tones did nothing to diminish the celebrations in the district round the Moonie discovery itself. When confirmation well Moonie No.2 came in during February 1962, the Tara townspeople had the news in a flash. Local station owners saw smoke from the flare rising above the trees and immediately phoned the information on. Publican Jim Murphy's Commercial Hotel was the nearest 'watering hole' to the field and most of the off-duty drill crew joined the locals there for a party that went on well past closing time. Having been caught short of beer when the first well was successful, Murphy had not made the same mistake again and during the evening the Commercial was gleefully dubbed 'Moonie No.3'.

Such was the enthusiasm of the time that plans were laid, and approved, to build a township close to the field which, it was hoped, would some day have a population of 100 000. A number of prominent citizens in Toowoomba, including the mayor and a member of State Parliament, were involved in the scheme. A marker was put at the crossroads of two tracks — one running between Toowoomba and St George, the other between Tara and Goondiwindi — and a motel on the spot was the first feature. The building was completed with good accommodation, swimming pool, a dance floor, restaurant, a garage and even a cricket ground. Water supply came from a nearby bore drilled by the oil rig. A school was added and for a while it looked as though the project might develop quickly. However the 'town' never grew any larger, principally because one well was drilled at a time and the developers made the mistake of thinking that other companies would be allowed to come into the area and set up drilling operations. That did not happen, and the final blow came when the drilling crew, who were the motel's first permanent tenants, were withdrawn to a camp on the rig site.[12] The appraisal drilling contract had been awarded to Richter Drilling (a company previously set up as the drilling arm of the Associated Group) because supervisor Roy Richter had been willing to meet Union's request for bids based on footage drilled. Most of the crew were either local hands living at home or Canadians billeted in town, and each man worked a solid eight-hour shift without a meal break on a 24-hour rotational basis. The aim was to drill 600 m during a shift and the record achieved was 594 m.

During the early part of the programme the rig sites were fairly spartan, comprising two caravans and a small mess hut fronted by three lounge chairs which the geologists used after dinner. A big shed nearby housed the bags of cement and other materials. The drilling sites were cleared from dense thickets of brigalow and sheoak and the area abounded with unwelcome reptiles — particularly death adders and black snakes.

This menace was so bad that the men were supplied with anti-venom serum and the field was named 'Death Adder Valley'. As many as 20 snakes over one-metre long were recorded killed during a shift in the summer months. In addition, the drillers put up with clinging, biting flies, scorching heat and the ever-present danger of bush fires, while floods were possible during the winter.[13] Even so, morale was kept buoyant by success and conditions were improved as the 17-well appraisal programme continued. Dennis Clow, a Cockney who had been in Papua New Guinea during the early 1950s before joining the Associated Group and then moving to Richter Drilling as administrator, visited the Moonie work on a regular basis.

> For the short moves round the Moonie field locations we didn't bother to break down the drilling mast each time. We just lowered it in one piece and set it up again half a mile or so away. There were several transport companies who became adept at this and it could be done quickly without any fuss.
>
> Up until that time the Richter drilling camps were all canvas, but when the company joined with Canadian Peter Bawden to become Richter Bawden Drilling, the incoming Canadians brought portable Atco cabins with them. They were widely copied in the industry after that. The Canadians preferred a woman in the camp, so often a married couple was hired to look after the chores like washing, cleaning and cooking. At Moonie we put up a proper mess hall too.
>
> We had one cook nicknamed Denny who was a real outback specialist. One time I suggested we should differentiate the cuts of beef, but old Denny reckoned that as far as he was concerned it was fillet steak from the tail right up to the animal's ears.
>
> Roy Richter himself came through early one morning in time for breakfast, so he joined the chow line and asked what was on the menu. Denny looked him in the eye and said, 'Steak or rissoles'. 'What's the difference?' Richter asked, gazing at them sceptically. Quick as a flash Denny replied, 'Three days', and closed any further discussion.[14]

Moonie was declared viable late in 1962 even though further drilling was needed to fully delineate the reservoir. In March 1963 Queen Elizabeth II lent royal recognition to Australia's first commercial oil field by unveiling a plaque at the mouth of the Brisbane River where Amoco and Ampol had already announced plans to build a refinery each, and where the soon-to-be-built Moonie to Brisbane pipeline would end. Moonie came on stream in mid-1964 and a few months later the Union group was successful again at the nearby Alton field found on Crown Land (a stock route), although this was soon recognized to be a smaller accumulation. Further north the Associated group continued to notch up gas discoveries round Roma, and gradually the proven reserves there were built up to a level where they could sustain a 20-year contract to supply a new ferilizer plant planned for the Brisbane area. Oil was found in AAO's acreage too, with discoveries like Sunnybank and Richmond, although unfortunately the size of the fields did not match the gigantic headlines they received in the local and national newspapers.

Surprisingly, for such a risk-laden industry and the rather casual 'cowboy' attitude adopted by drillers at times, there were few tragedies. Fatalities did occur when equipment failed, such as the time a bolt came loose from the derrick of a rig drilling a well in Grafton, New South Wales, and pierced the helmet of the only unmarried man working on the drill floor. On other occasions, particularly in the north of Australia, men were drowned after being washed off vehicles or trapped inside them during floods — accidents that could generally be attributed to a lack of caution or appreciation of the dangers involved. Plain stupidity played a role too, such as the accident in which a rig hand given notice for dangerous practice was subsequently killed on his last day when he overfilled a split rim truck tyre and it burst. Serious accidents such as lost digits or limbs, broken bones and burns were more common, most of them the result of flouting safety practices. The saying: 'Don't cut corners and things will be pretty safe' was often the only guide used during exploration operations. While this unwritten 'standard' was practised and policed as much as possible by most operators, it lacked the bite of the strict safety regulations and severe penalties which were enforced during the more aware 1970s and 1980s. Despite the casual appearance of attitudes in the 1960s, alcohol was rarely a factor in oil field mishaps because it was either strictly rationed and policed, or banned from operations altogether.

By the 1960s too it was standard practice in the remote locations to set up and keep regular radio schedules with operational bases or headquarters. Police and station owners were generally informed of the whereabouts and intentions of survey parties and drilling crews, while the flying doctor service was always as close as the nearest radio transmitter. The stock market was also very attentive during the years following Moonie, but for a different reason. The companies would always report discoveries as soon as they were made, but this did not stop people trying to anticipate the result. Some of the rig workers themselves were not above a little share speculation from the drill floor, but few attempts were profitable. Dick Dumbrell at ODE recalled one occasion when an American working on a Surat well came unstuck rather obviously.

> Working around the rig floor, he'd picked up a few comments from the geologists and reckoned that the well was a sure thing. Anticipating this success, he rang his broker to buy about $20 000 worth of shares. But he had to use the phone

> at a local farmhouse, so the farmer knew about it as well. He couldn't keep it quiet and pretty soon the neighbours had the story and everyone in the surrounding area bought shares.
>
> Anyway the well was not the discovery that everyone had put their money on. The American welshed on the payment to his broker and fled the country, leaving the farmers to carry their own burdens.[15]

A more serious 'scam' involved a hoax telegram to the Sydney Stock Exchange concerning Longreach Oil Ltd in February 1963. The message, purporting to come from Longreach directors, claimed the company had struck oil in its Alice River No.1 well about 1 000 km northwest of Brisbane. It added that a technical report from the well site was expected within 48 hours and that the directors would then issue a detailed statement about the significance of the find. The Stock Exchange sought confirmation of the telegram from the company but, due to the absence of the secretary, contact was delayed. In the meantime, trading in Longreach shares began the day on a high note forcing the price up two shillings. About 15 000 shares changed hands before any warning was received that the report might be a hoax. The Exchange suspended trading just before 11.00 a.m. after Longreach directors denied all knowledge of the telegram and of any oil find. When trading in the stock was resumed later in the day, shares immediately fell back to their overnight level and, although enquiries were made, the perpetrators of the hoax were not found.

The industry abounded with characters during this period, the more so as greater numbers were lured to the search after the Moonie success had finally dispelled the notion that Australia was too old to contain commercial oil. Somewhat surprisingly diviners were still abroad. One old fellow circled the Moonie No.12 location and declared there was no oil to be found on that spot. Moonie No.12 was subsequently found to be one of only two appraisals on the field which were dry. Some practitioners were straight out of the old 1920s mould like the character AOG geologist, Tony Wright, came across in the southeast of South Australia.

> This chap had a long brass rod with a metal lid and a box with numerous jars in it. Each jar contained a sample of the substances he was looking for and they were all quickly interchangeable on the end of the rod. He'd slowly move the thing to and fro while walking forward. I never saw him have any success, but it was a good show.[16]

On other occasions it was the rig workers themselves who 'hammed it up', particularly if there was an audience. Terry Watts, an Englishman who came to Australia during the mid-1960s to begin work with the Cundill Meyers consultancy, saw a number of these antics.

> There were many times in Gippsland and round Roma when the men took school parties on a tour of the rig. For the locals it was also a Sunday trip to see the derrick and machinery, and cars were often two and three deep along the perimeter of the drill site. So there was no shortage of audience.
>
> The roustabouts used to wander round the drill floor with twitching rods or mock-seriously study the ground close to the rig. Another favourite was to sit by the sump with a fishing line and loudly talk about how the sand sharks were biting.[17]

The 1960s was also a period when American entrepreneurs made their mark by taking up acreage in the inland basins where few companies had ventured to that point. Some made capital by broking permits to other explorers and retaining a small, but potentially significant, royalty or nonparticipating interest in any future discovery. Earl Noble, who brought Union and Kern to AOG, was one who benefited greatly from the latter arrangement. Another was a professional dealer and consulting geologist named Robert Kamon, who sold southern Surat Basin acreage to a Texan named Bruce Anderson, who in turn farmed out a large percentage to Beach Petroleum. A further example was Sydney Kahanoff, resident geophysicist in Australia for Union Oil till he left to begin Voyager Petroleum in Calgary. He took up areas in northern and western Queensland and had an office in Sydney for a short period. But perhaps the most flamboyant of them all was Herschel Albert Hackathorn, a Texas oilman, complete with boots and hat, who was president and director of Hackathorn Oil Australia Ltd. The *Bulletin* magazine in Sydney published a good description of the man and his motive.

> ... an oil drilling contractor... he has had wide experience in the US where he drills about 100 000 ft [30 500 m] of hole a year, operating in Wyoming, Colorado, Kentucky, Oklahoma and Texas, and owning altogether more than 40 producing wells.
>
> He's a hard worker — apt to start at 7.00 a.m. and end at midnight. His grandfather, formerly an Ohio riverboat captain, worked in the early days of the oil business in West Virginia and Pennsylvania and built the first gasoline plants for the Texas Co. in Oklahoma to supply aircraft fuel in World War I.
>
> Hackathorn states his aim definitely: to discover a major oil province, and he's prepared to spend his time and money realizing that ambition. He has taken up leases in the Murray River Basin, because, he says, small American oilmen are bottlenecked in their own country for lack of markets.
>
> Looking for a place where he could expand — where

> there are ready markets, a friendly government and friendly people — he chose Australia as meeting all those conditions. Up to date he's doing fine.[18]

The Murray Basin drilling was farmed out to Beach Petroleum, but was ultimately unsuccessful. Hackathorn also began the original Outback Oil NL which drilled two wells in the Great Australian Bight just off the coast of South Australia. However both hit basement at very shallow depth. Another of his involvements was Banner Petroleum, which had a permit interest in the Great Artesian Basin of western Queensland right through into the 1980s.

The first Australian equivalents of these free-flying American entrepreneurs were L.H. (Lou) Smart and John Fuller, both of whom were active participants in the exploration scene during the 1960s. Smart was already a millionaire by this time via his importing business in Sydney — mainly in ceramic tiles. A 'Bondi boy' and a member of the surf club, he was a big, fit man with a dynamic, likeable personality. It was Reg Sprigg, an oil entrepreneur of a different kind, who interested Smart in the oil business in Queensland. Ray Twist, a South Australian geologist initially employed by Sprigg, became one of Smart's first oil employees.

> Lou took to oil exploration immediately. He was an entrepreneur to his bootstraps and he'd go for anything if he saw a quid in it. But he used to play the government's subsidy system.
>
> Although he had plenty of money from his other business, he didn't use it for oil, preferring to rely on the drilling subsidies of the time.
>
> He got hold of a Hydromaster cable tool rig to use in southwest Queensland and the first wells were Orient No.1 and Orient No.2, both plugged and abandoned. But he kept acquiring interests and soon had vast areas stretching across western Queensland from Adavale through Quilpie.
>
> Smart never wanted to go public. L.H. Smart Exploration Ltd was always a private oil explorer. There were other shareholders, but Lou was the largest and it seemed he didn't want to lose control.
>
> When negotiating a deal he was a hard, shrewd businessman. He always had things in the fine print and never wanted to be squeezed out, preferring to keep equity for the future. Some jokers called this ploy 'the Smart Option'. Yet if he liked a person there was a lot of trust and very little questioning. It was an odd quirk of his nature.
>
> In his heyday Smart not only held the western Queensland acreage, but also permits in the Clarence-Moreton Basin off the northeastern coast of New South Wales and round Nowra

> in the southern Sydney Basin. Lou liked to make outrageous statements about the potential of his areas and I was for ever hosing them down to an acceptable professional level.[19]

John Fuller first became interested in the oil business within the confines of a Second World War prison camp at Kobe in Japan. Behind the barbed wire he had talked for hours with an American major who had been a drilling superintendent before the war and who had visited Australia and formed the opinion that the country contained oil. After the war Fuller went to Sydney University to study economics and accounting, which he regarded as a vital background to a business career. An ex-member of the Australian Wallabies rugby team and son of one of the partners of the Fuller theatre chain, he had a strong will to strike out on his own and succeed in the oil business. On meeting the American Gene Goff he became involved in Oil Drilling and Exploration, acting as Goff's alternate for a time. He formed his own Planet Exploration Company in 1957, also joining the boards of the Associated Group and Woodside (Lakes Entrance) Oil. Fuller added to his knowledge with a night course in geology at Sydney University and a three-month stay in the US 'learning the oil business'. He took Planet public in 1962, resigning his other board memberships to devote his full attention to the newly floated company. For the rest of the 1960s and into the early 1970s he presided over one of the most active explorers in the country, drilling in virtually every State, unfortunately without commercial success.[20] Equally as fit and strong a man as Lou Smart, the two Sydneysiders would often be playfully physical with each other and neither would miss a chance to take a rise out of the other in business or social life.

Less visible, at least to the public eye, were the large number of expatriate American drilling personnel who were brought to Australia by the US oil and contracting companies in their hundreds during the 1960s. American oil field technology led the world and the US companies were willing to pay for importing this technology wherever they went. Because of the high level of activity in Queensland, many of the companies based themselves in Brisbane — among them, Phillips Petroleum, which drilled 50 wells in the western part of the State during this period. Australian geologist Ralph Spinks joined the team towards the end of this stint but quickly came to know the 'old hands' very well.

> Two in particular — Fred Peden and W.O. Bumbard — left an indelible impression on me. They were a breed.
>
> Both were born in the American mid-west during the 1920s and their educations were high school at best. Neither was strong on paperwork or writing skills. They both wrote with that rather awkward capital letter American style.
>
> They started work in the 'oil patch' at perhaps age 16. For W.O. the war interrupted proceedings but, that apart, their

> careers had been continuous through to the time I knew them. Each had in excess of 25 years of concentrated experience in drilling oil wells.
>
> Their first jobs were roughnecking, and competition for jobs was tough. Only those with a will to work and learn survived, and it was common for men to work on the rig floor for 10 years before being promoted to derrickman.
>
> Fred and W.O. made that first promotion fairly quickly, then up to driller, and eventually toolpusher and company supervisor. Their knowledge when I met them had been accumulated through involvement in hundreds of wells all over the US. Each well presented a new challenge and the approach was not scientific or engineered as it is now.
>
> Solutions were developed by trial and error, and quickly, because time was money. And money was only made by drilling the well to total depth faster than the opposition. It meant the men in charge had to have persistence, patience, imagination and acceptance of ideas from the youngest to the oldest on the team.
>
> That's where Peden and Bumbard shone. They got anyone who had a contribution to make involved and interested. They were good motivators, relaxed and totally at home with the drilling rig environment. They liked an off-duty drink in town and a joke, but their life was on the rigs.
>
> They were also instrumental in passing on their knowledge to any Australians who were interested enough to learn. The teaching was informal and on the job but it was effective and had a tremendous influence in Australia where there was little experience or background in the oil industry.[21]

Phillips Petroleum itself began in Australia during 1960 in partnership with another US independent called Sunray Midcontinent Oil Company and headed for southwestern Queensland where it took up acreage in the region between Quilpie and Blackall. After initial survey work the first well, Buckabie No.1, was drilled on the Nickavilla Station about eight kilometres west of Quilpie township. In a typical inland operation, Delta Drilling Company's National 130 rig was road trained across for the job from the unsuccessful Delhi/SANTOS programme at Innamincka and Beetoota, while other heavy equipment, air-conditioned accommodation for 60 people and drilling materials were brought in from Brisbane. Two airstrips, one capable of landing a DC3, were cleared from the scrub and graded. Unfortunately, like so many of the inland Queensland efforts of the period, Buckabie No.1 was dry, and another three years of survey and drilling elapsed before the partnership made a significant discovery. This was heralded by a strong gas flow of around 140 000 m^3 a day from a deep 4 360 m wildcat named Gilmore No.1 during 1964.

There were a number of mechanical problems encountered downhole during completion of this well which resulted in damage to the formation. Subsequent production tests flowed at much lower rates than the original drill stem test and a lot of work was done to try to restore the reservoir permeability. Driller Bob McCulloch and his wife, who invariably accompanied him round the country, had their caravan a short distance from the rig.

> I went outside at 10.00 p.m. one evening to check on an acid fraccing [fracturing] job we were doing to improve flow from the formation. The men had apparently opened a valve to pump the solution down to the reservoir level when gas started escaping. All I could see was a white mist of gas enveloping the whole rig.
>
> I didn't have spark-proof boots on, but there wasn't time for delay, so I tiptoed gingerly onto the rig floor and turned off the valve. Later we found the friction of sand in the gas flow from previous tests had worn clean through the steel in an elbow bend of the pipework.
>
> Another time at Gilmore the Phillips VIPs had come out to look at the operation and wanted to see a test flow. That was OK because we were all set up for it. But, instead of opening the flare line valve gently, the 'big nob' spun the wheel round so the valve was wide open and, with all the released pressure, the line began whipping round like a mad snake. Luckily no one was standing too close.[22]

Although the next well, Gilmore No.2, gave a disappointingly weak flow, Phillips/Sunray persisted and confirmed the gas field with two further appraisal wells. However, despite the initial excitement and postulated plans of a pipeline to link with the Associated Group's growing reserves at Roma 650 km to the southeast, no further appraisal work was attempted. Realistically, the accumulation was judged to be too remote to warrant commercial development.

The remoteness of drilling operations from main centres could also be a problem for the equipment agencies, particularly those supplying the increasing need for rock bits. In the early 1960s the tri-cone rolling cutter bit did not have the hardened teeth of later models and the average life was between five and ten hours, depending on the formation being drilled. It was not till the 1970s that the 'journal bearing' bit extended life to as much as 100 hours. In the US the tool and bit companies divided the country into areas and had regional offices and field salesmen who called on the rigs once a week, often leaving bits with the contractors 'on consignment'. Queenslander Eddie Prackert found that this system did not suit the vastness of Australia and the general paucity of rigs. Prackert began working in the equipment business for local pioneer

Jim Bayliss during the early 1960s before switching to Ed Hoeppner, a New Yorker who was put in charge of the first Hughes Tool office in Australia during 1966.

> Hughes started with the tri-cone rolling cutter bit and had it under patent till 1957. After that, competition from companies like Smith Tool, Reed Tool and Dresser made the going tougher not only in Australia, but right round the drilling world. In Australia, Dresser sold through the ODE service company called Oilfield Supply Co in Sydney; Smith worked through Evans Deakin in Brisbane; and Reed, like Hughes, had its own people in Australia.
>
> Interestingly, it was the oil companies that bought the bits, not the rig contractors unless it was a turnkey operation [complete supply and equipment management of a job]. Even so, the superintendents, toolpushers and drillers in the field had their own preferences. Some of them would keep a record of drilling performances in various formations and then compare bit brands.
>
> I found visiting the rigs to get to know the men working out there was a good idea because, although those blokes did not make the decision to buy, they could make a recommendation to the oil company involved. It meant a lot of travelling, but it was worthwhile because in the end it all came down to personalities, sales and service.
>
> A lot of the old timers wouldn't have anything that was not manufactured in the US, which was OK by me because Hughes always fully imported its bits. Some others told a few fibs which made it difficult because I was never allowed to discount the price.
>
> At times there was the odd bit of bribery — less than in the US, but it was still about in Australia. For instance a driller would order a bit with a VO thread. That particular type came with a bottle of bourbon in the packing![23]

Spurred on by the finds at Cabawin and Moonie, explorers pushed further and further inland in the quest for more discoveries. Once it was shown that oil had been accumulated in commercial quantities in Australia, no one could believe that the Surat Basin was the only place in which it would be found. While the Queensland search moved west, South Australian teams moved north until work bordered and then entered the sand dune country of the Simpson Desert. Brian Fitzpatrick, who left Reg Sprigg's Geosurveys in 1960 and formed a working partnership with fellow geologist Bruce Wilson, put in a four-month reconnaissance survey in the Hay River/Tarlton Downs Station region for a company called Three States Gas.

> A lot of the country was shifting sand, as Fred James on Tarlton Downs was the first to lament. He told us that when he was buying the place it was described as being on the northern edge of the Simpson. By the time we met him in 1960 he reckoned he was right in the desert because the sand was drifting into the cattle yards at his south bore.
>
> Our aim was to track down and examine as much rock outcrop as we could. We had two four-wheel drive vehicles fitted out with long-range tanks and often worked alone so we could cover the whole area.
>
> It was the ideal spot for a workaholic. The way we lived left little option. We slept in swags and changed camp almost every morning. That meant rolling out at first light, cooking our own tucker, packing the swag into the vehicle and moving off before the flies woke up. Likewise at night we had to wait for them to go to bed before cooking anything.
>
> There wasn't a damn thing to do out there so we figured we might as well work seven days unless we were handy to a bore where we could get a wash, change clothes and rinse the dirty ones.[24]

Geosurveys, working for Beach Petroleum, also traversed through the Simpson during the early 1960s, while further south and east the Delhi/SANTOS consortium began measuring the stratigraphic section round the margins of the Great Artesian Basin. Bush treks were not confined to the oil explorers either, as no lesser persons than the Premier of South Australia, Sir Thomas Playford, and Mines Minister, Sir Lyell McEwin, joined company and government geologists on several four-wheel drive journeys between Birdsville and Oodnadatta to see the exploration programmes firsthand. Both men delighted in the outback life and were adept bushmen. The South Australian Government and the State Mines Department of the late 1950s and 1960s had thrown off the image of being nonbelievers in the State's oil and gas potential and were supportive of the search effort. In fact, it was on the recommendation of Heli Wopfner, Senior Petroleum Geologist at the Department of Mines, that the Playford Government insisted Delhi/SANTOS drill a second well on the Gidgealpa structure after the first had recovered just a faint 'sniff' of gas. Wopfner was familiar with the region through his earlier work with Geosurveys and was convinced that the down-flank well had not adequately evaluated the big feature. Delhi/SANTOS, on the other hand, felt the expense of a new well was not warranted.

> Gidgealpa No.1 was located on the eastern flank of the anticline. At that stage it had become customary to punch down the wells to the pre-Permian unconformity with little regard for the younger rocks. This resulted in some very untidy and sloppy holes.

> In Gidgealpa No.1 the Permian section was so badly washed out and caved that it could not be tested with any tools available in Australia at that time. This, I think, in the long run proved to be beneficial.
>
> As the logs of No.1 indicated excellent porosity in the Permian sands, the South Australian Government insisted that these sands be tested at any cost or another hole be drilled on the same structure.
>
> Delhi/SANTOS opted to drill another hole, Gidgealpa No.2, which on New Year's Eve of 1963 discovered gas at a flow rate of some 2 million cubic feet [56 000 m³] a day.[25]

The group confirmed the field with five more successful wells during the following 12 months and, after several dry holes elsewhere in the vast permit, went on to find the substantial Moomba gas field in July 1966. Reserves from the two reservoirs were estimated in mid-1966 to total 39.2 million m³ of saleable gas, certainly enough to supply the city of Adelaide 800 km to the south on a 20-year contract. Some long-range thinkers even then began toying with the possibility of the new South Australian gas fields also supplying the New South Wales market based round Sydney.

Just as the industry and potential consumers were digesting this turn of events, exploration in the southeast of the State revealed gas of an entirely different nature. The company Oil Development NL, which had been associated with the earlier Penola and Beachport wells and had changed its name to Alliance Oil Development Australia NL in 1963, began drilling a well called Caroline No.1 in the Caroline pine plantation about 16 km from Mt Gambier. During the last days of 1966 the well struck gas in Cretaceous-age sands which flowed strongly under test, but confused all concerned at first because no one could light the flare. The humour in these abortive attempts was not apparent until analysis showed that the gas was 99 percent carbon dioxide — an inert gas and a rare occurrence in such pure natural concentrations. Further tests indicated that the gas flowed from two zones and reserves were estimated at just under 28.3 million m³. Within weeks a feasibility study was begun which culminated 12 months later in an initial 15-year contract to supply Carba Australia Ltd. The resultant processing plant was unique to Australia and one of only four such operations in the world.

Flushed with this success Alliance remained in the news during the late 1960s as one of eight Australian companies that signed farm-in agreements with Delhi/SANTOS for entry to the permits in the border region of South Australia and Queensland. Although it was not the first time that farminees had joined these pioneering inland explorers (Total and Vamgas had done so several years earlier), with pledges totalling $12 million in proposed exploration programmes, such participation underlined the upgraded potential of the area following the discovery of gas at Gidgealpa, Moomba and more recently at Daralingie. The farm-ins also brought a number of relative newcomers to the spotlight — among them a company called Flinders Petroleum which, like Alliance, had begun with technical links to the Melbourne-based mineral and oil consultancy, Reg Hare and Associates. Managing director at the time was young Melbourne geologist Graeme Stephens, who was a consultant with the Hare group.

> Hare was mostly a mining man and his first major contact with the oil industry was in setting up a company called Farmout Drillers which he started with Minerals Securities chief, Ken McMahon, during the early 1960s. It should have been called 'Farmin Drillers' really because that's all the company did — take interests in permits operated by other explorers.
>
> Anyway Flinders was floated next and also managed by Hare through me. We then went on to absorb Farmout Drillers. I remember at the time there was a plane with 'Flinders Petroleum' emblazened across it. We thought we were sheiks of Australia.[26]

The plane was a twin-engine Aztec owned by the Melbourne air charter firm Convere (later called Pegasus Airlines) which was used to support the Flinders programme in the Delhi/SANTOS farm-out blocks, mostly on drilling crew changes in and out of Toowoomba in Queensland. A smaller aircraft, the Cessna 206, was used for 'hopping' between rigs in the different permit areas. John Considine was Convere's chief pilot during the six-year exploration effort.

> The inland was in the grip of drought for most of the time and there was a dust storm virtually every day. On long hauls I'd try to get away early in the morning to escape it. There was rarely much wind but, as the day heated up, the thermal activity created an updraft of air which raised the dust. More than once we had to talk pilots in to a desert strip by radio because they couldn't see a thing.
>
> It was that hot some days — over 50 degrees Celsius — that dingoes were just lying flat out under the sparse shade of the scrub with their tongues hanging out. It didn't take long for a man on his own to perish in that country.
>
> One time I went out with the toolpusher to a site the company was going to drill in Lake Callabonna, north of Lake Frome, so that I could pick a spot for the landing strip. We landed near a small waterhole miraculously still present in the bed of the Cooper Creek, and then walked out over the salt to the proposed drill site. After checking it out we

> turned back in what we thought was the direction of the plane. But our disorientation was so bad in the heat and glare that we went about half a mile before realizing we were actually walking away from it.[27]

The perverseness of the inland was such that another Convere pilot had a narrow escape from the same lake — this one during a flash flood.

> I was in Melbourne at the time, but Bill Fone was there in the Aztec. As the lake filled with water coming down the Cooper, he raced against the coming 'tide' to get the plane airborn. To make it lighter he took out everything that was not absolutely necessary including half the fuel and even his shoes.
>
> Bill managed to get the plane off the gluey strip by a whisker, but ODE's Baron aircraft which was also there at the time was not so lucky and it was swamped. When they got to it after the water subsided they found that rats had eaten the ignition harness and the wiring. It had to make four forced landings on the way back to Toowoomba for a complete overhaul.

After working in the area constantly for a month or two the pilots were able to navigate readily enough via landmarks, but it was easy to become lost if they strayed outside the flight 'corridors' between rigs and operational bases. During trips around the Flinders leases high-frequency radio communication was maintained with Broken Hill, and sometimes Mildura and Dubbo on the more wide-ranging flights.

> Generally we carried enough fuel to get to a rig and try a landing, plus some [extra] in case we were forced to go for an alternative. We only operated at night in an emergency because none of the rig strips had flare paths or radar. If possible we flew in the morning or late afternoon because often during the middle of the day the top of the thermals was 15 000 or 20 000 ft [4 600–6 100 m]. In a nonpressurized plane we couldn't get above that and it was far too uncomfortable at lower altitudes.
>
> The strips themselves were chosen carefully. Usually we took the Cessna low over the ground first to survey the land before picking a spot as near to the rig as possible. A lot of the time the drill site was in an inconvenient area and we had to find a ridge with the gentlest down grade.
>
> I never liked tearing up the ground to prepare a strip. Grading with a length of railway line was a lot better even if it did leave a few lumps.

> A clay base was best in the dry, but it didn't drain if there was any wet weather. Sometimes we had to shift the strip more than once during a well. Sand was better in the rain, but in normal dry conditions we formulated a special take-off technique — lift-off a.s.a.p. If we didn't the vortex effect from the props would suck sand into the engines.

Farmout Drillers, and hence Flinders Petroleum, were also interest holders in an even more remote exploration venture during the mid-1960s — a search that took place in the Amadeus Basin of the Northern Territory, right in the centre of the continent. Perhaps because of its remoteness the Amadeus work brought together some of the hardiest bushmen, geologists and oil hands in the country. It also accentuated the camaraderie that was often part of oil operations the world over. The first to take up leases in the region during the 1960s was US company Magellan Petroleum. It had set up in Brisbane and initially picked out some areas in western Queensland. However when Duncan McNaughton, the Canadian consultant retained by Magellan for its global operations, came to appraise them he was not enthusiastic and suggested the company look elsewhere. In Canberra, Alan Condon at the BMR suggested the Amadeus Basin, which had recently become vacant after the withdrawal of Frome-Broken Hill. Magellan accepted the advice and took the whole area in two bites during 1960, putting its name to one lease and that of an affiliate, United Canso, to the other. Neville Johns, a jackaroo, horse breaker and general station hand who had done a couple of jobs for Magellan in Queensland, was contracted to take a party of six or seven of the company's people and four vehicles from Brisbane to Alice Springs.

> I was just told that Magellan had a job for me in a new area out of Alice Springs. American geologist Roy Hopkins had not long come in from the Philippines to the Brisbane office at that time and McNaughton was in the country. They wanted to look at the permits. My job was to help with about three months worth of initial reconnaissance. I ended up staying with Magellan in the Alice for 28 years and I'm still going.[28]

The geologists quickly spotted several major features from the air. One in particular in the western permit was labelled Mereenie and another anticline in the eastern lease was right near Hermannsburg Mission overlooking Palm Valley. The next step was to examine them on the ground. Johns was again the transport organizer.

> The road for 50 km west of Alice Springs towards Hermannsburg must have been the worst in the country. Then it became a couple of wheel tracks and after that we were on our own,

bush bashing through the spinifex and navigating using the air photos to pick out the main landmarks. It took about a day and a half to get to Mereenie, which was about 240 km from Alice Springs.

We used Willeys jeeps for most of the early work, often with four or five vehicles spread out during the day and meeting at a designated air photo reference each night. By that time we'd have staked a few tyres and there'd be four or five punctures to mend of an evening. That's not to mention wrecked clutches, mangled exhausts and ripped brake lines.

A little later on Roy Hopkins and I were out there on our own, hitting rocks, mapping the outcrops and measuring bedding dips. One time we had a week of dust storms and could hardly see each other through the murk to do the theodolite and staff work.

Generally we'd go out for three weeks, maybe even six, at a time, camping under the stars. The routine was to be up at 4.00 a.m. and have a good, big breakfast with lots of tea to drink. Then we'd work through till about 1.00 p.m., break for a small lunch and a rest till fourish, and do a couple more hours before dark and a big camp dinner. We'd try not to drink anything after breakfast till mid-morning — longer if we could — because once we took a sip we'd be at the water bag all the time.

Once Magellan had done the initial mapping, gravity surveys and some seismic work, the company looked round for farm-in partners, and the first on the scene was Bill Siller, chairman of the newly formed Exoil, along with US consultant Grover Murray from Louisiana State University. They picked out three areas based round three big structures — Ooraminna, Alice and Mereenie — and made plans to drill them in that order to earn a 50 percent interest. The two largest structures, Palm Valley and Mereenie, were not on offer together. In any case Siller was not keen on Palm Valley as an oil play.

Although the Amadeus was miles from anywhere and the sediments were ancient (Ordovician and Cambrian in age), the positive thing was that they were of marine origin and the structures were huge. If any one of them had been full of oil it would easily have paid to put in a pipeline to the coast, even at the $US1.50 or so a barrel that oil fetched those days.

The other point was that the centre was a 'freight island'. It was as expensive to ship goods in as it would be to send them out, so owning a field, pipeline and perhaps a refinery in the area would have been pretty sound economics.[29]

Hopes were fired immediately when Ooraminna No.1, about 65 km southeast of Alice Springs, tested a small gas flow. It was not commercial, but it made geological history in being the first hydrocarbon discovery in the world emanating from such old rocks. The second well, Alice No.1, also found some gas along with a small amount of oil. Again it was not in commercial quantities, but the well caused a stir when it proved difficult to kill the flow. Increasing the mud weight only prompted loss of circulation into the porous formations and, in the end, the drillers gathered thousands of empty beer cans round Alice Springs to force down and kill the well. Similar problems, not helped by some short cuts in the casing programme, besieged the next location, Mereenie No.1. However the outcome was far more spectacular. Drilling contractor Dick Dumbrell of ODE recalls the incident.

> Gas inflow from the reservoir was tested at around 11 million cubic feet [308 000 m^3] a day and then drilling was continued for a while. But suddenly the mud weight got out of balance and she blew out. From then on it was a race to stem the flow and everything handy was thrown down the well, including blankets and shredded mattresses from the camp.
>
> Apart from the obvious danger at the wellhead, the problem was that the nine and five-eighth inch [24.4 cm] casing shoe was not set deep enough. It was only halfway into the Mereenie Sandstone, which was the main aquifer for the region and immediately above the Pacoota Sandstone gas reservoir. If too much gas bled into the aquifer it could have caused a drilling hazard everywhere else in the vicinity.
>
> Admittedly it would have been better to set the casing right through the Mereenie Sand, but time and money were short for all concerned in those days. It was also very hard rock and the supervisors thought it would be OK.
>
> In the end the well was killed and then cemented up completely, including the tools left down the hole, so the aquifer was effectively sealed off.[30]

Despite a narrowly averted disaster, the rig was quickly moved across to a new location designated East Mereenie No.1. Black tails on the gas flare during tests at the first well had indicated liquids were also present, and this was confirmed when cores bleeding oil were cut from the reservoir after East Mereenie 1 had come in with a roaring test flow of 850 000 m^3 a day, sending a flame 30 m into the air. The third well, East Mereenie No.2, put the matter beyond doubt when it recovered 55 m of oil in the drill pipe after the test. Throughout this programme and the drilling that followed Exoil and Magellan cut their own roads. Canadian toolpusher Jim Hodgkinson, who came into the area for ODE, joined Exoil as field supervisor after Mereenie was discovered. It was he who was responsible for many of the routes into the well sites and he was a great believer in straight lines.

The usual method was to fly over the area to plan the route, take bearings and then send in a bulldozer. There were a number of bare rock-spined ranges which others said could not be traversed in vehicles, much less 30 or 40 semitrailer rig loads. But Hodgkinson never failed to find a way through. Often it was pretty rough, but in most cases the road was only needed to bring the rig in and then out again. The principle was that roads could be upgraded if any oil or gas was found. In fact, some of the original exploration roads were later made into tourist routes, such as the one into King's Canyon between Mereenie and Ayers Rock.

The exploration at Palm Valley was operated by Magellan and, if anything, it was less accessible than Mereenie despite being closer to Alice Springs. The crest of the structure was high above the surrounding plain and the initial route up the steep incline was little more than a bulldozer scrape with larger rocks pushed to one side. It had been picked out by Neville Johns, who also became Magellan's troubleshooter, adept at setting up camps, scouting for water, lending a hand on the rig floor and generally fixing things. Roy Hopkins, who rose to become managing director of Magellan Australia, remembers Johns taking a couple of Sydney auditors up to the Palm Valley site.

> The way up was only a four-wheel drive track at the time and one auditor, hanging on to vehicle door supports to brace himself against the bumps, made conversation by remarking that he hadn't seen much wild life round the area. Deadpan, Johns replied, 'No, not along these main roads you don't'.[31]

The Palm Valley No.1 wildcat was drilled in March 1965 and immediately came in as a big dry gas discovery. That opened the way for a series of wells on other structures in both permits, including the interior of the curious feature called Gosses Bluff, believed to be a meteorite impact crater. For several years the Amadeus Basin became a hive of activity. Oil field characters abounded, not least of them being Magellan's consultant Duncan McNaughton. Then in his 60s, he had been a noted athlete in his youth, winning the high-jump gold medal for Canada at the 1932 Olympics. He had also been a bomber pilot with the Pathfinder group in the Second World War and was still at home in the cockpit. Roy Hopkins accompanied him on a number of early reconnaissance flights.

> One trip to Mereenie, Duncan wandered up and sat in the copilot's seat. He wanted to go down low for a real close look. It was as hot as Hades and the plane bucked all over the place in the thermals. I didn't worry too much as I thought there was plenty of experience in the two drivers' seats.

> Anyway we skimmed up and around the anticline with the rim rock just brushing the wing for a while before turning back to the Alice. Later on that evening I spoke to the pilot in the pub and he seemed to be a bit nervous and jumpy. Then he told me he was very glad his passengers knew what they were doing and where they were on the flight, because he didn't.
>
> It turned out he'd only had his commercial licence for three weeks. Before that he had been a taxi driver in Sydney.

McNaughton also became known as a 'big bang' specialist and there was a period during the late 1960s when, to increase the permeability of the Palm Valley reservoir, a nuclear device was suggested. Again Hopkins was very much involved.

> The proposal was known as 'Pacoota Boost', and it was about the time Lang Hancock suggested blasting iron ore ports out of the northwest coast of Australia. Duncan was all for it as it had been tried in the US in a project named 'Gas Buggy'.
>
> However on closer inspection the US work found that the explosion opened the reservoir cracks all right, but caused fusion of the rocks round the well and defeated the object of the exercise. It also produced unwanted strontium 90 and meant that the gas might not be useable till well into the twenty-first century or longer. So 'Pacoota Boost' never left the proposal stage.
>
> Putting that aside, McNaughton also thought of using nitroglycerine on our Palm Valley problem. It too had been used in the US to fracture wells, but only in small quantities. Duncan wanted to use gallons of the stuff.
>
> He got me to search round Australia for a supplier, and after a bit of research I found the only one was ICI. When I rang up and asked if they could supply 200 or 300 gallons [900–1 360 L] to central Australia there was a real shocked silence on the other end of the phone. Some moments later the guy told me they didn't let it outside their gates in larger quantities than a drop at a time in heart attack pills.

By the time Palm Valley No.2 was drilled late in the decade Magellan knew more about a problem that had cropped up with a number of wells at Mereenie and elsewhere — formation damage caused by chemical reaction with the drilling mud. The use of compressed air as the circulatory fluid had been pioneered with East Mereenie No.2 at the suggestion of Roger Planalp, Exoil's first operations manager, who had successfully employed the technique in his native Oklahoma. After that, gas from some of the neighbouring wells had been tried because it did not need compression. This latter method was used on the Palm Valley appraisal using gas fed in from the discovery well. BMR drilling inspector Harry Taylor-Rogers was on location to check details for the government's subsidy scheme when the No.2 well came in.

> A small gas flame about a metre long was lit at the end of the flow line to burn off the excess gas from the circulatory system. Without warning there was a roar as the bit struck the Pacoota reservoir and gas rushed up the annulus. Within seconds it shot out of the well and enveloped the rig in a heavy haze.
>
> All this happened before the blowout preventers closed, and the ground was literally shaking. The main worry was the flame at the end of the flow line. If that ignited the wild flow it would destroy the rig, the camp and everyone with it. There was a lot of scurrying about as the blowout preventers were finally activated and the well choked back.[32]

For the second time, the danger of colossal power being unleashed out of control from the Amadeus reservoirs had been narrowly averted. The later production test at Palm Valley No.2 indicated the huge pressures involved when the well flowed at 1.96 million m^3 a day, sending up a flare which was mistaken for a scrub fire by a commercial airline pilot passing over central Australia at the time. However it was typical of the drilling staff to down play the ever-present dangers they worked with. Roy Hopkins later asked the Canadian drilling supervisor, Lloyd Anderson, if he was scared when such a massive flow came from the Palm Valley well. Anderson replied, 'Hell no, but I sure sprinted past plenty of people who were'.[33]

Although about 15 structures were drilled in the Amadeus during the 1960s, only Mereenie and Palm Valley seemed to have any hope of economic viability. Even they had a number of problems to overcome before there could be any possibility of development, not least of which were the fields' isolation and the fragile chemical makeup of the Pacoota Sandstone reservoir formation. Nevertheless the programme went on as the list of farm-in partners grew. The entry of companies like Austram Oil, Freeport, Farmout Drillers (Flinders Petroleum), Krewliff Investments, Southern Pacific Petroleum, plus input from United Canso and two Exoil associates called Transoil and Petromin, all helped to fund new wells and surveys in the effort to delineate the huge permits.

This trend of seeking farminees to help defray costs and meet the work obligations set down by the various governments became noticeable throughout Australia and Papua New Guinea as the 1960s advanced. Few companies, no matter how large, could single-handedly carry out adequate exploration programmes in leases the size of those awarded during the 1950s. In Papua, the percentage interests in the APC consortium changed several times during the decade as BP's and Mobil's views of the prospects waxed and waned. Oil Search, however, seemed determined to continue whatever the cost. At one point that company's share in the consortium rose as high as 80 percent. To help spread the load, a farm-in agreement was signed with Burmah Oil of the UK and Murphy Oil of the US such that the two internationals could take up 50 percent in all the APC concessions. The condition was to carry out a compehensive study of the viability of establishing a local industry to use the natural gas reserves previously found at Kuru, Barikewa and, more recently, at Iehi. In 1963 Burmah/Murphy concluded that the plan was not economic and declined to take up their option.

Oil Search signed another agreement in 1967, this time with Esso, in which the Ini No.1 well was drilled in the Papuan Gulf country 320 km northwest of Port Moresby. The aim was to evaluate suspected Miocene-age limestone reef targets in the region, but the well was dry and Esso withdrew. Then Australian companies Interstate Oil Ltd and Endeavour Oil NL joined Oil Search in the Fly River delta region, while late in the decade Texaco and California Asiatic farmed into a new permit held by APC in the central highlands. The logistics of work in Papua New Guinea had been improved with the growing use of helicopter support, but high costs could still cripple a company. Although there was no commercial success during the 1960s and most farminees chose not to stay after the initial obligatory programmes, their input allowed the original explorers like Oil Search and BP to keep going. BP in fact chose to return to active participation in the search programmes during 1967 after taking a back seat for six years.

The company's revival of interest was prompted by experienced geologist Frank Rickwood, who had done of lot of pioneering traverses for APC in the Purari River area and the mountains behind the delta region during the 1950s. When BP stepped out of the operatorship in 1960 Rickwood stayed with BP and worked in other parts of the world including the North Slope of Alaska. However he still believed Papua had potential and persuaded BP's management in London that the company should look again in the region, particularly in the reef plays round the Gulf of Papua because reef reservoirs had been so successful in Canada. Unfortunately it was not repeated in PNG and the APC consortium, with BP back in the lead role, turned its attention away from the Gulf country to begin the first detailed surveys of the central highlands. Despite the lack of commercial success during the 1960s, the data bank had being enlarged and improved. Thanks to input from farminees, the group was stronger financially and there was a much better technical understanding of the complex geology in the region.

Back in Australia even WAPET's stranglehold on permits in Western Australia had begun to loosen. In 1964 agreements were signed with US companies Continental Oil Company (Conoco) and Sun Oil, and with French explorers French Petroleum (Total) and Australian

Aquitaine Petroleum for work in the Carnarvon, Geraldton and Fitzroy regions. A few years later these companies widened their activities, while Gewerkschaft Elwerath of Germany plus two American companies — Union Oil and Marathon Petroleum — also took part in searching WAPET's acreage. Not that WAPET itself was idle — far from it. Much of the company's onshore effort during this period was spent reshooting old seismic surveys using more advanced techniques than were available in the previous decade. Greater detail was being gleaned from seismograms, both old and new, through improvements in recording and processing. Western Australian Rod Hollingsworth already had experience with the BMR teams in the 1950s under his belt, as well as a stint in the Antarctic, when he joined contractor United Geophysical Corporation (later Seiscom Delta) in 1964.

> United Geophysical had just won a contract with WAPET at the time. The work was still with dynamite and shot hole drilling for the five years I was with the crew, but the charge patterns had become more complicated than in the 1950s. For instance, we might use hundreds of pounds of explosive in the one firing across a diamond-shape pattern of up to 36 shot holes.
>
> It was still tedious work and we'd be lucky to shoot three kilometres of line a day. But it did mean we could have multiple coverage — 3-fold, then later 6-fold traces — although it was still a long way from the 24-fold work and upwards achieved in the 1980s.
>
> Along with this we had processing and playback units in the field with the crew. That gave the advantage of being able to see a day's work very quickly and repeat it if necessary. Everyone was on the crew in those days — drillers, juggies [geophone handlers], computer processors and the interpreters. Later on, when the playback centres were kept back in town, the seismic crews really just became acquisition tools.
>
> The usual thing was to spend three weeks in the field and then take a week off. The whole crew, maybe 30 or 40 people, went off together. So, effectively, work ceased for a week every month.
>
> The field camps were never 'dry'. There was plenty of beer available, but no spirits. It was more tolerant than the drilling camps because we only worked in daylight. Parties could be held, but the rules were: no damage to vehicles, no noise after a certain hour, and everyone had to front for work at 6.30 a.m. each morning.
>
> Camps consisted of two-man tents, caravans for the office, playback centre and ablutions. The kitchen was also a caravan which had a big annexe attached for the mess.[34]

WAPET's survey work took the crews into unusual and sometimes restricted areas. Probably the most unusual was a gravity traverse through the main city streets of Perth. It was done in the early hours of the morning to avoid interference with daytime traffic. In addition, a marine seismic survey of the Swan River was completed at the Barrack Street Jetty, less than half a kilometre from the Perth Town hall. Restricted areas included the bombing range near Lancelin just north of Perth, and a remote section of the Canning Basin controlled by the Weapons Research Organization for monitoring Blue Streak rockets fired from Woomera in South Australia. Permits were needed for employees to enter such an area and these were withdrawn whenever a rocket test took place. An all-weather airstrip, constructed in the region by WAPET engineers and named Swindells Field after the job foreman, was a valuable aid to the government researchers as well as to the exploration drilling of the Sahara No.1 well during the mid-1960s. The region round the old RAAF base at Learmonth could be equally hazardous because live ammunition remained undetected for many years after the war. On one occasion when a scrub fire went through the Exmouth Peninsula, the Learmonth area was described as being like a battle field with ammunition exploding in all directions.[35]

Already regarded as the largest road builder in the State, WAPET had opened up more than 32 000 km of access tracks by the end of the 1960s for use by the seismic and gravity crews. In addition, the company had built over 2 000 km of new roads, many of which became public thoroughfares even before the rigs and survey teams had left the region. Others, such as the trails through the forests round Pemberton to the south of Perth, were available to assist Forestry Department fire-fighters. Yet, for all this activity in the mainland basins of Western Australia by WAPET and its various farm-in partners, the success rate remained low. The exception was a moderate gas find in a geologically complex reservoir at Dongara (plus nearby Yardarino and Mondarra), just south of Geraldton in the north Perth Basin, which was eventually declared commercial in 1969.

Although this find was recognized as an important milestone for the State, the major breakthrough for WAPET during the 1960s occurred 45 km off the mid-west coast on a windswept dot of land called Barrow Island. Not strictly classified as an offshore programme, but not wholly onshore either, the Barrow project issued a unique challenge to the Perth-based company and led to a phase of exploration which spread right down the western seaboard to the Abrolhos. This became known as 'island hopping'.

Chapter 11
THE ISLAND PROGRAMMES

Although it was not common practice, Australian explorers were no strangers to island drilling. The Humber Barrier Reef Oil Company had drilled on Wreck Island off the Queensland east coast in 1959. Prior to that WAPET had spent some time on Dirk Hartog Island in Shark Bay, and had run a brief geological survey across Barrow Island under the watchful eye of the Federal Government before a permit ban in the Monte Bello atomic test area was lifted. The practice was continued during the 1960s by several smaller Australian companies because it gave them a glimpse of the stratigraphic column in some intriguing offshore areas without the financial and technical worries associated with 'getting their feet wet'. Melbourne-based Oil Development NL, for instance, had negotiated a farm-in interest to a SANTOS permit covering Bathurst and Melville Islands in the Bonaparte Gulf off Darwin early in the decade. The permit was a logical extension to the two onshore permits the company held at the head of the Cambridge Gulf round Wyndham, which had been acquired on the advice of its consultant exploration manager, Leo Stach. Stach had a long felt liking for the Bonaparte's potential and had also pointed the permits out to the previous lessee, Westralian Oil, before it decided to quit the oil scene and change to mineral ventures.[1]

Westralian had drilled Spirit Hill No.1 to 900 m in the onshore Northern Territory sector without success, so Oil Developments decided to see if there was more potential on the seaward side of the basin. The company ran a three-month reconnaissance geological survey over Bathurst Island followed by a stratigraphic well (Bathurst Island No.1) during 1960. Unfortunately the drill rods became stuck in the hole at 253 m when a severe earth tremor rocked the region. A second well in 1961 reached 312 m, but drilled through Cretaceous-age marine shales all the way without penetrating a potential hydrocarbon-bearing reservoir sand. It was several more years before the company (by that time called Alliance Oil Developments) moved back onshore to the Western Australian sector and drilled two wells called Bonaparte No.1 and Bonaparte No.2, the second of which tested gas at 42 000 m^3 a day. Although it was subcommercial, it proved the presence of hydrocarbons in the Bonaparte Basin.

Another programme in Australia's north was carried out by the Delhi/SANTOS group on Mornington Island at the southern end of the Gulf of Carpentaria. The consortium decided on the northern venture because the Carpentaria Basin was believed to have a geological history similar to that of the Great Artesian Basin round the South Australian-Queensland border where the bulk of the group's permits were located. A National T-32 rig belonging to the Associated Group's drilling arm, Richter Drilling, was used for the programme and the job of arranging the two-well project went to Bob McCulloch, one of Richter's senior drilling supervisors. He arranged for the loan of a landing craft from the Australian Army to ferry the rig, equipment and supplies across the 80 km stretch of water from the mainland. Loading was done at the Gulf prawn-fishing port of Karumba and the entire stock of food, materials and spare parts to last the four-month programme was landed

Location of key places mentioned in text.

on the beach at Mornington Island during May 1961.[2] The operation went without a hitch but, unfortunately, the results of the wells themselves were disappointing. The first, planned for a depth of around 1 830 m, hit granite at 843 m and was abandoned. The second, 35 km away across a series of faults, gave a similar result. According to an account in Delhi's house magazine at the time, the only real excitement during the programme came from quite a different quarter. The article began by explaining that the Presbyterian Aboriginal mission on the island ran its own herd of cattle tended by Aboriginal stockmen.

> The island is 40 miles [64 km] long and 16 miles [26 km] wide... and some of the cattle have wandered to the end farthest from the mission and gone wild. Nobody goes to this part... and so the cattle, specially the bulls, know no fear.
>
> [To survey the location of the second well, two Delhi men]... started to traverse inland from the coast and were doing nicely when, crossing a grassy plain, a horrible bellow came from a point on the left... Over the rise came a huge red bull equipped with gleaming white needle-pointed horns and all 2 000 pounds [900 kg] of him breathing fire.
>
> The men headed for a line of mangroves bordering a tidal creek. They arrived there just ahead of the homicidal-minded bovine and crashed through to the other side. The bull thrashed about for a while, looking for them, but eventually gave up, much to their relief.[3]

Mornington Island was obviously not the place for oilmen, unless they were also budding toreadors!

Although carefully planned and well executed, both the Mornington and Bathurst Island programmes assumed the proportions of minor excursions when compared to the extensive work which took place during the 1960s on the flat, virtually treeless islands strung along the Western Australian coast. This programme began with Barrow, the largest of them all. Covering an area of about 260 km^2, it is 29 km long and 13 km wide. The Monte Bello Islands, where Britain conducted its atmospheric atomic bomb tests in 1952 and again in 1956, are 24 km north, while the port of Onslow, which WAPET chose as its supply base, is 100 km south. The visits by company geologists during the 1950s had been brief, but had indicated that Barrow contained the crest of a large anticline on trend with the features seen in the Rough Range–Exmouth region to the south. In fact, the island did fall within the original WAPET Carnarvon Basin exploration permits, but it was not until the early 1960s that the government of Sir Robert Menzies lifted the bans on access.

In 1962, WAPET landed a party of geologists equipped with a four-wheel drive vehicle to carry out a detailed survey and traverse the whole area. The men were surprised to find that extensive fires had burnt out the massive clumps of spinifex which had hindered the movement of previous parties. The new growth was much easier to negotiate, but the lack of water on the island was still the hardest problem to overcome. Following kangaroo tracks they eventually found shallow pools on the west side. Two refraction line seismic surveys were run in 1963 and these confirmed that the island was the surface expression of a subsurface anticline which had closure in potential hydrocarbon-bearing sediments of Jurassic age. At this point some consideration was given to farming out a percentage interest to Conzinc Rio Tinto in return for being carried through an exploration drilling programme. There was some idea that if gas was found, it could be used in the Pilbara area for power generation. The difficulty was that CRA was part of the Frome-Broken Hill group and the company would also have had to offer the deal to all the partners involved in the consortium.[4] In the end WAPET decided to drill an exploratory well itself during 1964. Jesse Haddock, by that time one of the company's most experienced field managers, was asked to work out the cost of putting a rig on Barrow for one well and then off again after three months.

> Although the island is only about 60 m above sea level at the highest point, shallow reefs and cliffs narrowed the choices of a landing site to a beach at the northeast corner. It was obviously going to be a barge operation, but we also needed an airstrip, so Bell Brothers built that for us along with a road up from the beach landfall we'd chosen.
>
> They'd just finished that when a cyclone came through and wiped out all the construction work, so it all had to be redone. During the blow the men took refuge in the cabs of the three heavy-duty vehicles and they were more or less confined to them for 36 hours.
>
> I based the Barrow project out of Onslow, loading up in Beadon Creek which was specially dredged to allow the barges easier access. About 35 loads of rig, camp and supplies were sent over to the island in April 1964. It was hot work and the mozzies and sand flies made life pretty unbearable at the Onslow end.[5]

Prince Launch Services won the initial contract and used a 60-tonne ex-US army wartime barge renamed *Jarrah*. At a maximum speed of about 8 knots, the vessel took 12 hours from Onslow. It was about 21 m long with a beam of six metres and a draught when loaded of one metre. Bevan Cook, who had also been with WAPET through the 1950s, was given some of the early logistics work because of his experience in water transport with the Citizen Military Forces, and with landing barges in particular.

> The vessel had a basic four-man crew plus the captain. There were only a couple of bunks on board.
>
> A small rig to drill a water well was brought across at the same time as they were building the airstrip. The first 'housing' was made up of one-man tents.
>
> Most of the material came up from Fremantle by State Ship to Onslow where it was transferred to the barge. At Barrow the vessel would discharge cargo at high tide and we developed a fork lift of sorts by taking the bucket off a front-end loader to help with the unloading.
>
> Apart from the cyclone there was only one mishap, when a sudden squall caused a load of drill casing to shift its position in the barge. With the cargo and vessel in jeopardy, they reached the lee of a small island. But the crew couldn't reposition the load, so the skipper put the men in a dinghy behind and kept going. He finally made it and negotiated the reefs to the beach at Barrow, but the barge was almost on its side by that time.[6]

The water well was drilled first, successfully finding an aquifer to cater for the rig and camp needs. Then, skid-mounted huts were set up for the main exploration well. The location of the wildcat, Barrow No.1, was changed a couple of times before drilling. First it was sited on the crest of the anticline, but then moved closer to the airstrip as a cost-saving measure. In the end the geologists prevailed and it was returned to the original crestal position. Years later a well was drilled near the old airstrip and gave a poor result. As with the planned Rough Range No.1 location, one can speculate on the course of subsequent exploration history if the airstrip well had been drilled as the initial attempt. Barrow No.1 was spudded on 7 May 1964 with the traditional gesture of blessing the bit being performed by Ampol's exploration manager and chief geologist, Sami Nasr. Nasr, a Palestinian Arab, had been engaged by William Walkley exactly two years to the day previously. An interesting character, he already had a wealth of experiences behind him before coming to Australia.

> I did my original geology degree in the US during the 1930s and then went back to the Middle East to work for Iraq Petroleum Company for 15 years, where I eventually became exploration manager and, finally, general oil field manager with responsibility for production of around 800 000 barrels a day.
>
> After that I became scientific adviser to King Feisel of Saudi Arabia, but after Feisel's murder in 1959 I was arrested by the revolutionaries and held in solitary confinement in Baghdad because they thought I was involved with counter-

> revolutionaries. I was eventually let go and given 24 hours to leave the country.
>
> The irony was I had had four nationalities, all by right — Palestinian, Turkish, British and Jordanian — but, at that moment, I was stateless. My wife was Irish so we went there. I did an MSc degree at Trinity College in Dublin and became an Irish national.
>
> About 1961 I met Lyn Noakes, head of the Australian BMR, who was in Britain looking for people to come and work in Australia. I came out on a two-year contract, but I'd only been in the country a few months when Ampol approached me. Apparently Harold Raggatt had spoken to Walkley and the government was willing to waive my contract with the BMR. That was 7 May 1962.[7]

Nasr's knowledge and experience lent new weight to Ampol's input to the WAPET consortium and it was deemed fitting that he should officiate at the Barrow Island ceremony.

> At the time I jokingly said that WAPET had not found commercial oil yet because the company had not let an Arab bless the bit. After all, most of the world's oil is between Libya and Indonesia, and they are all Moslem countries!

Whether influenced by Nasr's 'blessing' or not, Barrow No.1 flowed gas at rates up to 308 000 m^3 a day during June 1964. Early the following month oil was tested from a Jurassic reservoir at just under 1 000 barrels a day. Headlines screamed the news but, after the Rough Range experience, WAPET refrained from speculation about viability until it had run an adequate appraisal programme and, as with Rough Range, the results of further drilling did bring some surprises. Unlike Rough Range, however, the news was unexpectedly good. Geologist Robin Elliot was a member of the Barrow appraisal team.

> By early 1965 the fourth and fifth wells had been successful and completed as producers from the Jurassic. But after a bit it was noticed that oil was oozing out of the wells around the outside of the casing. We began to wonder if there was another reservoir above the Jurassic. When tests were done from a potential formation we then noticed in the Cretaceous, they flowed around 200 and 300 barrels a day.
>
> Although it was still company policy to drill for the Jurassic in appraisals during the rest of 1965, we did propose a recovery programme on the upper reservoir (the Windalia Sands) and we were able to run long-term tests from this formation in two of the crestal wells. One came in oil and

> the other gas, and they flowed for six months without a pressure drop.[8]

From that point onward the Windalia target became a major part of the appraisal programme and eventually replaced the Jurassic as the field's premier reservoir. It also helped WAPET decide the field's viability and, on 26 May 1966, the Barrow Island field was declared to be commercial. A telegram sent from WAPET's Perth office to the Petroleum Information Bureau in Melbourne gave the details.

> WAPET announced today Barrow quote economic unquote production expected to commence nine thousand barrels a day in May 67 estimated production rate in two years may be approx twenty thousand barrels day stop
>
> Area productive field twenty four thousand seven hundred acres [100 km^2] stop Average net effective sand thickness 44 feet [13.5 m] stop
>
> Sand very low permeability consequently low production rate conservative estimate eighty five million barrels will be recovered during life of field depending wells performance stop Recovery may be increased application new techniques now being studied stop

Seven months later the company was able to report favourably on the new techniques, which included a reservoir fracturing programme designed to force open the Windalia Formation and allow a greater flow of oil to the wells. The field's recoverable reserves were upgraded to 114 million barrels and development work began in earnest. The proposal called for the drilling of 246 development wells, and two rigs — one from ODE and the other from Richter — were brought to the island. Geological surveyors covered the structure with a grid system and pegs were driven in at regular intervals to mark the proposed well locations. Speed was of the essence and the competition between rig contractors was intense. Dennis Clow was with the Richter team.

> We got to the stage where we could drill to the reservoir down at 1 000 m in about 48 hours, then complete the well and move to the next location and be spudded there in another 24 hours. We were drilling wells for less than $3 per metre.
>
> Bad weather could slow us down a bit though. One time Roy Richter in Brisbane telegraphed wanting to know why there had been a five-day rig move. I sent back a telex which said, 'Genesis 7, 24'. He was apparently a bit nonplussed by that till someone told him the relevant Bible passage read, 'And waters came across the bosom of the land and flowed for 50 days'.[9]

Cyclones were a constant threat during the summer months and several times the drilling rig derricks had to be lowered and their tops chained to heavy earth-moving equipment as a safety precaution. Winds of more than 160 km per hour were recorded and up to 300 mm of rain could be associated with these disturbances. A permanent camp was set up as the drilling and construction work gathered pace and, generally, conditions on the island improved rapidly. Humidity was one of the most difficult things for some to become used to, but others felt the worst thing for work force morale was the prevailing wind, particularly in the spring months, which brought dust from the mainland in great clouds. The men soon dubbed the fine red grit 'Ashburton snow', because that was the nearest mainland region to the island.

With such a sudden influx of men and machinery to the island, there was concern in some quarters to ensure the protection of the natural fauna and flora on Barrow. The island had escaped much of the degradation seen at other parts of the coast over the centuries because it was a long way from the main trade routes and had little natural water to attract settlers. Consequently the plants and animals had not been disturbed by feral species and the habitat had remained intact. The island had been declared a wildlife reserve as early as 1908 and gazetted as a Class A reserve in 1910, largely as a result of the work of naturalist John Tunney, who had spent two months on the island during 1900. When the oil explorers arrived, well-known naturalist Harry Butler made several trips to catalogue the various species and investigate the possible changes likely to occur because of the field appraisal and development work.

In 1967 he was appointed consultant to WAPET in what was one of the earliest examples of environmental protection initiatives set up by the oil industry in Australia. Butler's report said that although changes would undoubtedly take place, it was possible to go ahead with development without doing serious or lasting damage to the delicate natural balance. Among his recommendations were a total ban on domestic animals, introduced plants, firearms and the building of jetties. He added that there should be immediate restoration of areas once use was completed.[10] At the same time, the West Australian Wildlife Authority began a series of independent periodic studies late in the decade to keep a check on the state of the Barrow Island environment. In addition, all WAPET employees and the contractor workforce were required to formally accept these strict measures. Bevan Cook, who became one of WAPET's field supervisors on Barrow, had responsibility for the day-to-day observance of the rules. But he was also the perpetrator of a light-hearted approach to the programme on more than one occasion.

> After Harry Butler had been a few times and production was underway we invented a horse and called him Vrabitsio after an Italian who was working on the island.

> Next time Butler came over someone in the lunch queue casually mentioned in his hearing that the horse had broken loose and was last seen in the spinifex. Harry was startled and immediately asked about it. The cook explained that they'd been presented with a horse from Mardie Station on the mainland in return for giving the station manager flying lessons, and that they were going to start a riding school.
>
> Well Harry was horrified and he commandeered a Land Rover and the rifle we kept in case any feral animal did sneak onto the island, and out he went looking for Vrabitsio. One guy said he'd seen it near a well at the far end of the field so Harry took off in that direction. When he arrived there he saw the well's beam pump [sometimes called a nodding donkey] all painted up to resemble a horse's head. There was even a bucket of water on the ground in front of it which the 'nose' dipped towards with each pump stroke.[11]

With the first commercial shipment of crude scheduled to leave Barrow in April 1967, WAPET turned its attention to the other numerous islands in the region. The drilling contractor ODE worked with WAPET engineering staff to select a basic rig design that would be readily transportable by barge from one location to the next, yet strong enough to reach the potential reservoir depths. The resultant unit, an Ideco H-40, was dubbed the 'Island Hopper'. That done, a list of more than 15 likely islands between North West Cape and the Monte Bellos was drawn up as drilling targets. WAPET's Jesse Haddock and retired naval hydrographer Lieutenant Jimmy Little then began to examine them in detail and draw a beach profile for each.

> There was Murion, Hope, Direction, Observation, Mangrove, Sandy, Long, Lesieur, Anchor, Tortoise, Pascoe, Thevenard, Airlie, Beagle, Ashburton, Sholl, the Lowendahl group (Varanus in particular) and the two in the Monte Bellos called Hermite and Trimouille.
>
> Some of them, like Long Island — the first to be drilled — were little more than sand dunes about 200 m across the centre at high water. The next, Tortoise, was just over half a square kilometre in area. Airlie was the same. We had to build small protective groins on these small ones.
>
> At Trimouille some of the north part of the island was still fenced off by the government and there were multilingual notices to say 'clear off'. However WAPET was given permission to explore outside the signs.
>
> On Trimouille too the survey crew had reported seeing rats as big as rabbits. They were thought to have arrived via the atomic test ships in the 1950s and maybe mutated to that

> size. No one knew really, but they'd be a menace to any drilling crew so we took some advice on eradication.
>
> In the end we bought cases of frankfurts and laced them with poison. We had to wear gloves and masks to handle the stuff. Then a team with knapsacks dropped them off round the island just prior to the drilling operation. The rig crew never saw a rat, so it must have worked.[12]

The rig, camp, supplies and other equipment, including desalination plants (operated by the heat of the rig's engines) to make fresh water, were moved from island to island on two landing barges, while a 14 m work boat was used in support. Much of the loading and unloading was done at night because of tide conditions and in some instances a passage had to be cleared through the reef to reach the beaches. At other times the water was just too shallow to make a landfall at all. The WAPET programme was emulated in small part further south in the Perth Basin, when BP Petroleum Development and Abrolhos Oil drilled a well on the table-smooth Gun Island in the Pelsaert Group off the coast of Geraldton. Another island of sorts was man-made by WAPET in Cockburn Sound just south of Fremantle to site the rig for the Cockburn No.1 wildcat. It was connected to shore by just under a kilometre of causeway. Of all these island wells the most successful was on Pascoe Island immediately south of Barrow, where oil and gas was encountered at three different levels. The best test flow was 625 barrels of oil a day from Pascoe No.1; a second well was dry, but a third flowed oil at 50 barrels a day.

Although no commercial accumulations were discovered outside Barrow, WAPET in particular gleaned a great deal of stratigraphic information from the island programmes during the 1960s. The rationale was that such wells cost around $2 000 per day to drill, whereas utilizing an offshore rig or platform could push costs to as much as $20 000 a day for the same information. For WAPET and some of the smaller Australian companies, island drilling was a stepping stone to the deep-water areas round the west and north coasts. But, by the mid-1960s, there were other explorers with leases off the country's southern shores who did not have the luxury of relevant island locations. Instead they took high costs and new technology by the horns and plunged directly offshore.

Chapter 12
THE OFFSHORE REVELATION

The Gippsland region of southeastern Victoria was certainly no stranger to oil explorers by the time the 1960s came around. Government and private concerns had been probing the onshore parts of the Gippsland Basin since the turn of the century, and there had been some minor production of oil, particularly from the Lakes Entrance shaft project. But the results had been more teasing than successful and a number of geologists, surveyors and promoters had wistfully glanced offshore at the stormy waters of Bass Strait. Some even recognized that the central part of the basin lay in that direction. But it was not until the Rough Range-led revival in oil exploration in Australia during the mid-1950s that much 'science' was attached to the theories. At the Victorian Mines Department, Dr Nicholas Boutakoff was one of the first to suggest that the oil and gas found onshore was likely to have migrated up to these basin-edge locations after being generated in the more deeply buried sediments to the south. He made strong pleas for exploration companies to look offshore. Although several companies privately agreed with him, including the Esso/Mobil/BP/Conzinc combine of Frome-Broken Hill, the prospects might just as well have been on the moon. Offshore exploration was in its infancy and, as the Victorian fishermen knew only too well, the treacherous waters of Bass Strait were no place for the unwary or the unprepared.

That said, it was the least likely explorer — the fledgling company Woodside (Lakes Entrance) Oil — that made the first seaward move by applying for a thin strip of water along the Ninety Mile Beach in the late 1950s. The application was the work of exploration manager Eric Webb. He wanted more, but the company's already precarious finances prohibited a larger lease. As it was, the company could do little with the permit for a number of years. By that time events had overtaken Webb's early initiative and another man was destined to take the plaudits for what followed during the 1960s. His name was Lewis George Weeks.

Paradoxically, the story of Bass Strait began in onshore New South Wales about 1954. At that time the newly formed Australian Oil and Gas Corporation had approximately 75 percent of the Sydney Basin under permit. The other 25 percent was in and around Port Kembla where the Broken Hill Proprietary Company had its steelworks. Professor Eric Rudd, who had worked for BHP soon after the war and then became a consultant to AOG, had suggested several times that it would be sensible for the big steel company to find out what was under its plant and around its coal leases in the region. Efforts to bring the two companies together in a joint venture exploration programme came to nothing but, realizing the fervour of the oil revival, BHP did take up the remaining Sydney Basin acreage. One of the company's managers, Murray Lonie, was asked to look after the new oil interests along with a renewed push into minerals.

> BHP's need for oil developed from an increasing demand for energy to fuel its steel industry. There was the thought during the mid-1950s that if oil could be found near the plant at Port Kembla it would cut the energy bill. At the time the steel works was the biggest single consumer of oil in the country.
>
> Allied to that there was the realization that sooner or later explorers would want to drill in the Sydney Basin and

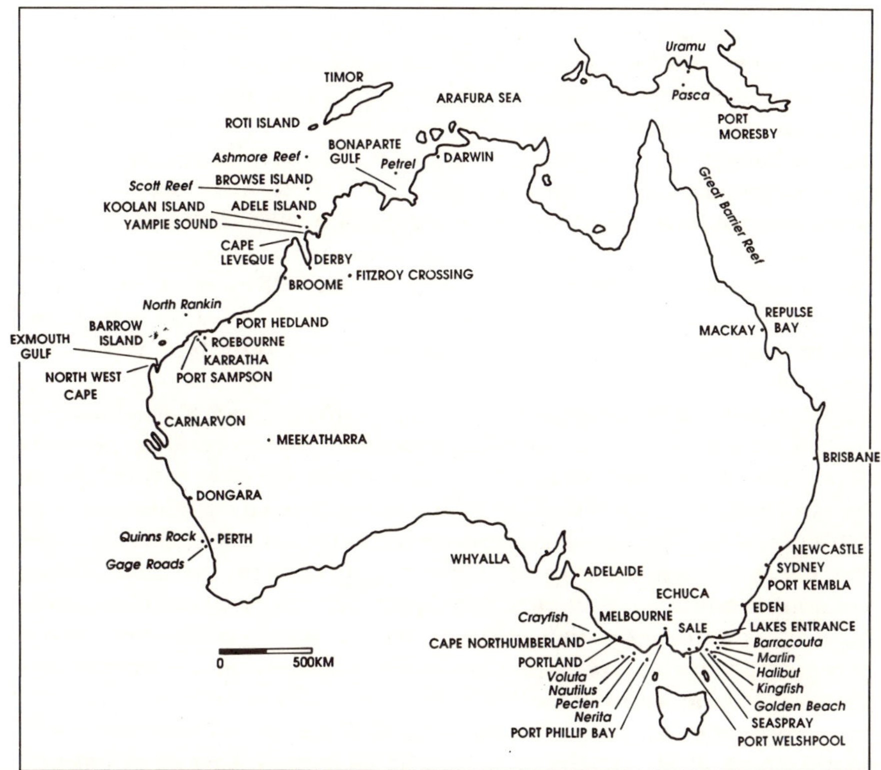

Location of key places mentioned in text.

> some of those wells would be in BHP's coal acreage. That might necessitate 'freezing' blocks of coal to act as a barrier to possible oil production and hence cutting the company's effective coal reserves. Some of our people had travelled to Pennsylvania in the US and seen it happen there.
>
> Set off by these thoughts, the company began a detailed examination of the coalfields with the twin aims of delineating the coal reserves and checking to see if there was any prospect of finding oil. In fact, the mapping did indicate oil potential and that moved the game on to the stage of seismic work and/or drilling.[1]

Even by that time (1959) there were few organizations who had the money or expertise to tackle drilling for oil. It was a big step, even for BHP, which spent $200 million a year on steel during that period. The company's chief general manager, Ian McLennan, decided that an expert opinion was needed, so he asked one of his executives, John Norgaard, to seek a petroleum consultant for advice during a business trip to the US. After establishing the company's bona fides with the US oil industry which, understandably, had never heard of BHP, Norgaard eventually met Lewis Weeks, a former chief geologist of Standard Oil Company of New Jersey (Esso, now Exxon) and the (then) president of the American Association of Petroleum Geologists. Described as a spry 66-year-old, Weeks was a world authority on sedimentary basins. Although he had never been to Australia, he was familiar with the geology and had a low opinion of the Sydney Basin as an oil province. Nevertheless, Norgaard was empowered to invite him to Australia for a first-hand examination. After a week of wandering around the BHP leases, the American's opinion was unchanged. Chatting about it with Lonie over a cup of tea at the Australia Hotel back in Sydney, Weeks asked whether the company really was interested in finding oil. If so he knew where to look. However the information would only be given to the top man.

On Lonie's recommendation Weeks was flown down to headquarters in Melbourne where McLennan received him, agreed on a commission of 2.5 percent on any production that might result, and was told to try Bass Strait. At first the BHP chief general manager was incredulous. While oil might lie under the seabed off the Victorian coast, he doubted that it was possible to do any exploration drilling in the turbulent waters, let alone produce anything that was found. Weeks, however, was unperturbed and pointed out that offshore technology was advancing rapidly and that production would be possible in such regions within a few years. The American's advice did not spring from a sudden whim. He was no stranger to the area, having studied it since the 1930s when he first learned of the small discoveries made round Lakes Entrance in Gippsland. He would have known of the comments by Sir Edgeworth

David and others that the Tertiary sediments seen in these areas probably extended offshore. This concurred with his own idea that oil occurs in the young sedimentary basins of the world. Perhaps even more relevant, he also would have been kept abreast of exploration through the 1950s via the Esso participation in Frome-Broken Hill.[2] Weeks never pretended to be anything other than a geologist. Certainly he did not profess to be a development or production expert. Nevertheless, standing in McLennan's office that day in 1960, he was the right man talking to the right company in the right place.

Born in Wisconsin in 1893, Weeks studied at the University of Wisconsin and Cornell University before beginning his career as a mining geologist and mining engineer in Mexico during 1917. A brief period with Whitehall Petroleum Corporation took him to the UK and India before he joined Standard Oil of New Jersey in 1924 where he stayed for 34 years until his retirement in 1958. Unwilling to step out of the industry, he set up on his own and by the time BHP brought him to Sydney and Melbourne his high reputation as a consultant and global geological thinker had preceded him. Australian oilmen, particularly the small band of local geologists, soon met Weeks and were impressed by his enthusiasm and his approach. Reg Sprigg was one.

> He was personally familiar with most of the oil fields of the world, and he literally thrived on geological facts. To him statistics told their own stories. He was also a man of strictly controlled imagination and rare geological insight. He saw things clearly in three dimensions, and [geological] time provided an important fourth.
>
> His publicly expressed philosophy came to be that in every field of knowledge each generation stands, pyramid-like, on the shoulders of generations gone before. His extraordinary perception carried far beyond the oil patch and included the revolutionary theory that some uranium deposits were in fact accumulated by sedimentary processes.
>
> After viewing the Athabasca Tar Sands in Canada he felt there must be liquid fields nearby, and this led to the discovery of the Leduc field and subsequently the vast Alberta Basin oil province.
>
> My first field encounter with Lewis was at his request to view the outcrop geology of the Otway Basin [again an area where Frome-Broken Hill was involved] and to see for himself the evidence of stranded coastal bitumen. It was a real plus when he and I stood on the beach at Cape Northumberland and witnessed five-cent-sized pieces of liquid bitumen stranding at our feet.
>
> I was happy enough just to collect it for analysis. Not so Weeks. He immediately produced a magnifying glass,

> viewed the oil and then announced that it was still degassing. Then he told me it was of very local origin. And he added, 'One day Bass Strait will prove to be a major oil province'.[3]

As a younger man Weeks was a polo player and swimmer, and even in his seventies he made a point of walking for 45 minutes a day to keep fit. Max Reynolds, then working for BHP, also came into contact with Weeks in the field.

> I escorted Lewis on a geological tour round Australia later in the 1960s. He struck me as a quiet fellow who took a lot of interest in what you told him. He had extraordinary stamina even though he was not particularly well at the time. The trip included Western Australia from Perth to the Kimberleys, central Australia and a number of points in the east.
>
> He did have a reputation for being a bit tight fisted and I certainly didn't see him pay for much. But he had a favourite saying :'Everyone hurries to wait'. He meant that people rush around so much that they end up arriving too early. He was always calm and collected himself.[4]

When McLennan took Weeks's advice the next step was for BHP to apply for the offshore leases across Bass Strait. As manager of the company's petroleum section Murray Lonie was intimately involved with this procedure.

> It was an unbelievable charade at first because the offshore boundaries had never been clearly defined, and certainly not scrutinized to the extent we needed. The Tasmanian boundary, for instance, was apparently five kilometres into Victorian territory. The South Australian–Victorian border was not taken as a natural continuation of the onshore north-south line. Instead it was made at right angles to the bit of coast where the land boundary hit the beach.
>
> On top of that New South Wales made a claim for inclusion saying that Victoria was originally carved out of New South Wales and therefore the two States should at least share the jurisdiction. In the end they had to go back to the declaration of the colonies to make a ruling.[5]

BHP, through its newly formed oil and minerals arm Haematite Explorations Pty Ltd (later Hematite Petroleum then finally BHP Petroleum), obtained exploration titles to the waters between Victoria and Tasmania in 1960, and to the western extension of Bass Strait from the South Australian Government in 1961. The grants covered a total area of about 166 000 km^2. At the time, geologically speaking, there was only scant knowledge of the boundaries of the Otway and Gippsland Basins through projections from onshore work, and no concept at all of the offshore Bass Basin. Weeks said that the first thing needed was an indication of the dimensions of the basins and the thickness of the sediments involved. That called for an aeromagnetic survey and he recommended that BHP call in Mike Reford, chief geophysicist for Aero Service Corporation of Canada, for the job. Reford and BHP's senior geologist, Brian Hopkins, joined forces to plan the preliminary programme despite the fact that a petroleum search subsidy was initially disallowed for the project. Hopkins, who had carried out the original oil survey in the Sydney Basin for BHP, had been brought back down again from a minerals exploration programme on Cape York Peninsula to work under Weeks's guidance on the geological aspects of the new Bass Strait work.

> BHP said we could have the money for a magnetic survey if we could obtain a Federal Government subsidy for it. The trouble was that the BMR said that magnetics would not give relative depths to basement and therefore there would be no subsidy.
>
> So with Reford's help we set off on a day's flying to prove we could get the accuracy we needed. We then resubmitted the subsidy application with this work as evidence and it was finally granted. After that Aero Service flew a group of eleven lines (2 000 km) across Bass Strait in December 1960. The lines were in various directions, spaced about 80 km apart. The aircraft (a DC3) was flown straight and level from one landfall to the next, using dead reckoning between points.
>
> The records were sent to Ottawa and a preliminary interpretation was phoned to Australia before Christmas 1960. The results were encouraging and they allowed areas to be outlined, indicating which basement was probably shallow and which, therefore, did not merit much attention. It was evident that a more detailed survey was warranted.[6]

The detailed work, using Shoran navigation (a radio signal distance-measuring system like that used by bats in the animal world) for more accuracy than the previously used dead-reckoning method, began in September 1961 and was completed in December the same year after 30 000 km of traverses were flown. The results clearly outlined the three basins and gave a good picture of the depth of the sediments in each. With exploration going to plan, Weeks next recommended that BHP embark on a marine seismic survey, and the US company Western Geophysical was called in for the work. New ground was again broken in Australia when two seismic boats (one for shooting and one for recording) arrived from Italy to begin the survey in November 1962, using 23 kg charges of gelignite as the energy source and Shoran as the navigation system. Altogether 8 560 km of lines were shot — 20 percent in the Gippsland Basin, 32 percent in the Bass Basin and 48 percent in the Otway Basin. When the survey was completed and results interpreted during 1963, it was clear even then that big structures of definite potential for oil exploration were present, as Brian Hopkins outlined in the company publication *BHP Review*.

> Each basin has a specific potential within the younger sediments, and the seismic surveys have outlined anticlinal structures and barrier sand ridges in the Gippsland Basin; stratigraphic 'wedge-out', reef trends and possible older sediments of interest in the Bass Basin; anticlinal structure and fault closure in the Otway Basin.
>
> The seismic surveys undertaken must be classed as reconnaissance only and, with the exception of Gippsland, more reconnaissance work is necessary. Although some detailed seismic surveys may be needed in Gippsland prior to drilling, studies of this basin are well advanced, and there is every likelihood of offshore drilling for oil beginning late in 1964.[7]

BHP had spent almost $2 million in building up this background data and, at the same time, it had received a rapid baptism in the preliminaries of an oil exploration programme. However at this point Weeks judged it time to bring in an experienced oil partner to run the expensive drilling programme that would have to follow. Initially, he wanted a different farminee in each of the three basins. Murray Lonie also realized the need for outside help.

> Although we had a lot of expertise in coal mining and steel production, we were still a bunch of kids in the oil business. We had sort of evolved into it, and even then no one had any real idea of how costly it was going to be. At BHP we were trying to understand every aspect, while the real oil industry had people specialized in all the different sectors.
>
> It was the same throughout Australia at the time. Most of the local geologists were good on structural theory, but they had nothing like the experience in the business of the US oilmen who came into the country in the 1960s. The people running the government subsidy scheme at the time were also conservative and limited in their oil experience.[8]

Nevertheless, with help and advice from several consultants, BHP put together a farm-out package that was far from an amateur effort. Its conditions for that period were tough. The incoming partner would have to pay the cost of all exploration, while BHP would take a 50 percent interest in anything found. Only at the start of production would

BHP begin to pay its half share and recompense the other party for half the exploration costs. There was also a clause inserted which said that if BHP did not want to become part of the production effort it could take a 12.5 percent royalty on production instead. The farm-out documents were sent to 10 of the world's major oil companies, a list narrowed from the original 20 that had expressed early interest. At BHP headquarters in Melbourne, Murray Lonie watched the replies come in.

> The offers were pretty half-hearted till Esso arrived and plonked a proposal on the table that was far above all the others. It suggested a $35 million initial programme, whereas all the others were barely $2 million.
>
> Esso's offer was obviously the product of a great deal of in-depth work and it transpired later that the Esso exploration manager in Australia, Al Caan, had been pushing hard to interest his headquarters in New York. What is more, when Weeks himself was with Esso in the 1950s, he also had tried to interest the company in Bass Strait based on his own knowledge of the region through Frome-Broken Hill.
>
> Caan was a protege of Weeks and it was not surprising that the Esso organization had followed BHP's movements closely.

Ken Richards, who had been with Frome-Broken Hill in both Gippsland and the Otway acreage onshore during the 1950s and early 1960s, confirmed the Esso interest.

> When Frome-Broken Hill withdrew from its onshore Gippsland acreage in the 1950s because the sediments were flushed, Caan (who was also part of the Esso group within the consortium) looked straight offshore.
>
> Then, when the decision came to split the Standard/Vacuum arrangement and the companies in Frome-Broken Hill were allowed to go their own ways, Esso set up several task forces in South-East Asia including a small group based in Sydney.
>
> John Hamlin was president and Caan was exploration manager, and their charter was to get a presence for Esso in Australian exploration and head it towards production. I was also a member of the team and, not surprisingly, Gippsland offshore was a major focus. We even did our own independent assessment which drew up the first sketchy outlines of what were to become the Kingfish and Barracouta prospects. Esso in New York finally showed some interest.
>
> When all the others saw our farm-in offer they thought we were crazy. But they had forgotten Esso's background knowledge in the area via Frome-Broken Hill.[9]

Naturally enough Esso's offer was accepted by BHP and the deal was signed on 12 May 1964. The proposed programme called for the running of detailed seismic surveys and the drilling of a minimum of five structures in the offshore Gippsland Basin. Similar agreements between the two companies were signed in February 1965 and then April 1967 for the Bass and Otway Basins respectively. The seismic work confirmed earlier interpretations and, by the end of 1964, the first location — a prospect named Esso Gippsland Shelf No.1 — awaited arrival of the contracted drillship, *Glomar 3*. Built in 1962 at a cost of over $US2.5 million, the vessel was 82 m long and had a displacement of 5 600 tonnes. Its drilling derrick, located amidships, was 41.5 m high and the rig could operate in up to 200 m of water, penetrating below the seabed to 6 000 m if necessary. Put beside some of the new drilling vessels of the 1980s, these dimensions are relatively small. But in 1964, when few people in Australia had seen a drillship, they were impressive. Certainly, the plans for Esso/BHP's first hole — 46 m of water and a subsea target of 2 440 m — were well within the rig's basic capabilities, although that was not the only consideration. John Evans was an operations manager for the US drilling contractor Global Marine when that company won the contract.

> Esso was looking for an experienced and capable offshore contractor and, as Global Marine had worked for the company previously, we were chosen. At the time the burning question was not the rig's depth capability, but whether it could handle the rough sea states in Bass Strait.[10]

But Global Marine was confident. The company had five drillships in operation at that stage, while two ore carriers were being converted into steam drilling vessels to add to the fleet. *Glomar 3* was the only rig available for the contract at the time, but it had previously performed satisfactorily in the Cook Inlet of Alaska where a nine-metre tide change created a 10-knot current. Evans himself felt the rig could do the Esso/ BHP job. A petroleum engineer by training, he also had a knowledge of geology and experience with machinery. He had served in the Korean War and then worked with Union Oil of California for a while in the 1950s before joining Global Marine, becoming involved with the offshore sector in 1960. In 1964 Evans came to Australia to oversee the *Glomar 3*'s work in Bass Strait.

> The rig sailed from the Gulf of Mexico early in October 1964 straight into a cyclone off Tonga. The crew were worried about the rig bearing away in the wind, but skipper Lloyd Dill had plenty of experience. The work boat *Pointe Coupee* was sailing in convoy with the rig, so Dill attached it to the drillship's bow to lengthen the total waterline and lessen the tendency to bear away.

> Some on board estimated that the waves were 15 m high during the storm. They were certainly coming over the drilling deck and that was 11 m above the waterline. But the vessel made it through. It was a pretty good test.
>
> It was loaded to the gunwales with equipment, including 3 000 m of drill pipe and enough supplies to last three wells. Esso wanted to save as much as possible on later transport costs for the programme. That was the reason the *Pointe Coupee* sailed out at the same time. It was fully loaded too.
>
> The rig crew members were classified as seamen for the voyage. It was the simplest way of getting round the red tape of insurance. Eighteen or 20 of them rode the *Glomar 3* and there were another five or six in the service vessel.

First Australian landfall was Eden near the New South Wales-Victorian border about 70 days after leaving the US. At the time it was the longest voyage undertaken by a rig of that type anywhere in the world.

> The rig cleared customs in Eden, although I'm pretty sure the Australian officials didn't know what they were looking at. Then it sailed round to unload a lot of the equipment and supplies at Port Welshpool in Gippsland. Lakes Entrance was unsuitable because of the sand bar blocking the harbour there.
>
> Three days later the rig was on station at Esso Gippsland Shelf No.1 and it spudded the well on 27 December. By that time too a second supply vessel — the Sydney-built *San Pedro* — had joined the fleet.

There was not long to wait for a result, but it was accompanied by some unexpected drama. In February 1965, after two months drilling, the bit penetrated a good gas zone in the target and the crew were directed to take a core of the reservoir. While this was being done gas apparently percolated into the drilling mud and, on resumption of drilling to check lower horizons, the well 'kicked', blowing the five-tonne drilling table straight up into the air. The first attempt at closing the blowout preventers failed when the vessel's heave caused the pipe to slip up and down between the rams. Finally the blind rams had to be activated, shutting off the well and allowing time for cement to be pumped down. Despite the near catastrophe, Esso/BHP realized the stunning fact that they had struck a gas reservoir with the first well in a completely new region — a very rare occurrence in the oil world, yet it had now happened twice in Australia's modern search history. In March, tests revealed that the gas was indeed likely to be in commercial quantities and a second well (EGS No.2) in June confirmed the reservoir.

Glomar 3's third contract well, EGS No.3, on a separate structure looking for the Lakes Entrance-type 'greensand' that contained proven oil onshore, was a dry hole. So was the fourth (Bass No.1), drilled in the Bass Basin in the Tasmanian sector of the strait. Then, in the early weeks of 1966 back in the Gippsland Basin, well EGS No.4, about 56 km south of Lakes Entrance, came in with a new, much larger gas discovery. The appraisal well EGS No.5 confirmed the flow and it was obvious to all concerned that the first two Bass Strait discoveries would support a commercial natural gas development for the State of Victoria. Surprised, but exalted by this early success, Esso and BHP realized that the basin had enormous potential and that a great number of both exploration and development wells were likely to be drilled in the coming years. To prevent any confusion in distinguishing between fields, the companies decided to change the well nomenclature. A joint statement was issued in August 1966:

> Indigenous fish names are being adopted for all prospects because of the absence in offshore areas of suitable geographic names such as those used on land. The names are assigned whether the prospective area is productive or not.
>
> Under the new designations, the EGS No.1 and EGS No.2 natural gas field will be known as Barracouta, and the EGS No.4 and EGS No.5 as Marlin. The prospect drilled by EGS No.3, which resulted in a dry hole, has been designated Cod.
>
> The new method of identification is one that is conventionally used in oil and gas exploration.

For BHP directors, standing on the threshold of the industry, the early gas finds effectively closed the door behind them. Barracouta and Marlin had become a reality so swiftly and unexpectedly that there had been no time to even consider a retreat. More importantly, as a further enticement to take a step forward and not back, the first Marlin well had also flowed oil from a thin zone below the gas-bearing sands. However the oil leg was difficult to interpret and the results of the second well indicated that Marlin's main attribute was its huge gas reserves. *Glomar 3* (retained by Esso under a second drilling contract) moved from there in April 1967 to drill a well on a large prospect named Kingfish. However Ken Richards, working with Esso, recalled that a technical problem nearly caused the consortium to miss what ultimately proved to be the biggest prize in Bass Strait.

> Kingfish highlighted a geophysical problem that was noticed back at Marlin. The structures as mapped on a time basis [that is, the subsea profile delineated from the times which the various seismic shock waves take to reach the rock layers

> and return to the recording microphones] were not an accurate picture of the actual physical position of the prospect. So when the geologists and engineers directed the rig to the high point on the Kingfish structure as delineated from the time map, the location was actually way down the flank.
>
> Luckily the feature was huge and the Kingfish No.1 well clipped the edge of what appeared to be a medium-sized oil reservoir, but nothing like what we expected going by the structure's mapped dimensions. We'd have probably returned to the structure for another go later, even if there'd been no hint of oil, but it might have been some time later. As it was, the engineers went back to the seismic interpretations straight away and planned the second well right on the physical crest.
>
> It took a long time [into the 1970s] before the geophysical problem was understood fully. Finally it turned out that a limestone layer distorted the seismic waves in a lateral direction and thus offset all the time maps to one side.[11]

While the engineers were grappling with the choice of location for Kingfish No.2, *Glomar 3* moved across to another large feature 20 km further north called Halibut. The first well on this structure set Esso/BHP, Australia and the oil world alight when it drilled straight through a 30 m oil column. The structure itself was so perfect in terms of reservoir parameters and performance that the oilmen declared the find commercial on the basis of this one well — something that is rarely done by the conservative exploration fraternity. Buoyed by that success *Glomar 3* moved to Kingfish No.2, and joy knew no bounds when the well tested a strong flow of oil. Shortly after, a third well put the matter beyond doubt. Esso/BHP had, in a matter of months during 1967 and 1968, eliminated most of the country's dependence on overseas supplies of oil. Headlines screamed the news and BHP shares, which had begun to move to the $8 mark when Kingfish No.1 alerted investors to the potential, fairly bolted up the boards to $17 at news from Halibut, and then $25 after Kingfish No.2 and Kingfish No.3. But the apparent ease of these discoveries belied the difficulties experienced offshore. Bass Strait weather had lived up to its reputation and taken its toll in the form of expensive delays and a fairly high turnover of personnel willing to work on the *Glomar 3.* John Evans and the Global Marine team toiled hard to combat both problems.

> To try and cut weather down-time we altered the torque converters on the anchor chains to compensate for the heave. The result was that the vessel's bow had a sort of rotating movement which was most uncomfortable. During one storm I was on a supply boat and I saw the whole keel of the rig at the bow

> end come out of the water. The anchors were spread as wide as it was prudent and it was enough to allow a change in the vessel's heading of about 20 points on the compass.
>
> We were worried about breaking something we could not replace readily from the stores we'd brought with us. It would have been a big expense to have to transport equipment over from the US, particularly if the rig was out of action waiting for it. The vessel hire cost was $100 000 a day (in 1962 dollars). It was better to be safe than sorry, so we erred on the conservative side and eventually that became the natural way of doing things.
>
> We did have a weatherman — a retired German naval captain who had been schooled in the Australian system. He was 67, but still came out to the rig bringing some of the meteorological equipment with him. We'd learnt that most of the bad weather came in from the southeast and a lot of the time our 'met' man was able to advise when a big blow was on the way.
>
> To help move about in bad weather there were guy wires strung along the deck.[12]

The rig crew worked 12-hour shifts, a practice new to the Australian oil patch, and something which the local hands were not keen on at first. The rotation was two weeks on and then two weeks ashore. Global Marine argued that there were less people on the rig when using this roster and it was easier to manage the operation. Nevertheless, John Evans agreed that the company probably overestimated the calibre of workers in the early days.

> Expectations were probably too high on both sides. For instance, some guys got pretty seasick and had to come ashore. I used to reckon it was best to leave them seasick for three days before sending them off. Some did get over it within that time.
>
> The work was hard, but not necessarily heavy. It was more a case of it being awkward, like trying to manoeuvre equipment round a heaving drill floor. I found the best people were farmhands because they were used to manual work and not afraid of getting dirty.
>
> In the early days all crew changes were by supply boat and in a storm the *San Pedro* could take eight hours for the trip out from Port Welshpool. The *Pointe Coupee* we used to call 'The Cork' and many's the time I was sick on a voyage in that. The fishermen used to think we were all crazy.
>
> On a crew change we'd try to leave the rig at night — it seemed to be calmest then on average — and try to get

> the men to the 7.00 a.m. train to Melbourne. If we had a good run they all went via the Welshpool pub and some of them weren't too flash by the time they arrived in the city. Others lived in the immediate Gippsland area.
>
> Other times the weather delayed the changes and the crews would be paid the extra days whether they'd finished a tour or were just starting one.

It was some time after the offshore programme began that helicopters were given clearance to fly to the rig for crew changes. There was apparently a wariness among the aviation authorities about flights over water but, once this was approved, Ansett was awarded the first flight contract. Alan 'Snoopy' Adams was one of the early pilots. Born in Sale, Adams was civilian-trained and spent the first half of the 1960s in fixed wing aircraft as a crop duster. He moved into helicopters when the spraying company bought one to use in forestry and orchard work round Echuca and Shepparton. In 1966 he joined Ansett's helicopter wing and came into contact with the Bass Strait work for the first time — ironically based back in his home town.

> Ansett had three choppers, Bell 47Js, which were piston machines with a capacity load of pilot and two passengers when fully fuelled. We moved the bases round a bit — Bairnsdale, Sale, Port Welshpool — depending where the *Glomar 3* was stationed.
>
> Soon after I joined, Ansett added two Sikorsky 62s to the fleet. They were single-engined turbine machines and a vast improvement in carry capacity because they could take six passengers with full fuel load, and 12 on shorter trips that didn't need full fuel.
>
> In the early days the flights were fairly ad hoc. The pilots were based in local hotels, although I was able to go home because it was so close, and the roster would be two weeks on and two off.
>
> There were a number of difficulties in those first years. For instance, the Bell 47s could only do 60 knots, so if the wind was 70 knots — which it often was — we had all sorts of trouble. I can remember one trip to Barracouta, which is 43 km. It took 15 minutes to get there and 75 minutes to get back. The Sikorskys were not much better in a headwind either.
>
> We stuck to the daylight hours if possible. There were some emergency night runs, but they were not popular. I went out in a Bell 47 to pick up the Strait's first fatality from one of the work barges in the late 1960s. They slung the litter onto the chopper skids, TV 'Mash' style, because there was no room inside.[13]

Ansett had the Esso contract till 1968 when it was taken over by a company called Helicopter Utilities (later Airfast) and, when Ansett closed its helicopter arm soon after, most of the pilots changed over to the new company. Helicopter Utilities also purchased Ansett's three Bells and one Sikorsky. The other had been wrecked offshore the year before.

> The machine, with pilot and eight passengers, had an engine failure as it approached the *Glomar 3* which was working on the Halibut wildcat at the time. It just dropped into the sea beside the rig. A workboat picked up everyone pretty quickly — only the pilot was injured — and took them to the drillship. Then it went back for the chopper.
>
> No one on board knew much about them and when they tried to lift it on to the deck, it just turned on its side. Thinking it would sink, the boat crew grabbed it by the tail section. Although this secured the machine, it effectively wrecked its flying future and we only used it for spare parts after that.

A worse helicopter accident occurred in March 1968 when an Esso/BHP-arranged press trip to the (then) new Barracouta production platform ended in tragedy. The 1987 Esso publication to mark Barracouta's 20th birthday tells the story:

> As history was in the making and a whole nation watched in earnest, all the good news concerning the Barracouta platform was shattered by tragedy.
>
> On March 22 1968, Esso/BHP had arranged with the media for an open press day to enable the journalists to report back to the general public the story of offshore oil and gas platforms. The journalists were flown out to Barracouta to be briefed on the production system and to take photographs.
>
> The [development] drilling programme was underway and was sure to win the appeal of the visiting press who would spread the word about the new industry. However, a terrible accident happened which meant another story was to be filed — a story that would involve the tragic deaths and mental and physical maiming of several men.
>
> At approximately 12.30 p.m. a Bell 204B, on approach to the platform, crashed onto the helipad. Three people died as a result of the crash and seven were injured.[14]

It later transpired that the helicopter had suffered a rotor failure on approach and some of the journalists, including an Esso staff man, were hit by the whirling blades as they waited (unfortunately in contravention of safety practices) on the helideck. For the pilots themselves, landing on a floating vessel was always more difficult than on the later fixed, and therefore stable, platforms. The helipads measure 18 m by 18 m and look very small from the air as the approach is made. Alan Adams said he found it hard to get used to.

> The rig pitch and roll was the most difficult to counter. There is a tendency to 'chase the deck'. The correct method is to watch the horizon, not the deck, and wait for a chance to set down when it is level in your vision.
>
> There was also a slide problem with the drill ships if it was very rough. We could handle a nine-degree pitch or a nine-degree roll, but if they occurred both at once even at five degrees it gave a corkscrewing effect that made landing tricky.
>
> In the early days, contact with the ship was by scratchy old HF [high-frequency] radio, and we'd work from actual weather — not forecasts. Often we would go out and assess it on the spot. We might land, but not switch off, so transfers of passengers and equipment had to be quick.
>
> Back then too, Kingfish was at the limit of fuel allowances. The *Glomar 3* had jerry cans of fuel on board, but only for emergencies, such as if a headwind sprang up while we were out there.[15]

Despite all the hazards, or perhaps because of them, the pilots kept their sense of humour. Admittedly sometimes it was at the expense of their passengers. Pat Washington, a deep-sea diver attached to the oil teams, was a frequent helicopter traveller and was used to their pranks.

> When Ansett had the contract the pilots didn't have uniforms, so 'Snoopy' Adams would wander out to the chopper mingling with the rest and take a seat in the passenger section. Everyone would strap in and sit there waiting for the pilot. Naturally he wouldn't turn up.
>
> After five minutes or so of inaction Snoopy would throw down his magazine and say, 'Bugger this, I can't wait any longer — I'll fly the thing myself'. He'd climb into the pilot's seat and fumble round for the flying manual. Then he'd start up and bump about the ground for a bit before taking off. By that time the passengers would be scared witless.
>
> I used to play up myself sometimes by taking my flippers into the cabin with me. Before long one of the greenhorns would ask me why. I'd tell him a lot of the choppers went into the drink and I was part of the salvage team. Or I'd just say they were needed for the long swim home.[16]

Washington was in fact one of another specialist group which soon became an integral part of the offshore oil scene in Australia. Originally a scuba diver with a part-time harbour job round Port Phillip Bay, he had no formal deep-diver training. In the beginning he made his own suits and tanks, the latter from converted gas bottles. His first oil work was in late 1966 attached to the *Glomar 3* rig.

> We dived in what we called 'heavy gear', which was the brass or copper helmet, canvas suit and boots plus 80 lb [36 kg] weight belt. Our air was pumped in from a compressor at the surface via air lines, and the pressure of this in the suit kept the water pressure off the diver. Later on we had the face mask and the hot suit.
>
> A lot of the early work was attached to a vessel called the *Nyhavns Rose*, a converted Danish trader operated by Wimpey (a UK company) which was conducting seabed coring round the projected platform sites for the two proposed Kingfish platforms. Using heavy gear and air we worked down at about 82 m. It had its dangers and one bloke was drowned when his hose was pulled in half.
>
> The accommodation on board was made up of caravans chained down in the forward hold. I remember one night a raging storm tossed the vessel all over the place and the caravans broke loose causing all manner of havoc below deck as they careered into each other.
>
> One of the most hairy jobs was working from the *Glomar 3* when it had a lot of trouble with its blowout preventer stack not making a good seal with the wellhead. The BOP would be lifted off the bottom about six metres or so and the diver, who couldn't fit inside because of the helmet, had to dispell air and go in upside down for as long as he could hold his breath.

Crew changes to the barges were usually by work boat from Lakes Entrance. Upon arriving alongside the barge the men had to wait for a wave crest and heave their suitcases onto the deck, wait for the next wave and leap at a knotted rope hanging over the side. There was little quarter given by the American skippers and contractors, a number of whom were accused of treating everyone as 'trash'. Even though many could hardly read or write, they had been brought up in the 'school of hard knocks' that the oil industry meted out in the 1950s and 1960s — particularly in the Gulf of Mexico. The divers were a small, elite group, mostly comfortable in the deep water. If they were not, the system would tend to weed the less suitable ones out. But there was a great camaraderie — no difference between the management and the divers in what was really a close, small world. It was, and is, a young man's profession with a lot of physically demanding work. When offshore the divers were on three-hour call, even if they were off duty.

> We could be dragged out of bed at 3.00 a.m. in a howling wind and cold water. Not everyone could stand up to that.
>
> Movement was slow in the heavy gear, but there was a knack to it. We'd lean forward to walk. We couldn't do things quickly, just move the body within the suit and be careful not to get upside down without venting. At 140 m we could spend between 25 and 40 minutes on the bottom.
>
> Sometimes in shallow-water work the job called for scuba gear. One time I was towed on a manta board behind a work boat looking for the moonpool doors which had dropped off the *Glomar 3*. I've never felt more like a piece of bait in my life. The area was notorious for White Pointers. Thankfully I spotted the doors on the second pass.
>
> There was one time when a tiger shark hit the landing stage I was on. The fish was about four metres long and came out of nowhere. Didn't touch me, but put a whacking great bend in the iron work and gave me a helluva fright.
>
> I'd look up and see seals hanging off the hose. They'd be friendly and curious, although in mating season the bulls didn't like the cows getting too close.

With the discovery of two gas fields within two years of beginning Bass Strait work, Esso/BHP felt confident enough to expand the exploration programme. A decision was taken to construct a new rig to enhance the search and the vessel, named *Ocean Digger,* was built at Whyalla in South Australia. More stable than the drillships like *Glomar 3,* it was a semisubmersible drilling 'platform' which floated on giant pontoons that could be flooded for drilling mode so that they gave support from below the wave zone — hence the term 'semisubmersible'. Unlike the *Glomar 3* it was not self-propelled, but was shifted between one location and the next using tugs. In 1967 *Ocean Digger* was towed to its first location, Crayfish No.1, to open Esso/BHP's programme in the Otway Basin. By that time several other companies had begun work in the waters of Bass Strait — one of which was Woodside, the first company to take up offshore acreage in the region back in 1959. It headed a group of Australian explorers including Planet and AOG.

Joined during the early 1960s by the British major, Burmah Oil Company, Woodside tackled its programme by drilling its first well in the sand dunes right on the coast between the town of Seaspray and a hamlet called Golden Beach. The latter gave the well its name — Golden Beach West No.1. Drilled late in 1965 it was dry, but the company resolved to try a similar structure seen within its acreage about four kilometres offshore. John Morgans joined Burmah Oil in Australia when he learnt that the company was looking for people with offshore experience. A toolmaker by trade, he had turned driller when working for APC in Papua and New Guinea before the war. He had also worked for Shell in South-East Asia for 10 years during the 1950s before returning to Australia to join the BMR's Muswellbrook coal drilling programme. He found, however, that 'big rigs' were more to his liking.

> Not that the *Investigator* was a big rig. It was really a dumb barge similar to a vessel that Shell had used in South-East Asia. Equipped to drill through a central moonpool in the hull or over the side, the drilling derrick was skid-mounted so that it could be moved to either position. *Investigator* began as a hull brought out from the USA to Newcastle State Dockyard for conversion to a drilling barge without self-propulsion and [was] then sent round to its first well at Golden Beach No.1.
>
> There were problems right from the start. Only a few kilometres offshore, it was a very shallow water location [15 m] so there was not much underneath the vessel in the wave troughs. We only had about 1.2 m of freeboard too and plenty of green water came in over the decks.
>
> On top of that the vessel had a very rigid marine riser, so in the rough weather it took a real pounding. The operation had a 40 percent down-time and the riser problem was the main reason the first hole had to be abandoned after a short time and respudded.[17]

Called Golden Beach 1a, the new well did find a small gas field, which tested at a rate of 126 000 m^3 a day. But in the face of the discoveries by Esso/BHP at Barracouta and Marlin, it was not considered to be commercial no matter which way the group looked at it. Burmah and Woodside shrugged off the disappointment and began what was to become an epic saga on the other side of the continent. The other company to show an interest in Bass Strait, if somewhat belatedly, was Shell. Along with a number of other majors the big Dutch-British company had scoffed at the high farm-in price which Esso had put on the table to win the BHP tender. It had been willing to offer two wells maximum, a ploy which was common practice round the world in those times to secure vast areas — often simply to keep other companies out, rather than with a serious intent to explore the areas.

Shell could rightfully claim to hold the largest exploration stake in Australia during the mid-1960s as it had interests in permits in virtually every State, including an area in the far eastern part of the Gippsland Basin (where it finally made the noncommercial Sole gas discovery in 1973) and a farm-in to Frome-Broken Hill's areas in the offshore Otway Basin. A preliminary survey programme was begun there in 1966. Reg

Thomas, a petroleum engineer, was headquartered in the company's Melbourne office under exploration manager Jack Spinks.

> Our Otway programme began in a bit of a rush. A new semisubmersible rig called *Sedco 135 E,* which was being built in Japan for the Pan American oil company to use off the north coast of the US, suddenly became available for long-term contract when Pan American decided to cut back its programme. Shell swapped a rig with them from another operation and the deal included fully paid mobilization costs for the semisub from Japan to Australia.
>
> It was a good deal. The trouble was the rig was just on completion, but we weren't ready to receive it because a drilling location hadn't been chosen.
>
> I went up to Japan to check on progress and found everything waiting on us. Even the tugs were on standby for the tow. So I sent a cable to Spinks back in Melbourne:
>
> 'Jack be nimble, Jack be quick
> Find a location mighty quick.
> Cannot wait another day
> Sedco rig is on its way!'.[18]

At the time no bottom surveys had been done, so Shell activated a quick programme using grab samples to check the seabed conditions. There were worries that the anchors would not hold the rig in the exposed Otway region. Eventually a prospect called Pecten was chosen and the vessel was duely towed down — a distance of 8 000 km which took 93 days and beat the *Glomar 3*'s voyage several years before. Once in Australian waters it unloaded supplies and equipment at Portland in western Victoria where it was a big attraction for the locals.

> There were no choppers so I had to board by personnel basket from the deck of a crayfish boat. That meant clinging on to the outside ropes while the rig crane driver winched me up. The rig was in full travelling mode so it was high in the water and I had to go up about 60 m by this means — not something I was keen on.
>
> Going back again was worse because the crayboat captain was not experienced in these things and he tried to manoeuvre the boat deck under the basket instead of letting the crane driver find him.
>
> Once out at Pecten location, 16 km off the coast, our fears about the anchorage were confirmed. The vessel was a three-legged design and the big triangular deck acted like a sail which took off in the wind. The anchors — nine of them

— were insufficient to hold the rig on station. In the end we had to drive piles into the seabed to secure them.

The seabed difficulties forced Shell to respud Pecten and the well was drilled to a depth of over 2 800 m without further trouble. Unfortunately it failed to find any significant hydrocarbons, and the story was repeated at two more wells in the region — Nerita No.1 and Voluta No.1 — before the rig moved away. Although disappointed, the explorers philosophically pointed out that the well data had given them a much clearer idea of the Otway Basin's geology. Further west, in the South Australian sector of the Otway, Esso/BHP had similar disappointments with three prospects called Crayfish, Argonaut and Chalma. The Victorian and Tasmanian Otway were equally frustrating for the two companies at Nautilus and Prawn, and a series of wells in the Bass Basin also drew a blank. It seemed that the Gippsland bonanza was not to be repeated — at least not easily — off the country's southeast coasts.

Returning to the scene of its early success Esso/BHP made a number of other, although smaller, Gippsland discoveries before the end of the decade. They included Dolphin (oil), Perch (oil), Tuna (oil and gas), Snapper (gas), Flounder (oil), Bream (gas and oil), Mackerel (oil) and Turrum (gas). In that short but concentrated burst of exploration activity between late 1964 and the end of 1969, Esso/BHP found 3 billion barrels of oil in the offshore Gippsland Basin and over 140 million m^3 of gas. Soon after the discovery of Barracouta and Marlin the two companies began to parallel their ongoing exploration activities with a development phase, and this accelerated when the Kingfish and Halibut finds were added a short time later. The first steel production platform, Barracouta, was put into position on the seabed during February 1968.

In other parts of Australia, however, offshore exploration was just breaking into its stride. One of those areas was in the diagonally opposite corner of the country — the North West Shelf. In recent times the name has come to be associated with the giant natural gas project centred on the North Rankin gas field and the onshore facilities on the Burrup Peninsula near the towns of Karrartha and Dampier. However, in its true geological and geographic sense, the term refers to the whole region from the Exmouth Gulf on Western Australia's central west coast right round to, and including, the Bonaparte Gulf and the beginning of the Arafura Sea north of Darwin. It was upon the area contained in this widest definition that the small Victorian-based explorer Woodside (Lakes Entrance) Oil focused its attention in 1962.

The story began eight years earlier and developed in some secrecy during the 1950s through the private work of Melbourne-based geologist Nicholas Boutakoff. Described as an academic, professorial type, Boutakoff was employed by the Victorian Geological Survey and was active in trying to interest explorers in the offshore Gippsland Basin a long time before the first permits there were taken up. Many of his ideas and theories were 'broad brush', but they offered a startlingly different view of the region's geology and its oil potential. Boutakoff spoke little about his own background, but would often expound at length and with great intensity on geological theory. One who was told a little of the man's personal side was American geologist Ed Durkee, who came to Australia with the Amoseas group.

> Nick was generous and hospitable, but also a dreamer and an introvert type who nevertheless would become excitable sometimes and get 'bug-eyed' as he expounded. He was a big man (about 190 cm tall) and an absent-minded professor character. As a consequence he would never win a prize for neatness in dress, or be 'Mr Fashion'.
>
> He once told me he was born in Washington DC before World War I. His father was Naval Attache from Czarist Russia to the US. With the collapse of Russia [the Czar's rule] the family returned to that country and his father was killed fighting with the White Russian forces.
>
> After that his mother moved the family to France where Nick spent much of his childhood. His first job was as a mining geologist in the Congo, and on one homeleave he was caught up in the Spanish Civil War and interned for a period.
>
> Then he apparently joined Shell and went to Trinadad. During World War II he emigrated to Australia and eventually ended up on the Victorian Geological Survey.
>
> The first I ever heard of him was when reading an article he'd written for *World Oil* magazine about the Lakes Entrance horizontal drilling programme in the early 1950s. At that stage of course I never thought I'd meet him.[19]

The seeds for Boutakoff's ideas about the North West Shelf were sown during a trip he made to the Exmouth Gulf region of Western Australia at the request of the Victorian Mines Minister soon after the discovery of oil at Rough Range. The Minister had decided that Boutakoff should acquaint himself with the geology of that area, and with the installations and operational methods used by WAPET, to see if there was anything to be gleaned for exploration in south Gippsland. While there, the geological and geographical relationship of North West Cape, Barrow Island and the whole North West Shelf with its shoals, reefs and islands first dawned on him as a working hypothesis. Once he had written his report to the Minister, Boutakoff followed up his initial thoughts in a private capacity by poring over a British Admiralty chart of the North West Shelf region. Using the depth estimates he drew contours of the seabed, and the more he studied it the more he felt that the obvious alignment of the shoals, reefs and islands was significant.

He developed a theory that Timor, in the Indonesian archipelago, was the active end of a geosynclinal basin filled with thick sediments that had been compressed and folded against the old rocks of the West Australian shield. As there were definite oil seeps in the region at Roti Island, it was possible that the aligned fold structures parallel to the Western Australian coast, as marked by the reefs and islands, could also be hydrocarbon-bearing. Excited by what he saw as great untried potential, Boutakoff gathered private capital and formed a company called Northern Holdings Pty Ltd. He then applied for a lease over the region and tried to interest others (including the giant Gulf Oil Corporation) in the exploration venture. These efforts failed and Northern Holdings was wound up in 1958 through lack of capital. Nevertheless Boutakoff kept all his maps and annotations.[20]

Throughout this period he continued to work on Gippsland geology and rose to become deputy director of the Victorian Geological Survey. He also did some consulting work for the fledgling Woodside (Lakes Entrance) Oil. Then, in 1962 when Eric Webb resigned from that company, Boutakoff left government employ to fill the vacant post of exploration manager. Up till this point he had told few people about his North West Shelf theories. However, upon joining Woodside, he presented the fruits of his eight years labour to the board. Duly impressed, managing director Rees Withers and chairman Geoff Donaldson agreed the company should make haste to apply for the area outlined in Boutakoff's map. Secrecy was still of the essence as WAPET was roaming round most of Western Australia and it had permits over the offshore areas to the south of Barrow Island. Woodside applied for the offshore region beginning at WAPET's northern boundary and extending up to the northwest corner of the continent, plus selected areas round some of the islands in the Timor Sea — a staggering total of 357 000 km^2, easily large enough to swallow the British Isles. Withers pledged an immediate £20 000 worth of exploration expenditure if the permits were granted.

Despite a leak to the newspapers in October 1962, there was no counter-application and Woodside was awarded the whole area in March 1963. Amazingly, the licence fee for this enormous tract of ocean was a mere £100.[21] The youthful company from Melbourne could afford that sum, but it had no hope of honouring its expensive exploration pledge unless it attracted farm-in partners — preferably some of the oil majors. Initially Burmah Oil and Shell were brought into the group, while prior agreements and cross shareholdings meant that Chevron and BP also became involved. Without really realizing the fact, Woodside had embarked on one of the most difficult, expensive, but exciting projects Australia, and arguably the world, had seen. For Nicholas Boutakoff, however, destiny held less buoyant plans.

Woodside's exploration manager believed his theories on the North West Shelf, worked out well before he joined Woodside, were his private property, and he had been under the impression that he would be given due recompense for divulging them to his new employer. When this did not happen he became bitter about the whole affair and parted company with Woodside before the decade was over. Whether it was a misunderstanding, or a wilful retraction of an oral agreement about his early work, has never been clarified. Boutakoff himself certainly felt the latter was the case and spent his last years trying to claim what he thought was his due reward. Painstakingly he compiled a series of documents (including his maps and notes from the 1950s) arranged in the form of 'exhibits' in the style of a court case, which he sent to Burmah Oil in London (as operator of the North West Shelf work) during March 1973.[22] His efforts were to no avail. In the 1970s his theories were proved right for the wrong reasons. The structures seen on the North West Shelf are in fact a function of the 'pull apart' mechanism of continental drift and plate tectonics, rather than compressional folds as he suggested. Despite this, Boutakoff deserved more than being relegated to the status of 'forgotten man' in the history of Australia's oil exploration industry. His death, in January 1977, went virtually unmarked in the obituary columns.

Field exploration of the North West Shelf began when Western Geophysical Company began running seismic surveys over the area during 1964. WAPET had just found oil on Barrow Island at the time and there was already some knowledge of the geology of this southern end of the Burmah Oil Company of Australia (BOCAL)-led group's giant permit. Also on the record were the oil seeps at Roti Island and Timor in the north, so the obvious plan seemed to be to begin the exploration work at either end of the region. Vic Garbin was working as a fisherman in the Lombardo company fleet when he was suddenly introduced to the survey work up the coast.

> I was working out of Dongara with the *Nelma* at the time and moored alongside the dock one night when an American voice said, 'You own this goddam rust bucket?'. I said yes and the voice said, 'How'd you like to work in the oil industry?'.
>
> Well it was near the end of the season so I said OK, not knowing much about it. Anyway we worked the boat out of Darwin for the winter at £120 a day just carrying stuff to and fro for the survey vessels. Then we did it again the following winter.
>
> Soon after, the fishing vessel *Andrew* went to auction in Perth and Lombardo bought it for £5 000. It was a 58-footer [18 m], a former Rottnest Island ferry, and we were sitting round wondering what to do with it when another Yank asked the same question. I was a bit wiser about the industry then so I jumped at the chance and we worked for GSI [Geophysical Services International] as a scout boat guiding the seismic vessel through the islands up north. That's how Lombardo Marine started in the service boat business.

> We were out in all sorts of seas, but one brush with high winds I remember was running into Port Hedland one night in heavy seas. The *Andrew* was bucking round all over the place and one of the crewmen was holding on in an open doorway on the lee side when the boat did a 40-degree roll and his back got wet as it brushed the sea. I've never seen anyone so white-faced as he was when we righted again.[23]

As the results of the seismic work came in it was seen that there were two obvious targets in the north (Ashmore Reef and Scott Reef), while in the south the Rankin trend offered a number of possibilities. Admittedly the two northern features were very isolated, but the surveys showed that the structures were so large that if they did prove to be full of oil there would be a good chance of commercial development. Another thing in their favour was that the reef atolls themselves offered shallow and sheltered waters in the midst of an otherwise very deep ocean environment. In the south the Rankin trend was covered by about 120 m of water but, as the 1960s advanced, so did drilling technology via practical experience in frontier areas like Bass Strait and the North Sea. Explorers were confident the Rankin area would not present a water depth problem when the time came to drill.[24] Nevertheless the remote locations under consideration presented a horrendous logistics problem, particularly in the north. Ashmore Reef was as close to Indonesia as it was to Australia, while Scott Reef was nearly 480 km over the horizon from the outpost town of Derby in Australia's far northwest. Both were beyond the reach of helicopter transport. BOCAL's newly recruited drilling and logistics man, John Morgans, was given the task of setting up the operation infrastructure.

> We chose Broome as our northern base. In the mid-1960s it was a sleepy little town where businesss closed at 5.00 p.m. every day, including the wharf. The first thing I had to do was impress upon the wharf authorities that the oil business ran a 24-hour a day, 7-day a week operation and the wharf would have to be available at all hours.
>
> Looking at the remote locations, we were lucky in a way that there were a number of islands off the northwest coast. We decided to use these as fuel staging points for the helicopters, and I boarded the State Ship *Cape Don* asking permission to divert it from its usual course up the coast to check the likely island sites. I used the vessel's dinghy for a close look at each one.
>
> Then I had a front-end loader sent up from Port Hedland and put on a barge to level a helipad area on the chosen locations. We went as far as Fitzroy Crossing for a suitable concrete aggregate for the pads, and also had use of the facilities

> at the iron ore set-up at Koolan Island in Yampie Sound to service the operation. Koolan was particularly useful for the base on Cape Leveque — not exactly an island, but pretty isolated just the same.
>
> I also remember Browse Island where the tide goes out 16 km. We had to work fast and coordinate with high water every 12 hours to unload materials. Our main tool was a small manual cement mixer but, as all the locations were tiny dots of land barely poking above sea level, all the fresh water had to be shipped in.
>
> Once each pad was set we left dumps of aviation fuel in 44-gallon drums. Sometimes it was buried. Other times we enclosed it in cyclone netting. But we had to keep checking because Indonesian fishermen got wind of this fuel source and pinched a lot of it. On Browse we also installed a 3 000 gallon [13 640 L] underground fuel tank.
>
> I had five crew with me on the job and I was cook as well as the company rep. For two months we lived in tents and were kept company by thousands of mice and booby birds.[25]

After a lot of preparatory work BOCAL chose Ashmore Reef as the first drilling location. An initial proposal to drive sheet piling into the reef and build a jetty to drill from was discarded as too expensive and impractical. So BOCAL decided to use the dumb barge *Investigator,* which had just completed the Golden Beach well in Bass Strait. The vessel had previously been stiffened for heavy-weather work, but in so doing it had become top heavy and, with its rounded keel, it tended to roll in big seas. However once it had completed the 6 000 km tow from Gippsland, the rig was quite stable in the 50 m deep, but calm, waters inside the reef. BOCAL's wildcat of wildcats, Ashmore Reef No.1, was spudded in during October 1967. For five months during the summer of 1967–68 the crews toiled in this remote part of the Timor Sea, 670 km from the Australian mainland and 1 800 km from the crew change point at BOCAL's headquarters in Perth. There was no other town along the west coast large enough to accommodate the number of people involved in the exploration operation. Geologist Doug Langton was one of many who regularly made the 13-hour trip to the rig.

> We'd board a DC3 at Perth airport at about 8.00 p.m. at night and eventually reach Broome at 5.00 a.m. the next morning. It was a plane with bus seats fitted and virtually on permanent charter to BOCAL from MacRobertson Miller Airlines [now Airlines of WA]. Often there was a stop in Port Hedland or Meekatharra on the way for refuelling.
>
> Then we'd have time for an hour's breakfast before boarding one of the 17-seater Wessex 60 helicopters run by the

> chopper contractor Bristow for the run to the rig. The four- or five-hour trip would be via one of the island fuel dumps — Browse, Adele, or maybe Cape Leveque. Sometimes there'd be food on these spots, but mostly not because the Indonesian fishermen often raided them.
>
> Other times they'd visit to barter seashells for old clothes. We had to have inoculations of cholera and smallpox because of the Indonesian presence and the injections were given in Broome.[26]

The Wessex helicopters were twin-engine machines and could fly on one engine for short distances with full load. If there happened to be helicopter engine problems, the pilot would make for the nearest of the islands, discharge the passengers and, with their weight removed, be able to lift off and fly back to base. Passengers would then have to wait until a supply boat came to pick them up. On one occasion though, the exhausting 'commuter' trip had tragic consequences of another kind.

> One of the passengers was a diver and when we arrived at the rig about mid-morning after flying all night this bloke had to go straight down because the previous team had run out of hours.
>
> He hadn't been below very long when he showed signs of the bends. They brought him up again, but he died there on board. By that time the chopper had left with the outgoing crew and no others would come out if it meant flying some of the way after dark. The atmosphere on the rig was pretty bleak that night.

Although the barge *Investigator* performed reasonably well at Ashmore Reef, BOCAL realized that it would not handle conditions outside sheltered waters. So the steam-driven drillship *Glomar Tasman* was brought in from the Middle East via South Africa for the next location — Legendre No.1 in the southern part of the permit. A base was set up at Port Sampson near the town of Roebourne. The attraction of Legendre over other prospects in the area at the time was a water depth of only 52 m. On the negative side, the seabed at the spot was billiard ball smooth, with no grip for the rig anchors. BOCAL decided to take a leaf out of Shell's book in the Otway Basin and pile the anchor chains into the seabed. Work began on this with the Wimpey vessel *Nyhavns Rose* before the rig arrived, but it progressed so slowly that the *Glomar Tasman* eventually tried to drill some of the pile holes itself. However a combination of strong currents and rough weather made this impossible and the *Nyhavns Rose* had to complete the job as best it could. The Wimpey vessel was on location for a total of 105 days before Legendre No.1 could be spudded. For half of that time it was inactive, waiting for the sea conditions to calm.

A mobile Failing 1500 rig used by the Frome-Broken Hill group in oil search programmes round the Portland region of western Victoria in 1958. Courtesy AIP Library, Melbourne.

Electric cable being wound up after a shoot during a seismic survey in Western Australia. Courtesy AIP Library, Melbourne.

'Thumper' weight at rest after its drop to the ground to create seismic waves during a survey in the Perth Basin of Western Australia. Courtesy AIP Library, Melbourne.

A geophone handler ('juggie') ready to lay out geophone and cable during a seismic survey in Western Australia. Courtesy AIP Library, Melbourne.

Seismic interpreter at work on data obtained during survey operations in the Otway Basin of western Victoria during 1958 for the Frome-Broken Hill group. Courtesy AIP Library, Melbourne.

Seismic explosion chamber (pan) which directed an explosive charge to a heavy plate on the ground to create seismic waves. Courtesy AIP Library, Melbourne.

An operator sets up a seismic record sheet in the field processing truck after a dynamite survey in the Otway Basin of western Victoria for Frome-Broken Hill group in 1958. Courtesy AIP Library, Melbourne.

A Sikorsky helicopter lowering equipment at a drill site in Papua for an APC programme during the 1950s. Courtesy BP Australia, Melbourne.

Cartoon of C.W. (Bill) Siller by Mac Vines (1987) depicting Siller's championship of the Queensland Petroleum Exploration Association since its beginnings in 1959. Courtesy Bill Siller.

Senator Sir William Spooner, Minister for National Development in the Menzies Governments of the 1950s and early 1960s. Courtesy *Canberra Times.*

Reg Sprigg, founding member and first chairman of the Australian Petroleum Exploration Association (APEA). Courtesy APEA Library, Sydney.

Alan Prince, founding member of APEA and publisher of the *Australasian Oil and Gas Journal.*

John Fuller, entrepreneur, ex-Australian Wallabies rugby team member, founder of Planet group of companies and founding member of APEA.

Lou Smart, businessman, oil entrepreneur and founding member of APEA who formed his own exploration company, L.H. Smart Exploration Ltd, with its own drilling rig.

A caricature of Texan oilman Herschel Albert Hackathorn who began Outback Oil in the 1960s. Courtesy the *Bulletin*, Sydney.

The oil discovery at Moonie as seen by cartoonist Rigby. Courtesy West Australian Newspapers.

Lowering the rig derrick at Moonie No.11 in 1963 ready for the move to the next location. Courtesy AIP Library, Melbourne.

Wet and muddy work roughnecking on the drill floor at Sunnybank No.1 well near Roma, Queensland, 1962. Courtesy AIP Library, Melbourne.

Union Oil field superintendent Ed McLeod (left) and Union's resident Australian Manager Doyle Graves at Moonie No.5 well, Surat Basin, Queensland in 1962. Courtesy AOG, Sydney.

Rural life continues as the Moonie No.6 well drills ahead in the Surat Basin of Queensland during 1962. Courtesy AOG, Sydney.

Working on the christmas tree of valves on the wellhead of Moonie No.11 in Surat Basin, Queensland, during 1963. Courtesy AIP Library, Melbourne.

Geologist David (Dave) McGarry examines cuttings from Moonie No.1 oil discovery well, 1961. Courtesy AIP Library, Melbourne.

Off-duty drillers from the Union Oil group's Surat Basin programme in Queensland quench their thirst at Murphy's pub, Tara, in 1961.

The Moonie motel was the first building of the proposed new township in 1962 after oil had been discovered in the region. Courtesy AIP Library, Melbourne.

Hughes tri-cone rolling cutter bit delivered to the field. Courtesy AIP Library, Melbourne.

Amadeus Basin explorers at Alice Springs airport in 1962. From left, Roy Hopkins, Ted Rannett, Duncan McNaughton, Grover Murray and Bill Siller. Courtesy Magellan Petroleum Australia, Brisbane.

Field camps during the 1960s (1) — WAPET's Meda No.1 in the Kimberleys of Western Australia.

Field camps during the 1960s (2) — The BMR's Beagle Ridge hole, Carnarvon Basin, Western Australia.

Field camps during the 1960s (3) — WAPET's Kidson No.1, Canning Basin, Western Australia.

Field camps during the 1960s (4) — Petty Geophysical's camp at Swindell's Field in the southeastern Canning Basin of Western Australia.

A corduroy road of saplings made easier access to survey bases in Papua where almost continual rain quickly turned dirt tracks into a quagmire. Courtesy AIP Library, Melbourne.

A sapling bridge across the Strickland River in Papua allows an APC survey party to cross. The river could rise and fall 10 m overnight. Courtesy AIP Library, Melbourne.

Sami Nasr, exploration manager of Ampol Exploration, blessed the bit during the spudding-in ceremony for Barrow Island No.1 discovery well.

Harry Butler, naturalist and environmental protection consultant, with a perentie (goanna) amongst the spinifex of Barrow Island. Courtesy WAPET, Perth.

George Swindells, construction and maintenance supervisor who had previously been foreman at WAPET's Swindells Field airstrip in the Canning Basin, displays some early Barrow humour. Courtesy AIP Library, Melbourne.

WAPET's Cockburn No.1 drilled at the end of a specially built 0.75 km causeway into Cockburn Sound, south of Fremantle, Western Australia, June 1967. Courtesy AIP Library, Melbourne.

WAPET's early base camp on Barrow Island in the 1960s.

Unloading casing for WAPET's well on Long Island off the Western Australian coast. Courtesy AIP Library, Melbourne.

The BP/Abrolhos Oil group's well on Gun Island off Geraldton, Western Australia. Courtesy AIP Library, Melbourne.

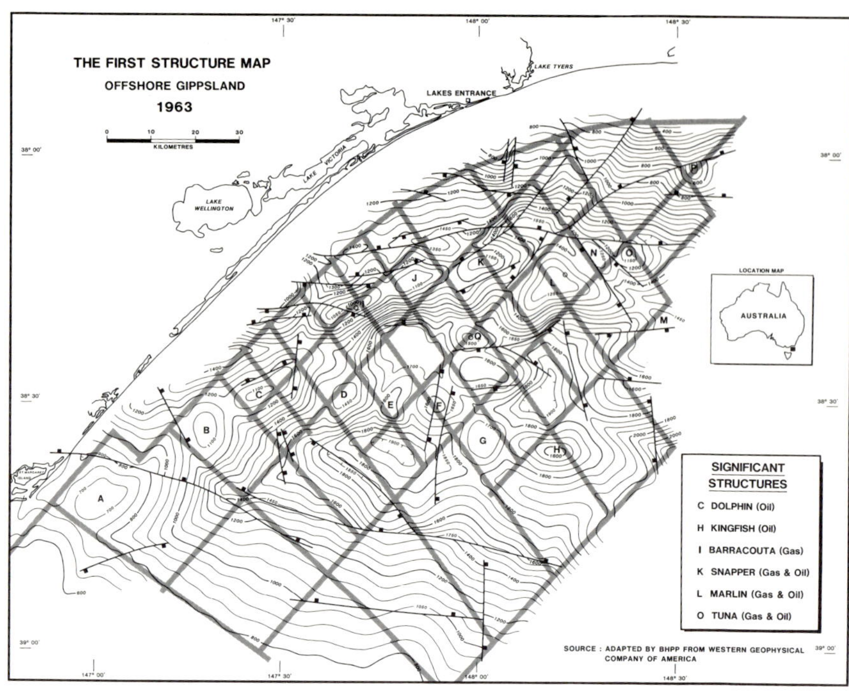

The first structure map of the offshore Gippsland Basin following a seismic survey run for BHP by Western Geophysical Company in 1963. Courtesy BHP Petroleum, Melbourne.

Lewis Weeks, the US consultant geologist who pointed BHP towards Bass Strait exploration in the early 1960s.

Supply/service boat *Pointe Coupee*, nicknamed 'the cork' because of its motion in heavy seas, made the voyage to Bass Strait from the Gulf of Mexico with drillship *Glomar 3* in 1964. Courtesy Esso Australia, Sydney.

Bass Strait discoveries were particularly exciting for Victoria. The drillship *Glomar 3* was depicted on the front cover of the 1966 Gippsland region phone book. Courtesy Telecom Australia, Melbourne.

The galley of drillship *Glomar 3*. Courtesy AIP Library, Melbourne.

Australian-built semisubmersible rig *Ocean Digger* in Bass Strait. Courtesy AIP Library, Melbourne.

Offshore workers home from a tour of duty in Bass Strait leave the helicopters at Esso/BHP's Longford heliport in Gippsland. Courtesy AIP Library, Melbourne.

Helicopter pilot Alan ('Snoopy') Adams at Longford heliport in charge of Esso/BHP flight operations. Courtesy AIP Library, Melbourne.

Semisubmersible rig *Sedco 135E* off the west Victorian coast near Portland. Courtesy AIP Library, Melbourne.

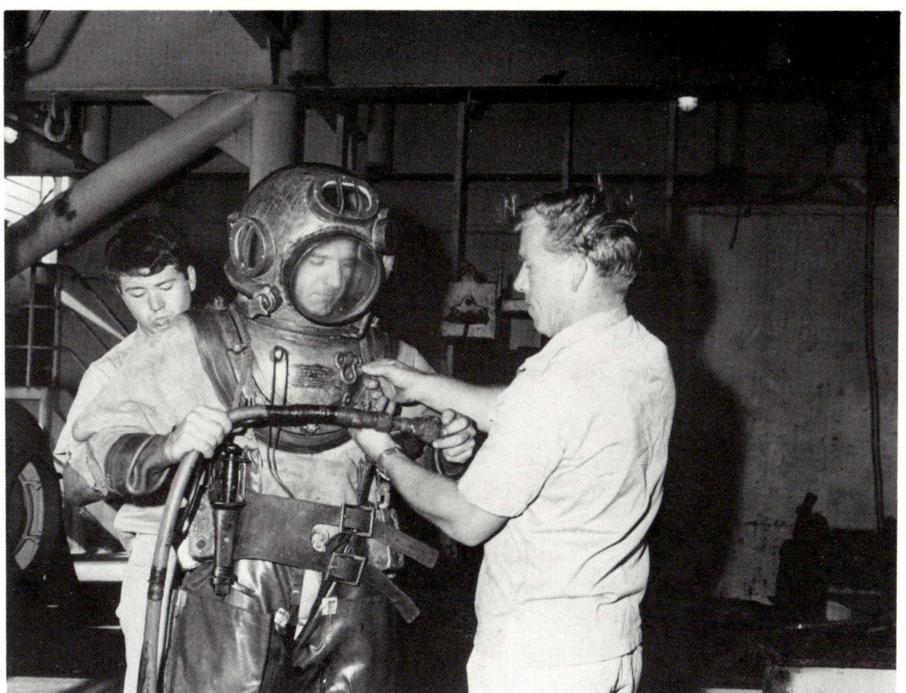

Diver Jim Duncan is prepared for Bass Strait work on board *Glomar 3*. Courtesy Esso Australia, Sydney.

In early Bass Strait work divers were lowered to the sea bed in a cage from the drilling vessel *Glomar 3*. Courtesy Esso Australia, Sydney.

Nicholas Boutakoff, geologist and exploration manager of Woodside (Lakes Entrance) Oil, initiated the company's move to the North West Shelf.

The drilling barge *Investigator* under tow off the Australian coast. Courtesy AIP Library, Melbourne.

The catamaran rig *E.W. Thornton* on location off the Queensland coast. Courtesy AIP Library, Melbourne.

The jack-up rig *Jubilee* in Western Australian waters.

Expenses began to mount even higher when it was found that the shallow formations were fractured and porous, causing circulation problems as drilling mud was lost through the sides of the hole. Flax, hessian bags and mattresses were stuffed down to counter the difficulty and eventually the drilling made more headway. Upon reaching the target Legendre did look promising, so a test was made. Initially it was disappointing as there was no flow, but when the drill pipe was brought out of the hole afterwards, the crew was surprised to find a small amount of oil. There was great excitement when the second test actually flowed oil to the surface at the rate of 400 barrels a day in the first offshore find on the North West Shelf. Unfortunately, however, the elation was short-lived because the flow rate did not improve and it was realized that the accumulation was small — certainly not enough for a commercial development so far from the coast. Nevertheless it did prove that oil was present in the region and, as the decade came to a close, BOCAL and its partners felt sure they were on the right track.

By this time too, WAPET had decided to try an offshore drilling programme. The company had first put its 'toe in the water' as early as 1961 with a one-line marine seismic survey down the west side of the Exmouth Gulf. The British company SSL, which had done some of the early onshore Rough Range work, was contracted for the job and geophysicist Daryl Johnstone looked after the WAPET interest.

> They started with a landing barge called the *Tarmona* and about half a kilometre of floating cable strung out behind. The crew lived in Learmonth and went out to the vessel in a small boat every day till they decided it was too rough. WAPET then hired a vessel called the *Lakanaukie* from Carnarvon which became the accommodation boat.
>
> After that they had two work boats, one for shooting and one for recording. Most of the time the shot boat followed the recording boat, but occasionally they'd run parallel and leave a gap in the geophone array to accommodate the arrangement.
>
> The results weren't too bad for quality — better than onshore in that area — and so the geometry did not have to be so accurate. I do admit that sometimes the distance measurements were by guess or by god.[27]

Once oil had been found on Barrow Island and the island hopping exploration programme was in full swing, WAPET decided to extend its coverage to offshore drilling in both the Carnarvon and Perth Basins. With the relatively shallow and sheltered waters of the Barrow Island area in mind, the jack-up drilling rig *Jubilee*, overseas at the time, was signed up in a two-year contract. Engineers had reasoned that this type of rig could be floated into a location in a minimum water depth of

16 ft [5 m]. Once in position the legs could be lowered down to the sea floor and the drilling platform itself jacked up above sea level to a safe operating level. This reasoning was sound enough. The only problem was that once the rig was on its way, WAPET decided to begin the programme in the Perth Basin where conditions were somewhat different. The rig was too wide to pass through the Panama Canal so it was towed from Texas across the southern Atlantic, round the Cape of Good Hope and across the Indian Ocean to arrive in Fremantle during August 1968 — a total distance of 22 525 km in 139 days. It was the first time such a vessel had been brought in to help an Australian exploration effort.

Back-up support for WAPET's programme came in the form of two Smit-Lloyd supply boats built to specifications in Adelaide, and two Bell Jet Ranger helicopters. Jesse Haddock, the company representative on the rig, was on board when *Jubilee* moved out to its first location — Quinns Rock No.1 — in about 40 m of water and 20 km off the beaches just north of Perth.

> It soon became apparent that the rig was the wrong type for this area. Persistent heavy seas and spring tides set up a swell that is peculiar to West Australian waters, and we spent a number of days either waiting for the seas to abate or trying to manoeuvre onto the location.
>
> We tried several times without success. In the end the contractor, Offshore Company, made experimental modifications and carried out tests on a scale model of the rig in a tank at Berkley University in California. Going on those results, modifications were made to the rig itself and it seemed to work. Still, we'd spent nearly two months trying to set up at the Quinns Rock location.[28]

The effort was to no avail. Quinns Rock No.1 was plugged and abandoned as a dry hole about two months later. The next, Gage Roads No.1 also in the Perth Basin, gave less trouble, but it too was dry. *Jubilee* was then towed north to the area for which it was originally chosen. Part of Jesse Haddock's job was planning the tows between each well location and watching over the jacking operation.

> Over the years I'd gathered a pretty good knowledge of the area between Barrow and Exmouth and I knew where all the sheltered havens were in case the rig tow had to be halted for any reason. There was one time when we were moving south past Thevenard Island when the rig suddenly developed a list. They jacked her up in the lee of the island and found a piece of timber had jammed a valve open.
>
> It was easily fixed, but the irony is that the rig had jacked

> up almost exactly over the centre of what is now the Saladin oil field. 'Course it wasn't found for another 20 years.
>
> The only trouble we had with bottom conditions up there was at Ripple Shoals. There was a deep sand cover over the location which didn't seem to affect the jacking up operation. But, jacking down again after the well was finished, the struts in one leg snapped. It must have slipped off a rocky ledge and it was aligned at a slight angle. The only thing to do was jack down slowly, welding each strut back into position as it broke.

WAPET's programme with the *Jubilee* continued till the end of the decade without any commercial success despite hopes of a repeat of Barrow Island. One well was even named Pepper No.1 because the geologists thought it might be a 'hot prospect'. It was unsuccessful at the time, but a small field (South Pepper) was found a few kilometres away in the 1980s.

Because of the seabed anchoring problems experienced by several of the early rigs working on the North West Shelf, a number of bottom surveys were carried out by diving teams at each proposed location prior to the vessels' arrival. Divers were also attached to the rigs themselves to help in the drilling operation. Pat Washington, who journeyed round from Bass Strait in Wimpey's *Nyhavns Rose*, found the conditions a lot different to those offshore Victoria.

> One trip went from Karrartha to Darwin taking samples all the way to test for anchor holding. It was pretty hot — even the water temperature was 82 degrees [28° C]. The sea was flat and the glare cut visibility down to about 20 m.
>
> Air temperature in the hold of the ship was 140 degrees [60° C] and it was impossible to sleep in the caravans they had down there. So we slept on deck.
>
> At one point we passed through a local tropical downpour and everyone rushed for the soap to have a good wash. Trouble was that by the time we were all soaped up the storm had passed, so the skipper turned about and steamed back into it.
>
> It was a pretty wild vessel that one. The crew was mostly Irish, except the cooks who were Chinese. The Chinese spoke little English so we called them simply One, Two and Three. They fished a lot and had hundreds of their catch strung up to dry round the boat.
>
> One Irishman became sick of the sight and the smell of them so he pitched the lot overboard one night. The Chinese retaliated by throwing a full case of beer over the side.
>
> Well that was a red rag to 'Mick' and he upped and

> threw Number One after it. Poor bloke couldn't swim either. But they hauled him back on board eventually. It was lucky too because there were always sharks about.
>
> Funny thing was I never had much trouble with sharks when I was in the water. Visibility was about 18 to 30 m, compared to six or nine metres in Bass Strait. Mind you I reckon you'd never see the one that gets you anyway. They can come out of nowhere.
>
> There was more trouble with sea snakes. On the *Glomar Tasman* there'd be six footers in the rig moonpool trying to get out. Even though their poison is deadly, their mouths are small and they can't get to you unless they grab hold of a finger. Their skins have terrific colours and we used to make belts out of them.
>
> I had more concerns off Darwin though. The worse thing there was the possibility of crocodiles looming up close. The big ones would come out to sea about six or eight kilometres, and I had heard of some which they reckoned swam across the straits to Indonesia.[29]

Another of Washington's experiences was being sent below to help with the big gas blowout at the Petrel No.1 well in the Bonaparte Gulf in 1969. With gas bubbles spewing out of the seabed, the divers had trouble staying on the bottom and a rope was fixed from the wellhead to the boat on the surface so they could pull themselves down. The echoing rumble of escaping gas was unnerving. On the surface the accident itself had been both dramatic and spectacular. Petrel No.1 was drilled by a third major group working in northern Australian waters led by Arco and Australian Aquitaine. The semisubmersible rig *Sedco 135 G*, a sister vessel to the *Sedco 135 E* in the Otway Basin, came in to spud the well in May 1969. On 6 August it penetrated a gas reservoir at 3 660 m, but the sudden flow could not be controlled and the blowout began. An anchor winch on one leg was started in an attempt to move the rig off location, and then left running as the crew was evacuated to the boats. Soon after that the gas ignited, sending up a fireball which enveloped the rig and apparently short-circuited the ballast pumps causing the rig to develop a list. However by that time the vessel was well clear of the gas boil and the fire had diminished enough for some of the crew to reboard, extinguish the fires and secure the vessel. John White of the BMR in Canberra inspected the scene soon afterwards.

> Once the fire was put out the rig was taken direct to Singapore for repairs. But the well continued to flow wild for another 18 months. The escaping gas made the sea 'boil' in an area 460 m across and there was great difficulty in igniting the gas so that it would burn off and not cause further immediate danger.

> First we tried drifting fire rafts through it, but they were too small and capsized. Then we used the Red Indian method of flaming arrows. That wasn't as easy as it looked either and it took a while before we could keep the flame on the arrow alight as it flew through the air. Eventually we were successful and a circle about 21 m in diameter caught fire. It was a strange sight with the sea churning up waves about a metre or so in the air and flames shooting straight up out of the ocean.
>
> No one was hurt in the actual blowout, but there was a tragedy later when the supply vessel *Sedco Helen* backed in to pick up a half-submerged buoy. It was a difficult manoeuvre because the gas had lowered the density of the water and the lack of buoyancy could sink a vessel that came too close.
>
> Anyway the marker became caught in the propeller shaft and wrenched open some of the plates in the boat's hull. She was abandoned when water flooded in, but then reboarded when it seemed that she might stay afloat after all. That's when the vessel suddenly turned over and the crew were drowned.[30]

American oil troubleshooter Red Adair's organization was called in to supervise the drilling of a relief well (Petrel No.1a) into the reservoir. But it was more than a year before the drilling rig was repaired and back on station for the job. The relief well finally killed the wild gas flow in January 1971. The experience had brought a fiery conclusion to the 1960s in that part of the country, but it had also proved the presence of hydrocarbons in the offshore Bonaparte Basin. Similar, although short-lived, encouragement came the way of explorers in the Papuan Gulf during 1967, 1968 and 1969. In an 18-month burst of activity, a five-company group led by Phillips Australian Oil Company drilled 11 wells in Papua New Guinea waters and made two substantial gas/condensate discoveries — Pasca and Uramu. However, of necessity they were rated noncommercial because of the lack of local markets for the gas reserves.

Ultimately it was a disappointment for Phillips and its partners, who had spent about $14 million on the project using the brand-new drillship from the Global Marine fleet, *Glomar Conception*. The swiftness of the programme — which averaged 51 days per well — belied a number of logistical and physical difficulties. Surprisingly, weather down-time was not one of them. Overpressuring in a shallow mudstone section and mud losses into a porous reef limestone lower in the sequence were the main culprits. Apart from slowing down the drilling rate, the frequent and copious use of mud chemicals, particularly barytes, to overcome the problems led to a supply shortage in the Port Moresby base. One well had to be prematurely abandoned when stocks were exhausted. Water depths varied widely between a minimum of 30 m at Uramu and a maximum of 98 m at Pasca. Ten metres was the absolute minimum depth in which the *Glomar Conception* could operate and, on those wells, Phillips took the unusual step of setting the blowout prevention equipment and wellhead within a caisson below seabed to provide sufficient clearance for the vessel.[31]

Further south, off the east coast of Queensland, exploration during the 1960s had been light. A group consisting of Australian Gulf Oil and the Australian Oil and Gas Corporation brought in the unique catamaran rig *E.W. Thornton* to drill two wells in the Maryborough Basin at the southern end of the Great Barrier Reef during 1967. The rig drilled a third well, for Phillips/Sunray, off Moreton Island near Brisbane in 1968. All were dry. Although plagued by bad weather, the programme was short and attracted little attention at the time, particularly as most of the news of the day was focused on the spectacular successes in Bass Strait. However, when Ampol Exploration and the Japanese explorer Japex announced plans for a well in Repulse Bay near Mackay during 1969, conservationists Australia-wide rose up in indignation. The Barrier Reef, they said, should never again be put at risk by oil explorers no matter what assurances were given about the safety of drilling techniques. As the decade came to a close the industry, the environmentalists and two governments were locked in a tussle destined to last several years into the 1970s. For the explorers themselves it was an important debate on two fronts. It was the first time in Australia that they had been asked to account for their treatment of the environment and brought face to face with conservationist opposition on such a massive scale. It was also the first real indication of unity between the local and foreign companies within their industry association APEA — a unity they had spent most of the 1960s sorting out.

Chapter 13

A QUESTION OF UNITY

At the end of 1959 about 5.18 million km^2 of Australian territory — nearly two-thirds of the continent, including offshore areas and Papua New Guinea — was either held by exploration permit or awaiting the outcome of licence applications. Queensland had the most coverage with around 2.2 million km^2, followed by Western Australia with 1.65 million km^2 and South Australia with 808 000 km^2. It dropped down then to New South Wales with 474 000 km^2, the Northern Territory with 260 000 km^2, and Victoria with 168 300 km^2. Papua was covered by 116 500 km^2 of permits. A glance at the tenement map showed that virtually all of Australia's sedimentary areas were involved, and the Federal Government was pleased with the obvious response to its subsidy arrangements. Yet, as the new decade began, there were still concerns about the country's true oil potential. Many of the basin outlines were only vaguely known, particularly in the inland areas, and numerous gaps yawned out from the stratigraphic profiles. What is more, most of the permits were huge tracts of land which individual companies, or even multi-company groups, could not hope to explore rapidly or thoroughly. The government needed assurance about the future more quickly than the oil companies seemed likely to provide. At the time, Australia was the fourteenth largest oil consumer in the world with demand moving rapidly towards 13.6 billion L per year. All of it had to be imported, and this imposed a heavy drain on foreign exchange. Discovery of indigenous oil would immediately ease the problem.

In March 1960 the Menzies administration engaged the acclaimed French body, Institute Francaise du Petrole (IFP), to carry out an Australia-wide survey of the country's oil prospects and related problems. It wanted a counterweight to the US presence and ideas of the time and, as Dr Raggatt had recently returned from a trip to the French-controlled Sahara regions in north Africa where physical similarities with much of Australia had been noted, IFP seemed the ideal choice of consultant. The government anticipated that the report would be a valuable 'second opinion' about Australia's prospects. A mission left Paris for Canberra early in April 1960 to begin a four-month review of the actual results of oil search to that time, and to study the main sedimentary basins before preparing a preliminary report on oil potential. At the outset it was realized that four months was too short a time to study all the country's basins in equal detail, so those chosen were the ones which presented the most attractive prospects because of their size, sediment thickness and the nature of the rocks themselves. The mission was led by Dr D. Trumpy, Director of the Bureau for Geological Studies at IFP, and included two other high-ranking geologists, J. Guillemot and M. Tissot. However Trumpy left the country for other business in mid-June and the work was continued by his two colleagues.

Most of the field trips were made during April and May and included whirlwind tours of the Amadeus, Georgina, Carnarvon, Fitzroy, Bonaparte, Bowen, Otway, and Sydney Basins. There were also talks and briefings in Canberra and the relevant State capitals with the companies and BMR geologists who had been working in each area. The preliminary report sent to the government in August 1960 was very pessimistic. The French researchers found that sandstone, siltstones and limestone predominated in most of the basins studied, while shales and clay played only a minor role. The two men felt that the absence of thick shale formations, which are usually the most probable source beds for hydrocarbons and also the most impervious reservoir seals, was unfavourable

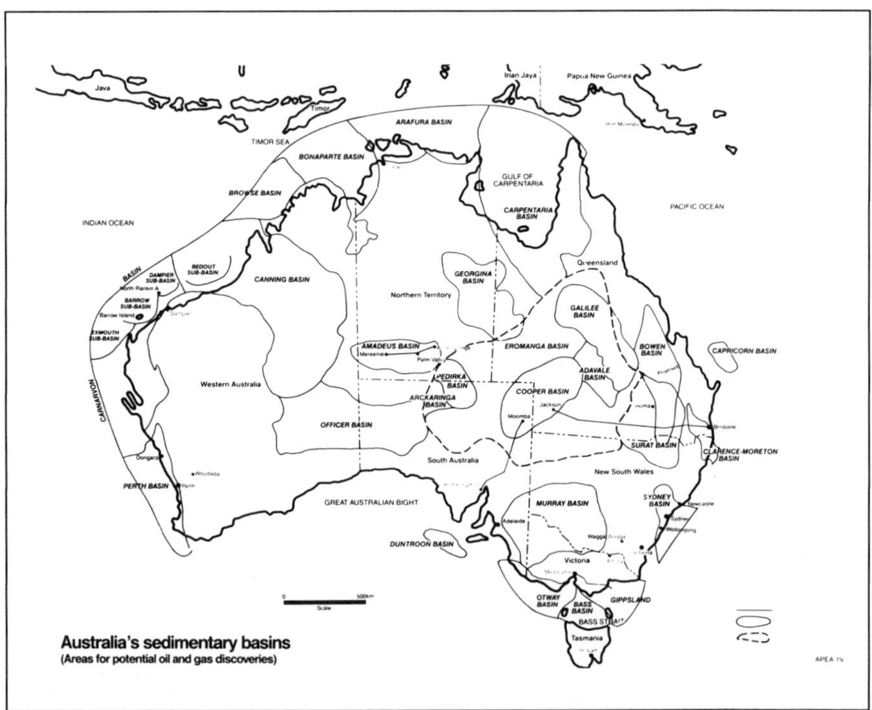

Australia's major sedimentary basins as mapped by the BMR following work which began during the 1960s.

for oil prospects in many Australian basins. They were equally disparaging about surface indications of oil.

> Apart from asphalt inclusions in the Antrim Plateau basalt and a doubtful reported oil seepage in the Proterozoic of the Bullo River [Northern Territory], no surface indications are known on the whole continent.
>
> Oil smell reported in outcrops should be better termed 'fetid smell'. They are not necessarily indications of source beds as they may never have liberated free oil.
>
> Oil shales have been reported in the mid-Cambrian near Camooweal, and perhaps in the Cretaceous of the north Artesian Basin, but they are only [seen] as facies which might indicate possible source beds in the general section.[1]

The report made detailed descriptions of the Western Australian Basins and the Great Artesian Basin, but relegated the Gippsland region to a section labelled, 'Others', along with the Eucla, Officer, Torrens, Murray, Carpentaria, Maryborough and Ipswitch-Clarence Basins. Overall the conclusion was pessimistic.

> In an effort to compare the Australian sedimentary basins with the productive areas of the world using only the information available at the present time, the choice would be made from facies and thickness of sediments on the Fitzroy and Carnarvon Basins.

Four other basins were listed as of possible interest — Georgina, Perth, Otway and Bonaparte — while not enough was known about the Amadeus and southeast Canning to venture an opinion. The IFP preliminary report also suggested that to clarify the assessment, a work programme of photogeology, geophysical surveys and stratigraphic drilling should be carried out by the BMR during the following two or three years with the aid of French expertise. Acting on this recommendation the Basins Study Group was set up within the BMR and much of the work was carried out as suggested. Dr Trumpy returned to Australia in 1963 to proffer a final report based on the mission's work in the interim. His revised list of prospective basins excluded the Georgina, but added the Surat (Moonie had been found by that stage) and the Gippsland Basins. However Trumpy was critical of what he saw as inconsistencies in oil company work practices in Australia. These included nonintegration of seismic work with drilling, poor attention to mud programmes during drilling, and a lack of standardization of logging in wells.[2]

The sum of both reports was a disappointing view of Australia's oil chances. Looked at with hindsight, it should be pointed out that oil was valued at under $2 a barrel at the time and the IFP therefore only considered the likelihood of finding large fields which would be economic at that price. Nevertheless, after receiving the preliminary version in 1960, the Federal Government hesitated before releasing the report for fear of reversing the upward trend in exploration activity begun by the subsidy scheme.[3] But the politicians need not have been concerned on that score because the local oil companies, under the banner of their fledgling organization the Australian Petroleum Exploration Association (APEA), had already taken steps to counter the pessimism. Its chairman, Reg Sprigg, led the way.

> Right from the beginning APEA was to be a technical association. It couldn't be a union, but it did need to be an industry voice. That included the need to produce a technical and scientific publication. I was particularly concerned that there was far too little known about Australia's huge sedimentary basins. What was more, the early councillors agreed that with the large permits being largely manipulated by the big foreign companies, not much progress could be expected in this regard.
>
> We all knew about the IFP study because some of us had talked with Trumpy and his associates. From this contact we had an idea that the report would be unfavourable about onshore prospects at least. What we needed was a publication to put an alternative view.
>
> I was a member of the Geological Society of Australia and the Association of American Petroleum Geologists, and I thought the best approach would be along the lines of the *AAPG Bulletins*, which have always contained thought-provoking papers on oil matters. Unfortunately the Geological Society of Australia didn't want to form a petroleum section, and an approach to the Australian Institute of Mining and Metallurgy was equally fruitless.
>
> Then we hit upon the idea of an annual oil conference where we could present papers and have them bound into a volume for distribution not only to members of APEA, but to a wider audience including government, the foreign companies and the press.[4]

The first conference was held in Melbourne's Hotel Australia in Collins Street for two days in March 1961 where 15 papers were presented to an audience of 70 delegates. The talks were primarily geological and technical, but it was the first time that information about Australia's oil exploration industry had been circulated so publicly. Nevertheless, the transition from speeches at the rostrum to the printed page of the *APEA Journal* was not without early difficulties. Alan Prince, a foundation member of the Association and also publisher of the *Australasian*

Oil and Gas Journal at the time, was instrumental in bringing the papers to press.

> We were very late with delivery of the journal in the early years, unlike the current system where papers are all organized and printed well before the event. The first conferences were put together by the Council itself and, with everything else to do, preprints were not possible.
>
> I can remember a number of times when I had to run to catch a speaker as he left the stage to obtain a copy of his paper, only to find he had spoken to notes. That meant I had to make sure he sat down later and wrote out his full text for publication.
>
> Martin Glaessner from Adelaide University helped enormously by editing the early volumes. Later on, when we had a secretariat headed by Ken Horler, this side of things became more streamlined.[5]

Admittedly the Melbourne conference, and the second held in Sydney the following year, contained a large element of propaganda to try to counteract the pessimism of the IFP report. The idea was to let the general public share the APEA conviction that oil was there to be found. As the organization settled into an annual routine, the speeches and technical presentations broadened, and did a great deal to fill in the empty spaces on Australia's geological maps. The published data also supplemented and expanded the more academic output from institutions like the universities and the Geological Society of Australia.[6] Despite this, the fledgling organization did not receive instant approval from the Federal Government. Far from it. In the first few years APEA literally had to fight for recognition, and officially the Association was regarded by the government and major companies alike as a band of rebels. Reg Sprigg and the Council had many a 'run-in' with Canberra.

> The majors were agin the organization and they told the government not to take any notice of us. They [the majors] said that Australia would be looked after from their international networks. If it came about that oil must be found in Australia, then the big companies would find it. They regarded the small local explorers, pretty well all of whom were members of APEA, as a 'bloody nuisance'.
>
> In fact, often if a contentious issue arose, the majors came down on a different side of the fence to the smaller companies. At times they even undermined the APEA position by arriving in Canberra a day early and hopping into the Minister's ear before we had a chance to state our case.[7]

Although judged a success, the first APEA conference was marred by the absence of any BMR or government representatives, all of whom had been forbidden to attend by Canberra. The stalemate was tinged with irony because both APEA and the government were working towards the same goal — indigenous oil discoveries. Clearly the impasse would not be overcome without some conciliatory moves from one side, or both. The first sign of change actually came from within the industry's own ranks. Despite a small attendance, the first APEA conference had shown that dissemination of geological information in such a manner was of obvious benefit to all companies, large and small. Soon afterwards, the discovery of Moonie field pointed strongly to the benefits of local and overseas explorers actually working together as AOG and Union Oil/Kern County had done. APEA members began to think again about admittance of the foreign companies to their ranks. The process was given further impetus by the Federal Government pointing out that APEA's opinion was hardly representative if it only spoke for half the industry in the country.

In May 1962 the Association decided to send membership application forms to the foreign companies with Union, Delhi, Arco and Phillips on top of the list. A subsequent resolution was passed stating that any company with a registered oil permit in Australia, as well as any company spending money to explore for oil in the country, would be eligible to join APEA provided there was a two-thirds majority vote of acceptance by the Council. Straight away the Association became a much steadier, stronger and more functional group. The change in status was marked in his own inimitable way by Senator William Spooner, the Menzies Government's long-serving Minister for National Development, at a dinner in Canberra held by the APEA Council. Rees Withers was one of the hosts.

> We'd just put out a brochure which detailed the shortcomings of the government in its lack of support for the industry and the possible ramifications if this persisted. It was included in a number of companies' annual reports and, as such, was not a direct lobby to government. Nevertheless it soon came to the Minister's knowledge.
>
> We then deliberately held a Council meeting in Canberra to make our presence felt even more strongly, and in the evening held a dinner for a carefully selected list of parliamentarians. It was a magnificent spread and everyone enjoyed the meal. Reg Sprigg then rose, as APEA Chairman, to welcome the guests and Bill Spooner was quick on his feet to respond.
>
> He opened with suitably complimentary remarks about the dinner, APEA's generosity and the kind invitations. Then, without actually referring to the brochure, he added that,

> notwithstanding his prior comments, he thought we were a 'mob of bastards' and as far as he was concerned 'the gloves were off'. He concluded by saying not to expect any favours, but that he would listen.
>
> From that day the ban was lifted on BMR and government attendance at the annual conferences. The industry greatly benefited from resultant government input.[8]

Spooner had developed a kind of grudging respect for the battles of the young, local explorers. But it was not until APEA became a more unified voice of the industry that he could give them evenhanded attention in his official capacity. Little did the explorers realize they would be back pounding on his door in anger and frustration before the year was over because of the government's move to cut the oil search subsidy.

The post-war Petroleum Search Subsidy Act had been a moveable feast from the time it was first conceived in 1957. At that point it provided for a government payout of 50 percent on all approved stratigraphic drilling costs. In 1959 a new Act widened this to include 50 percent of geophysical survey and well logging operation expenditure. In 1961 an amendment was passed to provide for the inclusion of detailed structure drilling and test drilling as subsidizable categories. This latter move was in recognition of the fact that in some areas of the country exploration had passed the regional stage and was homing in on specific targets. The amendments also covered detailed geophysical work but, at the same time, reduced the period for which information should be kept confidential from 12 months to six months. A further change gave operators the choice of being paid for drilling operations as a percentage of costs, as before, or as a fixed amount for each foot drilled.

In the wake of the Cabawin and Moonie oil discoveries in Queensland, both of which were made by subsidized wells, the government looked again at the scheme and decided that the payments need not be continued on such a generous scale. Exploration grants had increased markedly as drilling activity responded to the finds but, equally, officials felt that matching the pace of this activity with a 50 percent share taken from the public purse was not warranted. Choosing the August 1962 Federal Budget to make its announcement, the government cut the maximum amount of subsidy from 50 percent to 30 percent.[9] The move sparked bitter criticism from the oil explorers and, if anything, tightened the new-found unity among the companies within APEA. The Association's Council issued a statement saying that exploration programmes amounting to millions of pounds had been thrown into confusion and that the Federal Government must take full responsibility for cancellation of overseas deals then under consideration. An article in the *Australian Financial Review* gave full rein to APEA's protests.

> The reduction is 40 percent which means £200 000 in every £1 million programme which, in effect, must result in reduced drilling operations.
>
> Current activities by overseas explorers have been based on a consistent and reliable government policy. The government has dishonoured its undertakings and taken refuge behind the 'small print' of the Act.
>
> At no time has any alteration of this policy been indicated to the industry. The action is very badly timed, and confidence in the government's integrity is damaged at a time when a number of overseas oil visitors have come to Australia to a special conference dealing with oil matters.
>
> The Prime Minister, Mr Menzies, since the Liberal rout in the December elections, has repeatedly stated that the industry's opinion would be sought in matters of vital legislation. Without wishing to know the content of the Federal Budget, it would have been hoped that in a matter of such vital importance, the industry would have been given an opportunity of being consulted.
>
> A canvas of industry opinion by the Association has revealed a general reaction of disgust and alarm for government stop-go policies of which this oil measure is but one reflection.[10]

However the industry's protests fell on deaf ears and were in fact muted by the run of discoveries round the country during the following two or three years. Those same discoveries caused the government to review its policy yet again in the face of the steep rise in subsidy payments, particularly as offshore costs were much higher than those onshore. In October 1964 it was announced that operations would not receive a subsidy if they were within a 48 km radius of an oil discovery or a 32 km radius of a gas discovery. There was also a £200 000 limit imposed on the amount of subsidy payable for any one drilling operation as a further attempt by government to keep within the limits of the subsidy's allocated budget. Explorers sneered at what they called 'the government's silly circles', but they could do little to change the ruling. Just prior to the 1965 Federal Budget, Reg Sprigg, in his final year as APEA Chairman, called on the government to increase the subsidy budget from £5 million a year to £15 million. In particular, he asked that the 'circles' round existing discoveries be withdrawn.

> Many local exploration companies are facing crucial fund-raising difficulties which can only be overcome by increased subsidy... The restriction [of no subsidy round a discovery] has imposed a limitation on exploration in 'success' areas where discoveries have been made and which should be followed up with intensified drilling.

> The equity of Australian companies in exploration tenements is steadily being eroded because they have been compelled to surrender to overseas groups their interests in highly prospective areas, simply to raise funds for continued exploration.
>
> The oil share market has declined sharply despite an improved success ratio and the upgrading of tenements by widespread discoveries of oil and gas across Australia... But success is broadly a function of drilling intensity.
>
> Although the ratio of success to the number of wells drilled in Australia matched experience in North America, the rate of exploration is far too slow. The present drilling rate should be increased at least tenfold to discover and establish significant reserves.
>
> But the exploration industry must be given adequate long-range financial incentives and freedom from controls to achieve even 50 percent self-sufficiency [in oil].[11]

The government's Budget response — a price incentive of 75 cents a barrel over the parity price of oil imports to any successful producer — did help increase exploration activity a little during the middle years of the decade. But Sprigg's impassioned plea had also betrayed the beginnings of a crack in the solidarity forged between local and overseas companies only a few years before. It was true that some of the small Australian explorers had been included in the new discoveries, but the lead had been taken by the foreign majors in all cases. Once discoveries had been made, the big companies were quick to seize the initiative. The smaller locals were compelled to sell down some of their assets to match the pace and expense of the exploration programmes. What is more, as the discoveries continued, attention began to shift from purely exploration matters to those of development and production, and the financial commitments of the Australian companies involved were stretched to the limit. Understandably too, the government focused more attention on the foreign companies as the potential 'providers' of Australia's growing oil and gas reserve.

Despite the presence of Australian company BHP in the Esso/BHP joint venture in Bass Strait, the huge oil discoveries at Halibut and Kingfish brought matters to a head towards the end of 1968. The sudden success made Esso a major force in the country's future, while the Gippsland Basin became a bias of government and public attention. Esso, with BHP in tow, was keen to pursue its own line in pricing arrangements for Bass Strait oil production. It negotiated a fixed price, five-year contract with the Gorton Government set at a level above the (then) current world oil price. Although agreeing to that, the government withdrew the 75 cents a barrel incentive for successful explorers. The old rivalries between local and overseas companies quickly festered into an open wound and, for a short time late in the decade, there was a real possibility that Esso would withdraw from APEA. To make matters worse, it was considered likely that the other majors would also move out and start up their own association.

Alarmed at the way events were heading, the Council members cast around for someone who could weld the two camps back together. After Sprigg's long occupancy of the chair, Earl Abbott from the US major Conoco had been elected for a two-year term beginning in 1965, followed by Bill Siller of local company Exoil for a year. But the volatility of the situation in 1968 called for someone who would be acceptable to both foreign and Australian companies. The Council found him in the person of David (Dave) McGarry from Australian Oil and Gas Corporation. Because of his experience in working with Union Oil in the US as well as with AOG back in Australia, McGarry was able to mend the wound. Eventually, the majority of the industry realized that the era of subsidies for exploration was drawing to a close and that there was no point in fighting over the issue any longer. An amendment to the Subsidies Act in June 1969 had already limited its availability to the onshore areas. Its only possible application offshore was to the cost of an Australian contribution to an operation, and even then it depended on the extent of the Australian content. The footage basis of payment was also withdrawn.

Although the subsidy scheme was not officially terminated until 1973, to all intents and purposes its effectiveness as a stimulus to exploration had disappeared by the end of the 1960s. Even so, a total of $150 million had been paid out by the government during its 15-year operational period, the bulk of that going to Western Australian, Queensland and Papua New Guinean exploration programmes.[12] When introduced back in the late 1950s, there was no doubt that the scheme encouraged oil exploration and led to the first discoveries in the 1960s, and it may be argued that the actual finds then kept the momentum going during the later part of the decade. But undoubtedly the biggest single gain provided by subsidy legislation was the amount of geological information which had been made available to assist companies in immediate and future assessments of the country's potential. A good illustration was the decline in the total area of the country under exploration lease between 1962 and 1968. Geological consultant John Blumer explained what appeared at first glance to be a paradox in a paper delivered to the 1982 APEA conference in Sydney.

> The decline... is partly an expression of the rapid improvement in understanding of the regional geology of Australia over the period. This improvement was due primarily to the very large amounts of subsurface geological and regional geophysical data acquired in the course of exploration operations and disseminated widely as a result of the subsidy scheme.

> ... [in] 1962 very little was known of the extent, thickness, age or even the existence of many Australian sedimentary basins. Thus much of the exploration effort of the early 1960s was directed at determining whether or not a reasonable depth of section was present within a permit, and many areas were written off at an early stage.[13]

By the middle of the decade geologists had recognized 25 sedimentary basins in Australia covering a total area of 3.9 million km². Two of them, the Great Artesian and the Canning, could be further subdivided into six and two smaller basins respectively. Producing fields or potential producing horizons had been discovered in eight of these basins, while significant hydrocarbon occurrences had been noted in many of the others.[14] In Papua New Guinea, three main basins had been recognized covering an onshore area of more than 298 000 km². The offshore limits of each were less well-defined, but two of them contained numerous onshore oil and gas seeps and some discoveries.[15] The basins varied greatly in size, age, nature and structure, although most in Australia had remained relatively undisturbed by folding and faulting. The mid-1960s geologists saw that this stability did not necessarily detract from the country's oil-bearing potential. Nevertheless, they were gradually able to weed out the regions where sediments were too thin or at too shallow a depth, or where their original properties had been altered by intrusions of igneous rock. In PNG the task was more difficult, partly because the rugged terrain still made access a problem, but also because the tectonic activity which had thrust and folded the mountain chains made subsurface interpretation very uncertain. Thus much of the work in these regions was guided by the location of the same hydrocarbon seeps which had first attracted the attention of explorers 50 years earlier.

Offshore, the continental shelves round Australia and Papua and New Guinea also presented geologists with a relatively blank page at the beginning of the 1960s. But as the decade progressed the picture began to develop rapidly. In some ways, expense aside, it was easier to collect substructure data over the sea-covered permits because there were no physical obstacles in the way of the seismic or magnetometer surveys. The lack of outcrop material did present some difficulties, although seabed cores could be obtained at various points enabling some correlation with information collected along the adjacent coastlines and islands. Probing the offshore areas with numerous surveys during the 1960s, geologists found that the Australian continental shelf varied in width from about 20 km off southern New South Wales to more than 320 km in the north and northwest. It covered an area of approximately 2 million km² and much of it appeared to be prospective for hydrocarbons.

During this period the global theory of plate tectonics evolved with the world's oil industry in the forefront of its development. Associated with 'continental drift', it suggested that the earth's crust is made up of 'plates', the margins of which are marked by active seismic (earthquake and volcanic) zones. The plates 'float' on the earth's liquid inner magma and move apart in response to convection currents welling up from the molten core. As they do so, a new continental margin begins to form where the older continental cratons break up. The initial crack develops into a rift valley which is then filled with marine sediments. As separation continues, a new mid-ocean ridge is formed, the peaks of which are visible as many of today's islands, such as those in the central Pacific region. The theory became a valuable 'tool' in interpreting and understanding the development of continental margins and the evolution of continental shelves. During a series of lectures delivered round Australia in 1986, geologist Murray Johnstone referred to plate tectonics as the first revolution in modern geological thinking.

> Although the published accounts of the development of the theory give credit to the recognition of magnetic striping of the sea floor near mid-ocean ridges as the clue to understanding plate tectonics, we must remember that the massive acceptance of the concept was headed by the exploratory oil companies. This industry was also the one which gained most from it.
>
> The shooting of extensive seismic surveys over many continental shelves during the 1960s (including Australia) showed that the structural and depositional styles on the shelves of passive margins [where the moving oceanic crust contacts the the continental crust] were almost identical worldwide, and that even the timing of major unconformities was identical on both sides of a 'spreading' ocean, such as the South Atlantic.
>
> ... plate tectonics provided a mechanism to explain the unconformities observed in marginal basins on passive margins, and to help determine the nature of the sedimentary packages between these unconformities. It broadened our perceptions from our local shelf environment to similar shelves round the globe to provide analogs for traps, source sequences, and [oil] maturation history.
>
> Plate tectonics gave us reasons for the deposition of certain types of sedimentary sequences at specific times in the development of marginal basins. It furnished a simple answer to the enigma of thick sedimentation coming from an area which is now a deep ocean basin.[16]

In Australia, oil exploration, particularly offshore, went ahead in leaps and bounds spurred on by a growing understanding of the geology. The Shell Oil Company in particular actively pursued the theories of sea floor spreading and continental breakup, focusing on the Great

Australian Bight and Otway Basin as part of its worldwide investigations. In the process the company ran 10 000 km of aeromagnetics, 24 500 km of marine seismic and drilled three wells — unfortunately with little success.[17] So swift was the expansion of activity offshore Australia, however, that governments struggled to keep up and there was disunity between the States and Canberra over which would, or could, provide a sound legal framework for the continental shelf search. The sovereign rights of a country over its shelf were defined at the Geneva Convention in 1958 and, unless otherwise agreed, the limit of rights between two countries was set down as the median line drawn equidistant from each coastline. Internationally this only affected Australia's far northern areas. But locally, when oil explorers in Australia first began to glance offshore during the late 1950s, the individual States went ahead and issued oil search permits over their territorial seas (a distance of three miles [4.8 km] out from the coast) and beyond.

Some doubts arose, however, about the validity of this move, and the six States extended the operation of their onshore Petroleum Acts to the offshore sector, although no one was particularly sure of the legality of onshore rules being applied to offshore permits. The Federal Government quickly raised a counter-claim for jurisdiction over the continental shelf and raised the possibility of a claim over the territorial seas. Finally, in order to resolve the constitutional dilemma and draw up a suitable code for offshore operations, the States and the Federal Government began discussions in June 1962. Two years later (17 April 1964) a joint announcement was made to the effect that the individual States would administer offshore operations within their boundaries, and applications for permits would be granted by the States subject to approval by the Federal Government. Royalties were to be shared on an agreed basis. Harry Taylor-Rogers at the BMR worked on the technical side of the proposed new legislation.

> It was a very big job. I had John Livermore with me. He was a man who could pick out holes in an argument with clarity and accuracy and it was good to have him on the team trying to anticipate loopholes in the Bill. Ian Morley, the State Engineer with the Queensland Mines Department, was involved too. He was known as 'the Dictator' because of his hard-line approach to State regulations and determined attitude when his mind was made up. But a delightful man when you got to know him. He was always able to simplify things.
>
> We also had Bruce Lambert, the head of National Mapping, to help on the practical aspects offshore. For instance, Tasmania wanted to move its border with Victoria a little further north as the finds in the Gippsland Basin were close to the line.
>
> Clarry Harders, the Australian advisor to the Law of the

> Sea conferences, was involved and John Qualtrough Ewens, known as John Q, drafted the parliamentary Bill.
>
> There was much visiting, preliminary discussions and note taking. Everything then went to the Department head where legal questions were thrashed out before the individual States became involved.[18]

A Ministerial statement in November 1965 gave details of the objectives, but these were immediately attacked by the oil explorers as harsh and restrictive. In particular, the companies already working offshore objected to what they saw as a lack of security of tenure over permits, the failure to provide a fixed rate of royalty on any resultant production, and the possibility of enforced forfeiture of part of a discovery. Modifications were announced during June 1966, but the explorers were still dissatisfied. Nevertheless, Prime Minister Harold Holt announced in December 1966 that agreement had been reached between Canberra and the Victorian Government to enable Esso/BHP to develop its new discoveries in Bass Strait as soon as possible. The other States were pressured to fall into line and formal agreement between all parties was signed on 16 October 1967. The relevant legislation came into force on 1 April 1968 and the earlier State legislations asserting control over offshore areas were repealed.

Despite this there was continued dissatisfaction with the outcome. Some said Canberra had abdicated authority over offshore areas and others said the States had handed their responsibilities to the Federal Government. Legal opinion suggested the agreement between the Federal and State Governments was unenforceable by law and could be breached at any time. On the other hand this was seen as politically unlikely, particularly as the States would have strong economic reasons for abiding by it. The oil explorers had very little say in these proceedings and, to some extent, the fairly obvious bias towards ensuring a prompt start for Esso/BHP's Bass Strait production added to the friction between local and foreign explorers during 1968. On a purely operational front, however, the legislation set the ground rules for exploration offshore.

The basic unit of a permit was what is known as a graticular block (adopted from the system at work for the North Sea) and it measured five minutes of longitudinal arc by five minutes of latitude. In northern Australia the area equalled about 78 km^2 but, because of the convergence of meridians, a block offshore Tasmania measured about 60 km^2. Permit applications could be made for between 16 and 400 blocks in a single area, but had to be accompanied by a proposed work programme and an estimate of expenditure, along with the technical qualifications and financial resources of the applicant. The Designated Authority (State Mines Minister with approval of the Federal Minister) could then offer the permit subject to specific conditions, the payment of £5 000 security and a time limit for acceptance by the company or group of explorers concerned.[19]

During this period of political cut and thrust within and surrounding the oil industry, not all members of APEA were keen on being involved solely in the Association's lobbying functions. Early in the 1960s the Council conceived the idea of forming a more social Petroleum Club, which met once a month in the Airforce Officers Club premises in Sydney, often with a guest speaker. It might have continued in that form for longer had it not been for the American Association of Petroleum Geologists declaring its intention to set up an Australian branch, also in Sydney. The local oilmen felt this was unacceptable, so the Petroleum Club decided to head off the American organization's move by formalizing itself into a more technical body. As a result, the APEA Professional Division was formed during 1967 and went on to hold seminars and training lectures around the country. A branch was formed in each State including Queensland, where it moved in beside the still-flourishing Brisbane-based organization known as QUPEX (Queensland Petroleum Exploration Society). The Professional Division of APEA was also separate from the oil drilling companies which, late in 1964, had formed their own group called the Oilwell Drilling Contractors Association of Australia with Richard (Dick) Dumbrell of Oil Drilling and Exploration Ltd as first president.

The APEA parent body supported all the State Professional Divisions and they rapidly became an integral part of the Association — the annual conferences in particular. Nevertheless, as the membership grew, so did the general belief that the political wing of the organization should be kept completely separate from the technical aspects of the industry. Pressure for division grew as the decade came to an end and eventually, in 1974, the professional wing was hived off as a self-funding and independent body with the new name Petroleum Exploration Society of Australia. Although the two groups remained closely linked, from that point on membership of APEA was limited to companies only, while individual employees joined PESA.[20]

There was certainly no shortage of new and interesting technical innovations to occupy the professional division during the 1960s, particularly in the realm of geophysics, where a number of advances in acquisition and interpretation of data had been made. Nowhere was this more dramatic than in the running of seismic surveys. Instead of a simple dynamite explosive being used as the source of energy, the industry developed a choice of sources which operated with similar explosive force. They included compressed air, gas explosives, electric sparks, dropped weights and pneumatic or hydraulic machinery, the latter of which literally 'shook' the ground. Trade names like dynapulse, aquapulse, vibroseis and hydroseis became common to the geophysicist's vocabulary. In addition, the signal detector (the geophone) had evolved from the 1950s-style delicate but cumbersome package weighing about 9 kg to a durable miniature. Amplifiers to magnify the signal picked up by the detector could, by the mid-1960s, handle a much wider range of natural frequencies. What is more, they could eliminate much of the distorting unwanted 'noise' picked up from other sources, such as wind onshore and wave action offshore. The data itself, formerly recorded by pen or light-sensitive paper, was now recorded on magnetic tape which could be replayed many times. As the decade moved forward, the first use was made of digital recording instruments allowing accurate capture of large amplitude and frequency ranges. This naturally led on to digital processing of the information and, towards the end of the 1960s, operators were able to go a step further in improving the signal-to-noise ratio by introducing digital filters into the processing system. These could be mathematically tailored to the particular problems of a given area.[21]

With the advent of offshore work, there was a need to accurately know the position of the recording vessel at all times. Added to this, the geophysicists needed to feed the ship's velocity (both speed and azimuth) and the relative positions of the traverses into their calculations. Navigational aids thus became a vital part of the seismic operation and three types of systems were in use during the decade. Sonar doppler measured the shift in a sonic signal transmitted to and reflected from the ocean floor and calculated a ship's velocity to an accuracy of better than 0.1 knots. Also, very low-frequency radio systems made use of a worldwide network of radio stations transmitting on frequencies of less than 20 kilocycles per second. Geographical accuracies of half a mile were achievable at certain times of the day. More localized medium-range systems like Decca, Lorac and Toran used a three-station system (one master and two slaves) with position fixing done on the triangulation principle. Also in this class were the pulse systems like Loran and Shoran which measured the time taken for a radio pulse to reach a shore station and return and then converted it into a distance reading. But by far the most accurate 'fixes' were obtained by use of satellites which continually radioed time and orbital position back to earth. Signals could be monitored on board ship to locate the vessel to within 30 m if necessary. By the late 1960s there were sufficient satellites in orbit to ensure a position fix every 1.5 hours.

Through its annual conferences and the individual State meetings of the professional division, APEA was able to air such topics along with other advances in the fields of photogeology, geochemistry and drilling technology. The Association, for all its political machinations, was fulfilling the technical role that the original Council had intended. By 1969, David McGarry's leadership of the Association had healed, or at least soothed, the jealousies and conflicts within its ranks. Success in the field had come both on and offshore and the country was in the grip of a heady resources boom which spanned the mining sector as well as that of oil. The main murmur of disquiet as the end of the decade approached was a concern that Australia had not trained enough geologists, geophysicists and petroleum engineers of its own to cope with what had suddenly become a bustling industry. It was a concern voiced by McGarry himself at the time.

> Those few Australians who have worked with the industry during this time [the 1960s] can be considered pioneers and have been supplemented by some returning nationals who have worked overseas. However few graduates of Australian universities could be considered as adequately trained for petroleum exploration and development at this stage.
>
> It will be noted that 201 geologists are employed in Australia in petroleum exploration, and of these, 103 (or 51 percent) are from overseas. Indications are that around 50 percent of these overseas personnel would be replaced immediately or within a short period of a few years by new Australian graduates if these people were available to take their place.
>
> The position in respect of geophysicists is even worse. Ninety-nine geophysicists are employed in Australia and 77 percent are from overseas. Indications are that 27 of these overseas people could be replaced within a short period by graduates from Australian universities.
>
> Seventy-nine petroleum engineers are employed in Australia and only 24 Australians, predominantly trained men returned from overseas, are contained in this grouping.[22]

McGarry lamented the fact that there was a disproportionately low rate of graduates moving to private industry jobs seemingly because a large number remained at university to take higher degrees and eventually pass into research and teaching positions. Others went into the State and Federal Government Geological Surveys. He added that the industry needed people to come to petroleum exploration with a sense of optimism, inspiration, a knowledge of technical matters and also of the financial aspects of various projects. It was up to industry as well as the educators to promote the desired interest and standard.

So ended the 1960s, a decade that became known as the golden era of Australian exploration. With nearly three billion barrels of oil discovered in a four-year period, predictions of complete self-sufficiency within a few more years were heard from all sides. Training Australians to take the reins was not regarded as an insurmountable problem, and even the likely phasing out of the subsidy scheme was met with a shrug from many quarters. Overall, the country's explorers faced the 1970s with confidence, unaware of the major hurdles that would soon stand in their way.

PART FOUR

THE 1970s

A FEAST IS NOT FOREVER

Chapter 14

BOTH FEET ON THE TREADMILL

In 1970, with the oil successes of the 1960s still fresh in their minds and the nickel-fed mining boom raging round them, most Australians felt sure the country's resources would lead to unparalleled prosperity in the decade to come. After all, the explorers only had to poke the ground to find riches awaiting. The nation, many believed, had finally found its feet and was off and running. Certainly many of the finds were real enough but, in the case of oil, they had come with such suddenness and apparent ease that few people recognized that such riches might be a seductive trap. Not understanding, or choosing to forget, the exceptional good fortune surrounding the discovery of Bass Strait fields, expectations were raised for more of the same. Plans for the economy were made based on the establishment and maintenance of long-lasting self-sufficiency in indigenous crude supplies. Before long this developed into an unrelenting need. Australia was indeed off and running, but several years elapsed before an awareness developed that the country had landed squarely on the broad, endless belt of a treadmill.

Within the industry itself, some did try to put exploration and success back into perspective. Addressing the annual APEA conference, held in Surfers Paradise during March that year, Assistant Director of the BMR's Petroleum Exploration division, L.W. Williams, said that Australia's annual expenditure on oil exploration would have to double if self-sufficiency was to be a serious goal. At the time, the search budget was running at about $90 million per year. Production from three regions — Bass Strait, Barrow Island and Moonie — was judged capable of supplying around 65 percent of the country's oil requirements in the near future. But Williams claimed that to achieve and maintain complete self-sufficiency it might be necessary to find up to 10 billion barrels of oil reserves during the 1970s and 1980s. Moreover he was not confident of reaching that target.

> At the present average finding cost of 35 cents a barrel this will require an average annual expenditure of about $175 million during this period, which is about twice the present rate.

> It is difficult to see this type of expenditure being forthcoming unless further major discoveries are made, even with continuing government incentives... It must be emphasized that in Australia's present situation even one year without a discovery increases the average finding cost of oil by 5 cents per barrel.[1]

Williams took the statistics and probabilities further by suggesting that the $90 million to be spent in 1970 and the 35 cents a barrel finding cost would theoretically result in the discovery of 250 million barrels during the 12 months. Although that was greater than both indigenous production and local consumption during 1970, he stressed that Australia's consumption was increasing at a substantial rate — some estimates putting it at 9 percent per year. It would soon outstrip the production volumes. Nevertheless the pace of discoveries continued at a reasonably fast clip during the first years of the 1970s, particularly onshore, as new companies floated in the late-1960s boom began to carry out their

Location of key places mentioned in text.

farm-in obligations and take advantage of what was left of the Federal Government's subsidy scheme. Much of that action was directed to permits in the Surat/Bowen and Cooper Basins of Queensland and South Australia.

1970 saw the AAR group active in the Bowen Basin — round Roma, and further north with a gas discovery at Rolleston in an area now known as the Denison Trough. In addition, the Queensland Geological Survey began a series of about 20 holes in the Denison-Surat regions to stimulate commercial interest, deliberately selecting sites which would advance the geological data base of the area. Also in 1970, Hartogen confirmed an old Union Oil gas discovery at Kincora to the south of Roma and more gas was discovered by Delhi/SANTOS and its farminees in the South Australian Cooper Basin where Delhi found Toolachee in late 1969. Alliance drilled Merrimelia No.5 in 1970 and Pursuit found Strzelecki and Della gas the same year. In 1971, Delhi discovered Big Lake gas field and followed up with gas at Brumby, Brolga, Epsilon, Burke and Dullingari in 1972, while Vamgas drilled Kanowana early in 1973. The best prize went to Cooper Basin farminee Bridge Oil when it found Tirrawarra oil field in 1970 and followed that up with nearby Moorari oil in 1971. It then drilled the smaller Fly Lake oil field the same year.

The appearance of new company names on the permit maps during this period was matched by the disappearance from the scene of the longstanding Surat duo, Union Oil and Kern County Land. Kern had departed in 1967 as a direct result of unwanted takeover action in the US by Armand Hammer's Occidental Petroleum. To keep Occidental at bay, Kern solicited the help of another US heavyweight called Tenneco Oil Company, which then came in as a 'white knight' and made a more acceptable offer. However once the dust had settled, Tenneco was more enthusiastic about Australia's mineral potential than it was about the Surat oil inheritance from Kern and eventually onsold the permit interest, including its share of Moonie field production, to a local company called International Oil in 1970. International Oil was part of a stable of Australian companies run by entrepreneur Alec Dodson including Amalgamated Petroleum, Timor Oil and Apollo Minerals, and part of the deal was for Tenneco to take a position in Apollo.

The new regime brought an American, William (Bill) Wells, to Australia as manager of the International group. A reservoir engineer by training, he was no stranger to the Australian scene, having evaluated the Moonie field for Tenneco in 1967 before going on to become vice-president of the company's mineral division prior to his transfer. When he arrived to look after the International group, Union Oil was still in charge of the Surat programme, but Wells quickly saw that its management was beginning to despair of ever becoming a major oil producer in Australia.

> At the time Union was tired. Although it had Moonie, Cabawin, Alton and Bennett to its credit, they were small fields

> and the company had drilled over 140 holes elsewhere in the Surat without success. As big companies were wont to do, Union continued a policy of drilling anticlines in preference to the more stratigraphically controlled buried river channel sands in the region. The trouble was that the anticlinal 'bumps' were of ever-decreasing size and there was less and less chance of a big discovery.
>
> Union had found Kincora gas field in the 1960s, but didn't think much of it at the poor gas prices [around 25 cents per thousand cubic feet] of the time. Yielding to Tenneco's influence in the consortium, Union farmed out the northern parts of the acreage, including Kincora, to Pexa Oil [which in turn was bought by Hartogen in 1970].
>
> Bridge Oil struck a farm-in deal for the southern sector eventually finding Silver Springs gas/condensate field in 1974 and Boxleigh field a couple of years later. But Union held on to the discovered oil fields until finally the company decided to withdraw from Australia altogether around 1973. Both Australian Oil & Gas Corporation and Bridge Oil made offers, but International Oil stepped in to buy Union's interest in Moonie and the satellite oil fields to make its ownership up to the full 100 percent. AOG retained a 20 percent Net Profits Interest. At that point I suggested International's name be changed to The Moonie Oil Company Ltd.[2]

Another, more personal, link with the Surat pioneering days of the 1960s was severed with the death in October 1970 of T.W.H. (Bill) Dee, who had retired as managing director of AOG 12 months earlier. Friend and colleague Eric Rudd described him as a quiet and dedicated person who avoided the limelight to such an extent that few Australians knew of the contributions he made to the country.

> His talents were many but he was by training and profession a surveyor. In 1922 he went to the site of Canberra as a cadet with the Department of Interior and he stayed there for nine years. During that time he passed his Land Surveyor's examination and was attached to the newly created Federal Capital Commission. In this capacity he participated in the very extensive, detailed and difficult survey work for the layout of Walter Burley Griffin's plan for Canberra.
>
> In 1931 he joined the staff of Oil Search Ltd . . . and became the focal point of that company. [He] never lost his faith in the possibility of oil in Australia. Of all the team, he was the one who stayed with it through the difficult years of the late 1930s.
>
> As managing director of Australian Oil & Gas Corporation he played an important part in the ultimate successful

> discovery in Australia... Largely because of his own personal knowledge of the Roma district he had been chiefly responsible for the move into the Surat Basin. Throughout all the excitement of Moonie, Bill Dee worked quietly in the background ready at any time to explain or discuss any aspect of the oil business or his company with anyone who cared to ask... He treated all people with the same thoughtful consideration.
>
> He made many friends in the oil business and it was noticeable that oil men who are always on the move were never too busy to look up Bill Dee after his retirement to Rose Bay.[3]

Dee was probably something of an exception with his quiet ways in what was still very much a rough and tumble industry. His conservative approach was in marked contrast to that of aggressive entrepreneurs like John Fuller, Lou Smart, Bill Siller, Reg Sprigg and more recent arrivals like Paul Strasser, the property developer who piloted Bridge Oil's early years, and Pat Burke, the chartered accountant who took over the reins at Hartogen in the early 1970s. There was no quarter given in the field either, as Bill Wells testified.

> Out there in the Surat, when I arrived to look after the International group, the drilling crews were the meanest, toughest bunch of bastards I have ever come across. Some were Australians but there were also Americans and Canadians.
>
> There was one Canadian toolpusher who used to meet the plane bringing in a new load of roustabouts and roughnecks. He'd stand on the strip near the aircraft door and eyeball each one of them as they stepped down. Making a snap judgment, he'd have no hesitation in ordering some of them back on the plane and out if he decided they would not make a go of work on the rig floor.[4]

It was no place for the faint-hearted and, as far as company managers in town were concerned, certainly no place for a lady. But the late 1960s and early 1970s was a time of change in the industry, with women beginning to join previously male-dominated professions like geology and engineering. In 1971, geologist Ingrid Campbell became the first woman in Australia to step into the daily working environment of oil exploration drilling. Surprisingly, there was more opposition from head office than from the rig floor.

> The well was Horrane No.1 in the Surat near Dalby in south-east Queensland. At the time I was working for Amalgamated Petroleum and there was a feeling among some of the managers in the company that a woman would not be able to cope

> with supervision of a well on her own. The silly thing was that when I did convince them to give me a try, the men on the rig itself accepted me with very little fuss.
>
> They were initially interested to see what sort of person I was and what sort of job I could do, but only in the same sense as they would be curious about any new workmate.[5]

Completing her geology diploma at the Royal Melbourne Institute of Technology in 1970, Campbell had been the only woman in the course, so she was used to working in a male-dominated environment.

> For that well the crew had accommodation at the local pub and I stayed at a nearby homestead. The family there were good to me even though they couldn't really understand what a woman was doing in such a job. The fact that I played classical music on their piano and smoked small cigars raised their eyebrows even more.
>
> I had to fight the same battles when I joined Hematite Petroleum [BHP] the following year. There was no way they'd let me offshore in Bass Strait for more than a day trip, but eventually the company did send me up to central Queensland for a couple of wells round Winton. Lovelle Downs No.1 in 1972 was one of them.
>
> I stayed with the Elliots at Lovelle Downs Station and drove 20 km or so to the rig each day. They were a lovely family. The kids used to tease me by putting goannas and other crawlies for me to 'find' on the gate posts after dark. They taught me how to use a stock whip too.
>
> I loved the country up there. I can still remember the first sight of it — red earth stretching to the horizon, no green anywhere, tin roofs standing out as stark outlines against a clear sky. But I could never capture the clarity of the scene on film.
>
> Apart from the oil drilling work, I was fortunate to be there as a geologist because Lovelle Downs is where the fossil dinosaur remains were found and it's an important museum site now.
>
> The men on the rig were fascinating — all the types you hear about and see portrayed in outback films: the hard drinker; the strong silent type; the sly, sleezy one; the wag; the extrovert, and so on. They all had appropriate nicknames for each other like 'Jelly Belly', 'Sludge' and 'Mouse'.
>
> I had my own names for them and we all got along pretty well. I was there to work — and that meant being on call at any time of the day or night. Everyone accepted that I was the company representative and there was no nonsense.

Few disagreed that the best training for an exploration geologist, male or female, was at the wellsite itself, where an understanding could be gained of how data is collected and what weaknesses might lie in reservoir theories hatched back at headquarters in town. Nevertheless, the exploration programmes of the 1970s showed a marked change from the individualistic efforts of the few pioneers who tramped the outback locations during the 1950s and 1960s. There was much more of a team effort involved in planning seismic surveys and choosing well locations. This was seen to best effect in the Cooper Basin work, where SANTOS and Delhi began to second geologists from the farm-in partners that had joined the two original permit holders. Companies like Vamgas, Total, Alliance, Pursuit and Bridge brought in new ideas, and some programmes, like Bridge Oil's drilling at Tirrawarra to the north of Moomba, brought immediate success. The well, Tirrawarra No.1, was South Australia's first oil flow but, more importantly, it was found in a basin which, until then, had only hosted gas discoveries. Geologists had even begun to believe the region was not capable of producing oil.

Newspaper headlines screamed the news that Bridge 'magic' was turning the desert into black gold — an excess of reportage that had become common during the discoveries of the 1960s. Yet despite the fact that Tirrawarra proved to have more modest reserves, it did lead directly to further oil discoveries at nearby Moorari and Fly Lake, and help rejuvenate the Cooper Basin exploration programme overall. Another spur to the expansion of exploration was the potential for finding enough gas to supply the New South Wales market as well as the already-confirmed supply from Moomba and Gidgealpa fields to Adelaide. A rider to this was that some of the newer gas discoveries contained high percentages of condensate and, if a gas contract could be secured with Sydney consumers, it would allow the Cooper Basin producers to extract the liquids in a separate profit-making operation.

Such incentives were behind the expanded Cooper programmes of the early 1970s, but they did not make the logistics of the operation any easier. Working in untracked country on the edge of the Simpson Desert, the explorers negotiated ground alternating between gibber strewn plains and 23 m high sand dunes running for up to 130 km in unbroken parallel lines. As the *Petroleum Gazette* for December 1971 wrote, Moomba, itself a man-made oasis carved out of the desert, became the nerve centre and supply headquarters for the renewed exploration effort.

> Giant multi-wheeled trucks loaded with drill string and casing trundle in and out between the scattered rig sites; geologists and field hands in smaller vehicles add to the omnipresent pall of dust on the dirt tracks. Overhead the drone of light aircraft is commonplace.
>
> The value of aircraft can hardly be overstated. A well only 30 miles [48 km] from Moomba as the crow flies may

> easily be 150 miles [240 km] by the winding route that has to be taken through the numerous dunes in the parallel ranks. Even lightly laden four-wheel drive vehicles find difficulty breasting this sandy surf of the desert.
>
> This means a tough life for the drilling crews. Sheer distance makes daily commuting to the air-conditioned Moomba base impossible. So for up to three weeks at a stretch they are isolated on site. Isolated with no escape from temperatures consistently over the 100 degrees F [38° C] mark, nor from the swarms of bush flies.[6]

And yet there was always the possibility of a more daunting problem — flood. The desert receives an average of only 100 or 130 mm of rain a year but, when combined with the tide of water running down creek beds fed by the monsoon rains over the Channel Country in Queensland, the result inundates a vast area of the Cooper flood plain up to depths of one and a half and two metres in places. In the low-lying terrain the water cannot run off, so it lies for months until dissipated by soakage and evaporation. History indicates that one big flood occurs every 50 years or so, but exploration records to the end of the 1980s show there has been major disruption to the oil and gas programmes at least once every decade. Rising water requires the dyking of airstrips and all low-lying rig sites. Often a rig move is either impossible, or trucks are forced to travel up to 1 130 km on a circuitous route to reach a new location only a short distance away under normal conditions.

Jean Michel (Mike) Imbert, working with Aquitaine on Delhi/SANTOS farm-out permits on the Queensland side of the border, had firsthand experience of wet-weather problems during this period. An Algerian who had migrated to Australia at the age of 18 after being made unwelcome in France following the uprising at home, Imbert knew some geology and obtained a job with Forasol/Geoservices during the 1960s as a mudlogger. He then moved to Shell, working on the drilling side of the business for that company's Otway offshore programme and then to Aquitaine in the Cooper Basin.

> Usually we had 36 hours notice if the Cooper Creek was coming down. I remember on one well we'd just finished the programme and we raced to move out all the expensive equipment first — the Haliburton trucks, logging cabins and Slumberger vehicles. Even so we had to run a few of the creeks to make it through.
>
> There was no hope of pulling the rig out as well, so we moved the camp and as much of the unit as possible, including some of the valuable deviation drilling equipment, to the nearby dune top. The water came right up to the substructure.

> About the same time there was a blowout on the Della No.5 well across the border in South Australia. I delivered a load of barytes across to them, but they needed much more and army choppers were used to ferry the stuff in because all the roads to Adelaide were cut. That was an expensive operation.[7]

In March 1972 another blowout occurred, this one at Big Lake No.2 southwest of Moomba. Fortunately there was little chance of gas from the well igniting because it was also flowing a strong stream of water. Nevertheless it was a spectacular sight as the estimated flow of 700 000 m³ a day engulfed the derrick, virtually obscuring its top from view for two weeks. 'Boots' Hansen from Red Adair's blowout control organization in the US eventually cut free the drill pipe still in the well, skidded the rig to one side and fitted a new wellhead assembly to end the drama.

Imbert believed part of the difficulty in the early 1970s Cooper programmes was that there was never time to learn anything about the country — particularly the lie of the land.

> We were only drilling 8 000 ft [2 500 m] wells and they were finished very quickly for the most part. We'd send in a site clearer and water well driller first to prepare the location. Then we'd move the rig in with enough fuel and supplies for two weeks and keep replenishment orders running two weeks ahead.
>
> Drilling conditions were reasonable. There was a crumbly Cretaceous-age shale and a difficult Transition zone into the Jurassic where the dip of the beds used to kick the drill bit off line. By contrast, the mid-Jurassic Hutton Sand was pretty hard and that used to chew the bits out. But we came to know the drilling pattern and put most wells down swiftly.
>
> The French seemed to engineer the wells pretty thoroughly and they'd take advice on the geology when they designed a well. The Americans though were more inclined to do things the way they'd always done them in the US, and that did not always apply in Australia.

French geologist Maurice Cadart, working for Alliance during this period, recalled that there were frequent problems with electricity on the rigs.

> Many of the units were brought straight out from the US or Canada and had 110 volt/60 cycle AC current systems. Australian fittings and accessories were 240 volts and 50 cycles per second.

> Often during the day the airconditioners would overload the system and there'd be a voltage drop to contend with. The logging and recording equipment had to be kept going and the electrician was a key man on site. He was forever balancing equipment.[8]

The renewed activity and expansion of interest in the Cooper Basin exploration programmes during the first years of the 1970s brought a number of new international faces to the management of the major companies involved. Charles Easley, a Louisiana-born oilman/lawyer whose trademark was a splendid stetson he was rarely seen without, was head of the Delhi organization in Australia. He was also prominent on the APEA Council as well as on the boards of several South Australian-based companies including the Natural Gas Pipelines Authority. In addition, he was the US consular agent in the State and vice-president of the Australian–American Association. Geologist John Zehnder, managing director of SANTOS, had travelled widely in oil circles since his pioneering days traversing the highlands of Papua and New Guinea during the 1950s. Working for BP and then Union Oil he had been in East Africa, the US, Libya and Indonesia before returning to Australia. Zehnder was an enthusiastic proponent of the programmes to find Cooper Basin gas for Sydney and the setting up of a liquids scheme for the region.

Canadian born engineer/lawyer David Floyer was appointed chief of the UK-based Burmah Oil Company activities in Australia which included the Cooper Basin following the takeover of Sydney explorer Reef Oil. Floyer was well qualified to take a role in the gas development programmes, having spent 16 years with the Burmah group in Pakistan where he was involved in the production programmes of nine gas finds on the Indian subcontinent. Another American to come to Australia during this period was William (Bill) Diggs, who took up the position of general manager with Alliance Oil Development Australia in 1972. Previously a geological consultant in Arkansas, his new role involved the financial and administration aspects of AOD as well as the exploration programmes. Diggs was confident the company's interests in the Cooper Basin would enable it to share in the coming development schemes.

Partly through the influence of these people and other international companies that arrived in Australia during the early 1970s, local explorers began to contemplate expanding their own horizons by travelling overseas themselves. Hematite Petroleum farmed in to permits in Malaysia, Pakistan and South America; Crusader looked at New Zealand; Beach Petroleum followed suit, but added the major step of taking up permits in Turkey; Alliance began looking in the US; Hartogen went to Canada and the Associated Group acquired a 100 percent interest in a producing field on the Indonesian island of Seram. The Australian drilling contractors also began to expand with ODE sending a rig to Thailand and another to New Zealand, while Richter Drilling picked up contracts in South-East Asia — particularly drilling prospects for geothermal projects in the Philippines.

But Australia's closest overseas ties were still with Papua and New Guinea where, in 1969-70, the first serious attempt was made to run seismic in the highlands region and begin a drilling operation there. Previously the highlands had only been subjected to random geological traverses, although in the mid-1960s geologist Paul St John and three fellow PhD students conducted a gravity survey along the highland spine in conjunction with a geodetic survey being run by the Australian Army and the Department of National Mapping.

> Our PhD project was under the guidance of Professor Sam Carey at the University of Tasmania and, through his influence, we were able to 'hitch' rides on the helicopters being used by the army and the government in their geodetic work. That meant we could observe all the trig points they were using and do some other sightings as well. We even borrowed an army landrover and traversed up the Highlands Highway to Mt Hagen.
>
> Anyway our data gathering took about 13 months and from it we produced the first gravity map of PNG. It was published in 1967 and then presented as an APEA paper in 1970. What it showed was an interesting deep 'low' which I called the Southern Highlands Basin. It also indicated the Papuan Ultra Mafic belt and the massive overthrust where a 10-15 km thick slab of the earth's mantle had been wedged up and over itself. The survey indicated the whole geological sequence in the highland region more clearly than any previous work.[9]

St John then joined BP, one of the companies in Australasian Petroleum Co, and went straight back to work in PNG. However during the late 1960s the main concentration was in the coastal and offshore regions and it was not until 1970 that exploration switched to the highlands.

> The first seismic survey in the Kagua region was reasonable as far as access went because we could get in by road and run traverses down the valley, but the results themselves were fairly useless. The data was fair in the synclines, but hopeless over the anticlines which, of course, is where we needed it most. However all was not lost because we studied the surface geology and did more gravity work at the same time.

There were a number of oil and gas seeps known in the region and, based on this new work, Australasian Petroleum Company concluded negotiations with US major Texaco Overseas Petroleum to drill a farm-in well called Cecilia No.1 in the western highlands during 1971. Following this, APC itself drilled a well called Mananda No.1 to the southeast during the same year. Both were dry, although looking on the positive side, they did prove there was an overthrust belt in the region. Even so, the difficult logistics pushed the cost for the two wells to more than $6 million — easily the most expensive onshore programme in the history of the Australian onshore oil search to that time. Texaco ran some more seismic work in the region adjoining the border with Indonesian New Guinea, but shortly afterwards the company decided to withdraw from Papua to concentrate on Australia — and they did so with good reason. Despite the activity onshore Queensland and South Australia during the first two years of the 1970s, it was the offshore sector that had really captured the imagination of the big players. By 1973 Texaco (via Caltex), along with partners Shell, BP, Burmah and local explorer Woodside, had discovered vast volumes of gas and condensate on the North West Shelf.

The offshore regions of Western Australia and Northern Territory had seen their share of exploration activity during the late 1960s but, except for brief and isolated incidents like the oil find at Legendre and the blowout at Petrel, the work was overshadowed by the size and frequency of commercial discoveries in Bass Strait. The first half of the 1970s, however, told a different tale. Esso was having trouble maintaining a continuous drilling programme in the Gippsland Basin and its run of success in the region had dwindled by the end of 1969. In July 1970 the company called for a period of consolidation to evaluate the accumulated seismic and drilling data and plan the next move. Upon resumption of work it was soon shown that the early success rate in Bass Strait would not be easy to repeat. Bernard (Bunny) Brown joined Hematite Petroleum's (BHP) exploration team during 1971, in the middle of the data studies.

> The break lasted just over 18 months, with drilling resuming in March 1972 at the Mackerel No.2 appraisal location. Although oil had been discovered in this structure during 1969, the oil column found was only 15 m thick and was not commercial. Evaluation of data during the drilling hiatus indicated the discovery well was down flank and this was proved correct when Mackerel No.2 came in with a 90 m oil column.
>
> Yet there were still surprises because the oil reservoir in Mackerel No.3 came in much lower than expected. It was not until Mackerel No.4 in mid-1973 that the field was declared commercial, with 400 million barrels of recoverable reserves — about half the size of Halibut.
>
> By that time too, the 1968 Tuna oil/gas discovery had finally been declared a viable development project, but the

oil found in Cobia prospect was disappointing. The first well there, drilled in 1972, penetrated only 5 m of oil in a 37 m reservoir sand. It was not until much later (1977) that Cobia No.2 confirmed the field as a commercial development prospect, albeit much smaller than the original fields.[10]

Most of the Bass Strait support services during the early 1970s revolved round the production platforms. The helicopter link between Longford and the offshore operations had virtually become a scheduled airline with daily departures timed for 7.30 a.m., 11.00 a.m. and 3.00 p.m. Passenger numbers averaged 140 per day on these regular flights. But the pilots, many of whom were Vietnam War veterans from the Australian and US armies, also handled the crew changes to and from the Esso/BHP contracted exploration rigs like drillship *Glomar Conception*. A familiar figure on these flights and on the platforms themselves during this period was diver Henri Bource. Born in The Netherlands of Dutch/French parentage, he had begun diving during a two-year stay in Indonesia during his teens before emigrating to Australia and setting up a company specializing in underwater film-making. Bushy moustached, broad-shouldered and quietly spoken, he walked with a slight limp, the legacy of a shark attack in 1964 during which he lost a leg. He was also a gifted saxophonist, forming his own rock and roll band, and at one point contemplating a music career. But the diving business won out and he gave up music altogether in 1969.

Bource was not introduced to the oil industry until 1971, when Esso contracted him to do a photographic survey of a proposed gas pipeline route across Port Phillip Bay near Melbourne. Successfully completing that job, he found other underwater work soon followed, much of it taking him into the more technical field of platform maintenance and inspection. However there were also tasks associated with the exploration rigs. At the height of activity 11 divers, including Bource, made up the company payroll. A team consisted of a superintendent, three divers and a tender, and the below surface work was always done in pairs. At that time they concentrated on shallow diving tasks in 25 m of water or less. Even so, Bource's men were restricted to four hours below at one time because of the cold temperatures. Strong currents were also a major concern and all tools and materials had to be tied down at the work place. Continually fighting against the swirl round the platform legs or the marine riser of an exploration rig made any underwater operation a tiring one.[11]

Several other oil exploration companies tried their luck round the outskirts of Esso/BHP's discoveries during the first years of the decade, but the only 'sniff' came from Shell, with the noncommercial Sole gas find in the eastern sector of the basin during 1973. Others, like Endeavour Oil's Albatross and Gannet prospects, were dry holes. The most spectacular failure was John Fuller's Planet group with a well called Sailfish

No.1 drilled on a farm-in to Magellan Petroleum's Tasmanian permit T/1P just northeast of Flinders Island. The prospect, 10 km long and 3 km wide, was interpreted from seismic and geological data to be a potential buried biohermal reef structure, the like of which contained prolific oil fields in other parts of the world. Still an aggressive entrepreneur despite his lack of commercial discoveries during the 1960s, Fuller typically threw all his energy and resources into the project. Planet mounted a publicity campaign, including a film, to gain support in financial circles and from its own shareholders. As a result, all eyes were on the drillship *Glomar Conception* when it spudded the well during the second week of October 1971. A month later the rig delivered an anticlimax that shook many of the country's oil explorers and virtually all the investing public. Instead of being a buried reef, Sailfish prospect proved to be a volcanic dome with not the slightest chance of hydrocarbon accumulation. Although the news rocked Planet to its core, Fuller met the disaster the only way he knew how — head-on. He rationalized the companies within the Planet group and managed to continue for a few more years, even retaining an interest in the Gippsland Basin permits along with others in the Otway Basin further west. But the damage had been done and in 1976 the company's board conceded that the only course was voluntary liquidation.[12]

The Sailfish saga symbolized the changed fortunes in Bass Strait. Although Mackerel and Tuna were declared commercial in the early 1970s, Cobia was the only new oil discovery in the Gippsland Basin during the first half of the decade and even that was considered marginal at the time. In addition, the Otway and Bass Basins continued to yield disappointing results. Consequently, both Esso and BHP separately began to widen the scope of their Australian operations. BHP, through Hematite, considered the onshore Perth and Cooper Basins before drilling a couple of wells in the Canning and then moving to Queensland. Esso, through the work of its basin studies group based in Sydney, became interested in the offshore regions from Darwin right round the West Australian coast to the Great Australian Bight. Geologist Chris Curnow gathered the data for this far-flung survey.

> I was given virtual carte blanche by Esso, so I began with a field trip right round the coastal strip. I also managed to obtain the navy's bathymetric map of the areas in the study, but it was not contoured, so I spent about three months doing that.
>
> Through this work we found the Bremmer Basin off Esperence in southern Western Australia and we also became interested in the Duntroon Basin off South Australia. Esso applied for and won a permit in the Bremer area, but the Duntroon was already under licence.
>
> We did seismic in the Bremmer, but found it had water depths of around 1 800 m and was subjected to very rough

> weather conditions which put a dent in the economics of operations.
>
> In the Duntroon I saw the potential for a trend stretching from the Ngalia, Amadeus and Officer Basins in central Australia down through the Polda Trough in South Australia. It was likened to a river 'funnel' with the delta sediments postulated to be in the Duntroon Basin offshore. At the time we had to keep the theory quiet because Shell was in the area, but in the end it did not come to anything either.
>
> Through contacts in the industry I also managed to keep a watching brief on the exploration programmes of other companies in the study area, particularly off the north and northwest coasts. There was no possibility of a farm-in to the Woodside group acreage on the North West Shelf, but Esso did negotiate a deal with Aquitaine in the Bonaparte Basin.[13]

The permits in question were operated by Arco and the joint venture was still in the final throes of drilling Petrel No.1a, the relief well designed to kill the gas blowout in the Bonaparte Gulf that had been flowing wild for more than 15 months. Despite this mishap, which had cost a total of $13 million, the group was in high spirits. Arco's president Thornton B. Bradshaw, who visited Australia early in 1971, expressed the view that the Bonaparte Gulf could supply half the country's oil needs in the not too distant future. He reasoned that the joint venture had already found plenty of gas (indicated by the unabated wild flow from Petrel), and the hope was that oil discoveries would follow. He went on to pledge another seven wells in the region with Arco's share totalling $21 million. Indeed the group did continue to make discoveries. Apart from Petrel No.2 confirmation well in 1971, more gas was found in Tern No.1 during the same year and again at Penguin prospect during 1972. Yet in such a remote region with scant prospect of an immediate market, gas had little commercial value. Oil was the real prize and management hearts began to beat when the Puffin No.1 well in the Timor Sea registered oil shows in 1972.

For a brief period in 1974 Bradshaw's fond hope appeared to be near fulfillment when semisubmersible rig *Margie* tested a flow of 2 700 barrels of light oil per day from the Puffin No.2 well. But despite big headlines in the press, the explorers knew that the reservoir sands were thin, thus limiting the potential reserves. When added to the great distance from shore, the odds were stacked against commercial development. The disappointed consortium stayed together for a little longer before Esso pulled out in 1975, followed by Arco several years later. Aquitaine elected to stay, although it reduced its permit areas to those surrounding the Petrel, Tern and Penguin discoveries. After the Arco group had spent $70 million on 25 wells without a commercial discovery during the first half of the 1970s, it seemed plain that offshore exploration — particularly in remote regions — could only be born by the world's major oil companies. Yet flying in the face of this conclusion, albeit by the seat of its pants and the help of the large companies, was Woodside Petroleum on the North West Shelf. And with both hands firmly grasping the helm stood company chairman Geoff Donaldson.

Aged in his late 50s at the time, Donaldson was a man of imposing stature. From his towering height of 203 cm, his stare was enough to impale all but the strongest willed and his determination to succeed had already become legend. He was impatient with incompetence and certainly with dishonesty, and reportedly gave no quarter in the board room. Yet he was a courteous, nonaggressive man of the old school. He had taken up the reins at Woodside in 1956 when asked to underwrite a second share issue only two years after being instrumental in the company's initial capital raising through the family broking partnership Donaldson and Lynch. His style of chairmanship was very much a 'sleeves up' and 'open door' affair where everyone from the field hands to the office workers could feel themselves part of the team.

Donaldson had matriculated in geology from Scotch College in Melbourne in 1930 but, perhaps influenced by the fact that his father and two uncles were stockbrokers, he did not go on with the earth science, preferring commerce instead. Nevertheless he began in the lowly rank of office boy at Perpetual Trustees in Melbourne, rising in the ranks to be a Trust Officer at the outbreak of World War II, whereupon he joined the army. Coming home on a brief leave pass during the war years he found seats on the Stock Exchange were selling for £900, so he bought one and went back to the front. In 1946 he returned to take up his seat and join the family firm. Although remaining a stockbroker till the early 1970s, the added responsibility of piloting Woodside from 1956 increased his interest in developing companies. Yet it was not till the North West Shelf burst into the headlines in 1972 that Woodside business became a consuming role. An article in the magazine *Australian Business*, written by journalist Charles Wright on Donaldson's retirement 13 years later, best sums up the man and his ideals.

> What Donaldson is, essentially, is an engine man, not so much a buccaneer as a builder. He sees himself as a risk-taker, his gift that of calculating, and minimizing, the risk. He is careful to make a distinction between risk-taking and gambling. 'I don't mind taking risks, but I don't believe in gambling. That might sound strange. But for me there's a very big difference between taking a calculated capital risk, something that will achieve a worthwhile goal, a contribution to the wealth of the nation and of course one's own, and gambling', he says.
>
> 'That's the challenge, taking something that's valueless, and planning and developing it. Once you've found the cash

you've got to find the technical ways and means, and the markets. All that's worthwhile because you're doing something. You're not just throwing money on a horse and hoping to make a profit out of it.'[14]

Woodside (Lakes Entrance) Oil, as it then was, began work on North West Shelf exploration with Burmah Oil, Shell, BP and Chevron (Caltex) as partners in the 1960s. In the face of daunting costs Donaldson's leadership kept the small Australian company holding tight to its one-third equity position, although the company's name was changed to Woodside Oil during 1968 in view of the fact that it was no longer active in the Gippsland region. By the start of 1971 even the majors within the consortium were beginning to despair of making a commercial find in the huge permit and operator Burmah Oil Company Australia Ltd, better known as BOCAL, suggested some farm-outs. At the time the US mining company Amax was interested in setting up an oil arm and its management became interested in the idea of North West Shelf acreage. Geologist Max Reynolds had joined that company's Australian operations from the BMR in 1970.

> BOCAL put up seven areas for farm-in bids but, using the precedent of BHP's deal with Esso in Bass Strait, the terms were tough. The farminee could go in and explore one of the areas on offer, but it would only earn an interest if it was successful.
>
> Amax agreed to that and made up a consortium with Japanese companies Mitsui and Sumitomo to tackle two of the permits. Another company, Beaver Oil, was part of the group only for the very early stages. Seismic was run and a prospect picked. The well name — Wamac No.1 — was a rather unimaginative choice. It was an abbreviation for Western Australia and Macgregor, the man who was chairman of Amax in the US at the time.
>
> Anyhow it drilled in 1973 and was dry. There was some survey work done on the second farm-in area also, but no drilling took place. So in the end Amax came out of the exercise without any North West Shelf interest.[15]

However by that time the BOCAL group itself had broken its long run of dry holes. The change of fortune began in 1971 when the drillship *Glomar Tasman* drilled into the huge gas/condensate reservoir at Scott Reef 320 km out into the southern Timor Sea. The location, like that of Ashmore Reef drilled further north during the 1960s, was a coral atoll marked by a tiny island bare of any vegetation. Because of this the waters were relatively shallow and calm, although spikey coral heads presented a navigation problem when bringing the drillship onto location. Scott Reef No.1 was an undoubted technical success. For one thing, BOCAL had achieved an Australian offshore drilling depth record when the bit penetrated to 4 734 m. Apart from that the explorers knew they had found a very large field, but unfortunately it was too far from shore to be a commercial proposition. Had it been oil the remoteness may not have been such a handicap. Some schemes to expand the island perimeter with rock fill and build a processing plant on the spot were bandied about at the time, but they were more futuristic dreams than economic reality. Nevertheless the find did lift drooping spirits and the BOCAL group turned its attention closer to shore. Burmah Oil geophysicist George Edmond rejoined the team during the Scott Reef programme after a brief transfer outside Australia.

> There were three new prospects to consider. We'd named them Rankin, North Rankin and Goodwyn, and they sat about 190 km off the coast level with Dampier and Port Hedland. The seismic was unclear under a major unconformity in the region at the Cretaceous-age boundary, so there was no way of predicting the complex horst and graben features in the Triassic sediments which we found with drilling. Still, the structures were all obviously big ones and we thought at the time that they might be reef formations.
>
> Anyway we pushed this line to management, many of whom were Canadians familiar with such structures at home, and they were amenable to the programme proposed. We began with North Rankin because it was structurally the highest and the most prominent feature. The company could not believe its eyes when in May 1971 a huge gas/condensate discovery was made with the first well. Then came Goodwyn field later the same year followed by another prospect called Angel early in 1972. The only disappointment was Rankin, which had nothing like the volume of the other three.[16]

Success in the North Rankin No.2 and No.3 appraisals during 1972 confirmed the viability of a gas/condensate development based on the domestic gas market in Western Australia and the potential for export of liquefied natural gas (LNG) to overseas customers. By that time Geoff Donaldson had already sprung into action. He saw the potential for Woodside almost as soon as North Rankin 1 reached its target, but realized that the only logical way his company could raise finance for development was to combine Woodside and Burmah forces into a single Australian-based company. Flying to London later that year he put the proposal to the Burmah Oil board and came back with approval to set up a new entity called Woodside-Burmah Oil NL in Melbourne with a 50 percent share in the proposed gas project. The corporate arrangements and 50 percent equity were extended to exploration on the rest of the North

West Shelf permits with the principal operations base remaining in Perth. At that point the exploration programme literally took off, as George Edmond recalled.

> At one point we had three rigs drilling and the seismic team was flat out providing drilling locations. I remember one year over 16 000 km of seismic lines were shot and we had Shoran navigation stations set up on islands all over the area. Generally the stations had an operator who just pitched a tent and monitored the equipment in two- or three-week stints on his own, although on some islands he was accompanied by rats and snakes.

The accelerated search programme also meant a great deal more work for the service companies including the small, but rapidly growing interests of local firms like Lombardo Marine. Using its converted Rottnest ferry *Andrew*, Lombardo was kept busy setting up and retrieving Shoran stations all round the North West Shelf acreage. Skipper Vic Garbin recalled losing equipment only once during that hectic few years after 1971, and that was because they broke a golden rule and tried to retrieve a station at night.

> We'd arrived at Adele Island and anchored about three kilometres off to avoid the surrounding reef. Normally we'd have waited till daylight, but it was a dead calm evening so we lowered the dinghy and the blokes set off to ferry the Shoran station back.
>
> There was a lot of equipment involved and it took more than one trip. One bloke stayed ashore packing up while the other two rowed back to the *Andrew* with the first load. That was OK. But on the second trip a stingray tipped the dinghy over and they lost their load. They managed to right the boat, but water had saturated the engine and it wouldn't start. That was when their troubles really began.
>
> We didn't know any of this on the *Andrew* until the bloke on the island started sending morse messages with a light — badly I might add — to say he wasn't being picked up. No one on the vessel could decipher beyond the fact that something was amiss. But we couldn't do anything and spent a long worried night until at dawn we spotted the dinghy in the distance rowing towards us.
>
> Apparently the two men had tried to row to the *Andrew* after the capsize, but fought a losing battle against the tide. They watched as the light on Adele disappeared into the distance. Then the tide turned and they were able to make headway back again by rowing all night. They were very tired but relieved men when they clambered back on board.[17]

The expanded drilling programme also multiplied the problem of logistics, with supply and crew change lines stretched to the limit. David (Dave) Agostini joined Woodside in 1972, initially spending a lot of time in the petroleum engineering division which, among other things, meant involvement in well testing. In this capacity he experienced the extended travel routes first hand.

> The mid-1970s was a period when Woodside-Burmah was drilling wells closer to Timor than Australia and the company actually set up a base at Dili, the capital of Portuguese Timor then. But it was only part of the 'endless' journey to the rigs. Troubadour and Sunrise wells way north of Darwin were the main reasons for this and both actually made gas/condensate discoveries.
>
> Not forgetting our passports we'd leave Perth at midnight and be in Broome at 3.00 a.m., then on to Dili in a Citation aircraft, arriving there by 6.00 a.m. It was the first jet the locals had seen and it made a helluva show landing on the coral strip with dust and coral chips blown everywhere.
>
> There'd be a bit of a break to clear customs and then we'd be off again to Vequeque on the south side of the island where Woodside had a staging post for the Wessex helicopters. We couldn't use the choppers from Dili because they did not have enough lift to clear the central mountain range. The trip back over water to the rig just on the Australian side of the sea boundary took another hour or so. We'd change nine crew members at a time via this route and the flight would run five days a week.[18]

Woodside-Burmah had a preference for using drillships during this period because they were able to make the long-distance location moves around the North West Shelf permits more quickly than other rigs. The semisubmersibles and jack-ups only made about 3 knots under tow. Deep water locations also ruled out jack-ups for most of the time. An additional bonus with the drillships was their supply storage capacity, an important cost factor when supply runs often entailed a five-day round trip for a service vessel. But equally high in the early thinking was a drillship's ability to handle the bad weather during the summer cyclone season. Having its own propulsion enabled such a vessel to move out of the path of an approaching storm provided there was enough warning given. In fact, right through until the late 1970s the rigs themselves were used as the advance weather monitoring and warning posts for the whole of Western Australia's northern coastline. Data on wind and sea states was regularly sent back via Woodside's bases in Broome and Karratha to the Meteorological Bureau in Perth. Bureau meteorologists visited the rigs occasionally to make their own analyses and were amazed to see millpond conditions change to four-metre seas in the space of half an hour when a squall came through.

From the 1980s, weather satellites took much of the onus of forecasting away from the explorers in the front line, but there is no doubt that the well-polished cyclone alert procedure of today is a product of the (at times) bitter experience built up during the drilling operations of the 1960s and 1970s. Most of the rigs used by Woodside-Burmah (and WAPET further south) were buffeted by heavy seas at one time or another, but none gave the crews more awareness of the power of a storm at sea than the ship-shaped drilling barge named *Big John.* Used by Woodside-Burmah as the third rig in its fleet of three during the first half of the 1970s, *Big John* was a 'dumb barge', meaning it did not have its own propulsion despite having the hull shape of a drillship. The vessel began life as a railroad boxcar ferry plying between Anchorage and San Francisco on the US west coast before it was bought by American drilling contractor Atwood Oceanics and converted into a rig. In all, it drilled a dozen wells for the Woodside-Burmah group along the North West Shelf acreage. Dave Agostini found the vessel a comfortable enough work place — that is, until he was aboard when the barometer fell and seas began to rise.

> We were conducting a well test at the time and you could see the waves building up. The wind wasn't all that strong — maybe 60 knots — but the swell was heavy. A supply vessel nearby actually kept disappearing from the rig's view in the next wave trough.
>
> We'd shut in the well and hung off by that time, which is just as well because one anchor cable broke with a sound like a rifle shot. Then three more cables parted and the rig swung side on to the sea. It was a bit hairy for a while, particularly since *Big John* was a flat-bottomed vessel. The spray came up as high as the helideck.
>
> There was nothing much we could do except stay indoors for a couple of days. There were 80 people on board at the time, although that wasn't too bad because *Big John* had better accommodation than most of the rigs. Some people were a bit nervous though, and a few were sick. After about 48 hours the seas subsided and we went back to work. She'd shipped a bit of water, but otherwise there was no major damage.

Agostini experienced a severe storm, but during the first few years of work on the North West Shelf there was little knowledge of what to expect if any of the rigs were caught by the full force of a cyclone. Consequently, the procedures for dealing with one had not been set down in any detail. However in 1973, Woodside-Burmah operations were disrupted by not one, but two severe cyclones in swift succession. Between

January and February that year cyclone Kerry gave the explorers their toughest experience ever. Forming in the Timor Sea north of Broome, it immediately veered towards land, cutting a swathe through the drilling operations before crossing the coast at Dampier and continuing inland as far as the Meekatharra region. Immediately on its heels another cyclone, Maud, ran parallel to the coast between Derby and Dampier. Although its track was shorter and it did not turn towards land, this cyclone prolonged the disruption to all oil exploration on the Shelf permits. The experience clearly demonstrated that there had not been sufficient warning, and hence not enough flying time, to secure and evacuate the rigs before the storms hit. W.R. Whitley, Woodside-Burmah's operations superintendent at the time, described the maelstrom which surrounded the company's three rigs in a paper delivered the following year. *Big John* once again suffered the brunt of the violence.

> When the warning was received, the crew began pulling up the drill string... [and] the drill pipe was laid down on the deck to reduce the stress in the derrick and to increase the stability of the vessel. This was a slow operation in the face of deteriorating weather... and, due to rapid intensification of the storm, it could not be finished. The [remaining] drill pipe was hung off in the lower pipe-rams in the blowout preventer with the bit still in open hole at 1 372 m... There was no time to bring the subsea equipment to the surface... but the marine riser was disconnected and... using the anchor winches, the vessel was moved away from the vicinity of the blowout preventer stack on the sea floor.
>
> The barge was moored with eight 400 kg anchors. Thirty personnel had been evacuated in two helicopter sorties, but high winds and seas prevented the remaining 38 men from being taken off. By the time the rig was winched a little way from the wellhead, the storm (with winds coming in from the southeast) prevented any other human action. Movement outside the accommodation area was impossible.
>
> One eyewitness estimated Kerry's winds to be between 70 and 110 km per hour and waves 14 to 17 metres. The vessel itself was rolling to 21 degrees and pitching 15 degrees. The Master on the standby supply vessel reported that 18 m of the rig's keel was visible between the waves. Waves were breaking over the vessel.
>
> Number 5 mooring line parted first, followed shortly after by the parting of numbers 6, 4 and 3. The vessel swung round to hang on its forward moorings. However, numbers 8 and 7 cables parted soon after this and the vessel moved again to hang on the remaining two moorings. These lines were crossed when *Big John* swung round to face into the

> storm. Those six anchor wires all parted within a period of four and three-quarter hours. At no time could adjustments to tensions be made during the storm because the anchor winches could not be approached.
>
> At the height of the storm heavy seas broke open a watertight door on the sea-deck and within minutes the mud pump room (power room) was flooded. The control box was shorted out and all six DC generators were made inoperable. Both pairs of electric rig pump motors were also made inoperable... [but] the two 800 kW AC generators and the emergency 100 kW AC generator remained in operation.
>
> The moonpool well cracked and the pipe-racking system was damaged. The marine riser was severely buckled and the slip-joint and top pup-joint were also damaged by waves. A number of items of equipment were lost. During the height of the storm, when it was thought that only one anchor held the vessel, a 'mayday' call was issued to shipping by the master of the barge [even though] two supply boats were on hand to offer assistance... The eye of the cyclone actually passed over the rig, which experienced several minutes of relative calm.[19]

When cyclone Maud passed by, the barge was still held by mooring lines 1 and 2. The effect of this second storm was felt from the southwest and the vessel swung around into the weather. However the rig was held without further trouble. Luckily the only human damage in the whole drama was one broken collar bone, but lost operation time amounted to 36 days with a down-time cost of $876 000. The damage and repair bill came to a further $138 000, putting the total cost arising from the two cyclones at just over $1 million. Following this incident the cyclone warning 'three-stage alert' system was initiated where bulletins of a storm's progress were given at frequent regular intervals as soon as it came within 2 000 km of a drilling operation. Steps to progressively shut down any rigs in the storm's path and evacuate personnel (if a semisubmersible) or move the vessel off station (if a drillship) would be taken as the cyclone moved closer. A number of companies involved in the Pilbara mining industry also began to take an interest in weather surveillance in the northwest and, for several years during the mid-1970s, oil and mining companies clubbed together to finance a Navajo aircraft to do the job. Data was forwarded to Perth via the aircraft until weather satellite station links came into 24-hour service over the region.

By that time, however, the Australian exploration industry had run into stormy weather of quite a different character. In Canberra, the Federal Labor Party surged into government during December 1972 for the first time in over 20 years, promising sweeping reforms on a number of fronts. Realizing there would be changes, but unsure of exactly what to expect, the explorers held their breath. They did not have long to wait. Within months of taking office, the new Minister for Minerals and Energy, Reginald Francis Xavier (Rex) Connor, had made perfectly clear both his dislike for all foreign oil companies and his intention of setting up a government-owned body to take a major share in the running of the country's exploration and production programmes. By 1975 activity had fallen to its lowest level since the 1940s and large numbers of jobless explorers fled the country seeking positions overseas.

Chapter 15
LABOR DEALS ITS HAND

Although Australia's explorers experienced their darkest hours under the nationalistic fist of Energy Minister Rex Connor between 1973 and 1975, it is not true to say that the Labor Government of Prime Minister Gough Whitlam was the sole cause of their woes. As early as 1970, exploration activity had begun to recede from the heights of the late 1960s. This was partly in response to the scaling down of the subsidy scheme by the Liberal-Country Party Governments of John Gorton and then William McMahon. It was also caused by local oil prices falling behind the world price as a result of the fixed-price agreement for indigenous oil sales that had been insisted upon by the Bass Strait participants back in October 1968 when the world price was in decline. In addition, during 1970 and 1971 there was a re-emerging debate about Australia's prospectivity for oil. Despite continuing finds round the country, no major oil discovery had been made since 1967, and the doubters who had been silenced by the discovery spree of the 1960s began to return with predictions that the country would soon run out of the oil it had found. Even APEA began to push the line that production from existing fields would reach a peak of 63 percent of the country's needs by 1972 and then drop rapidly to about 25 percent by 1980 if no new big discoveries were made.

The fight for increased subsidies was maintained principally by the small explorers because the majors, including BHP, saw little need for government assistance in this form once Bass Strait had provided such a bonanza. The small local companies viewed it differently, claiming that the 1969 amendments to the Act, tailoring subsidies for offshore work to the proportion of Australian equity involved in the individual programmes, had in fact discriminated against the small explorers by cutting the amount paid for each programme. Rees Withers, managing director of Mid Eastern Oil and its associate, Woodside Oil, was particularly vocal on the subject.

> Amendments to the subsidy Act... have penalized Australian companies which have been sufficiently conscious of the need

> to link up with large overseas companies. The affect of the Act on Mid Eastern and Woodside has been to penalize them by reducing their subsidy by 66.66 percent because of their partnership with overseas associates.
>
> The subsidies paid for work on the North West Shelf are based on the proportion of Australian equity in the enterprise. The efforts of local partners to maintain their equity and raise risk capital has been penalized by the government's decision to reduce the subsidy of Australian companies engaged in offshore exploration. That is not the way to encourage exploration work by Australian companies with the objective of providing self-sufficiency in oil reserves.[1]

The question of price, however, was tackled by the whole exploration industry in Australia. Although producers were the ones directly effected by the selling price, the argument ran that unless the Australian oil price was allowed to catch up to the world price (by being priced at 'import parity' as the catch phrase had it), explorers would have no incentive to go out and search for new reserves. In other words, the cost of exploration would outweigh the rewards of discovery. World prices had been rising slowly but steadily since 1970 and, by the middle of 1971, Australia's price had fallen about 20 percent behind. APEA chairman in this period, Eric Avery, spearheaded a concentrated industry campaign beginning in September 1971 to try to make the government aware of the Association's concerns in this regard, casting aside the fact that the existing fixed-price regime was a 'monkey' which industry had put on its own shoulders.

> To reach and maintain self-sufficiency over the next 20 years, Australia needs to discover reserves equivalent to all those of the Bass Strait fields every two years. Past experience indicates that a substantially increased level of exploration activity will be needed to find these added reserves.
>
> Price is the most important incentive because it means that the most efficient earn the greatest reward, with the result that efficiency is encouraged.[2]

Avery claimed that the exploration risk in Australia was greater than in any other country in the world and that it was at least six or seven times greater than in the Middle East oil countries. Any one discovery had to carry the costs of many previous, and perhaps future, dry holes. The theme was continued at the 1972 APEA conference in Sydney, where it was pointed out that the only way of escaping from the Middle East domination of the market was for Australia to become self-sufficient in oil. However to do so, and thereby maintain the level of indigenous reserves, it was estimated that the country's explorers would have to spend between $2 billion and $4 billion before the end of 1977. The difficulty was that even the lower figure was over three times the total spending on Australia's oil exploration programmes to 1972 — and that was only the finding cost. A great deal more funding would be needed to develop and produce the discoveries that were made. Apart from the call for government help in the form of subsidies and tax incentives, suggestions to aid the explorers were few. The small companies' problems were compounded by the general disintegration of the stock market mining boom, as an article in *Finance Week* remarked.

> The plethora of small, inexpert and sometimes downright shady exploration companies in the new issues market is an impediment to raising the massive capital required.[3]

Fresh from his company's gas discoveries on the North West Shelf, Desmond Dewhurst, an executive director of Burmah Oil in London, suggested to the APEA conference delegates that a number of the smaller Australian explorers should band together to form Australian consortia of sufficient size to command the necessary finance and to buy, where necessary, the technical talents.

> One major reason for the difficulty Australian companies have experienced in fully participating in the Australian oil search is that most of the best exploration opportunities probably lie offshore where exploration is expensive. In most cases it puts the relatively small Australian company at a distinct disadvantage by comparison with his overseas rival.
>
> Perhaps APEA could play a role here and provide a really rewarding contribution towards meeting Australian aspirations. Something like this should be attempted because it is important for us all, for Australian and for overseas self-interest alike, for Australian participation in future oil discoveries to be as large as possible.
>
> It would be more than gratifying if all the funds necessary could be raised in this country... [But] however unpalatable, the gap will have to be filled partly by welcoming overseas risk money to work on your behalf.[4]

As Dewhurst represented one of the foreign majors, it would have been surprising if he had concluded otherwise. Nevertheless it was becoming obvious that Australia's oil search had developed into a sharp pattern of 'rich' and 'poor' areas. Robert Murray, energy journalist with the *Australian Financial Review*, made this observation in an article published on 10 March 1972. He defined the rich areas as those where exploratory and confirmatory drilling were relatively active, pointing to the onshore Cooper Basin of South Australia and the offshore regions of Bass Strait/Great Australian Bight and North West Shelf/Bonaparte Gulf. The poor areas were almost everywhere else. To support his observation Murray used the rig deployment figures issued by APEA at the end of February 1972. Of the 11 rigs active onshore Australia, seven were employed in South Australia; three of the others were in Queensland and the last — WAPET's faithful National unit — was in Western Australia. The figures showed that five of the six offshore rigs were active and dispersed evenly round the country's southern and northwest coasts.

Overall, the predictions of the time were that the discoveries on the North West Shelf would keep offshore activity at a relatively high level, with four or five rigs operating in permits there during 1973, and perhaps one in Bass Strait. But the onshore outlook was for little or no improvement in either drilling or seismic work for the year ahead unless there was a marked improvement in exploration incentives. The task of renewing APEA's quest in this direction fell to new chairman, Desmond Wittwer, the general manager of BHP's oil and gas division, Hematite Petroleum. But he had no sooner taken the role when the Labor Government came to power, and into the exploration ring on the government side stepped the burly figure of R.F.X.(Rex) Connor.

The press quickly dubbed him 'The Strangler', an allusion to his formidable bulk (190 cm tall and 114 kg) and his dour, sometimes volcanic manner. It was not long before the explorers who tried to deal with him added the words 'rude', 'arrogant' and 'paranoid' to the description. Connor came to the Minerals and Energy portfolio at the age of 69 after being shadow spokesman between 1966 and 1969. He had been brought up in the coal mining region of Wollongong in New South Wales, a background which had stamped him with a strong sense of anticapitalism and a consuming nationalism. When he took up the resources post in the Whitlam Government, his single-minded aim was to ensure that only Australians benefited from the country's mineral wealth. This attitude soon put him, and hence the government, on a collision course with the resource companies — the foreign oil explorers in particular. Openly calling top management officials 'mugs and hillbillies', he made it plain that he did not trust them. He ignored most of their letters and submissions and often kept them waiting in the corridor outside his office for an hour and more for appointments. The reason for this behaviour may have been connected with Connor's view of himself as a shrewd financial wizard who knew best how to develop the country's oil and mineral resources, and his determination to have his own way. Geoff Donaldson, chairman of Woodside, recounted his first meeting with the Minister to journalist Charles Wright during an interview in 1985.

> The first time Keith Wilson (managing director of Burmah Oil Australia) and I went to see him, he was wearing braces [a mode of dress that soon became Connor's trademark]. He

> told us roughly to sit down, stood up, thumped the table, and told us bluntly that the first thing we had to remember was that he was in power, not 'that other sleazy mob'.
>
> He warned us never to forget that, and he then told us precisely what he was going to do [with the North West Shelf]. He was going to buy the gas at the wellhead, ship it across Australia by pipeline, and sell it to America. Then he told us to get out.
>
> We were naturally very polite to him, thanked the Minister for his views and said we would take note of his aims and we would consider what he'd said. Then we went back to Melbourne and battened down the hatches for five years.[5]

Because of its high profile in the Cooper Basin and the North West Shelf, Burmah Oil came in for the severest treatment, so much so that the company managers simply shook their heads in disbelief. Esso too was targeted for 'hiding' a billion barrels of oil in Bass Strait. Connor presumably imagined that the company's studies into the potential of stratigraphic traps in encapsulated sand lenses below the main structural reservoirs were definite discoveries. The tragedy was that in his efforts to bludgeon the foreign majors, he effectively hit the whole exploration industry into a pit too deep for most of the small local companies to crawl out of again. The Minister's fundamental aim was to keep the price of oil down because he believed Australia would benefit from the lower prices of other commodities the country would then achieve. So when he saw exploration activity continue to fall, he vowed to establish a government-owned body which would do its own exploring followed by production, transport and marketing. Failing in his attempts to set up a government exploration group within the Bureau of Mineral Resources, he decided to create a new entity — the Petroleum and Minerals Authority — to achieve his aims. He then announced that all remaining drilling subsidies and tax incentives for explorers would be withdrawn and the money thus saved by the government would be used to fund the PMA.

Coming on top of Connor's repeated refusals to raise the price of indigenous oil in the light of rapidly increasing world prices, the industry immediately labelled the move as a disaster. APEA executive director Ken Horler pointed out that the government had effectively put control of exploration into the hands of the new PMA and a few major companies with sources outside Australia. He predicted it would be the death knell for many small explorers.

> The government seems intent on crippling Australian private enterprise in oil exploration and development at a time when it is crucial to find more oil with the energy crisis the world is facing. To eliminate taxation incentives to investors at a time when it is already extremely difficult to raise risk capital

> will result in a massive cutback in Australia's already inadequate exploration programmes.
>
> It is hard to understand the government's decision in the light of its previously stated policies of foreign investment and the need for greater Australian equity in mineral and petroleum natural resource development. The small Australian exploration companies have contributed much to the development of geological and geophysical knowledge of Australia and they have also been responsible for the country's early petroleum discoveries. It is a poor reward for years of endeavour . . .[6]

The chairman of Alliance Oil Development, Thomas Webb, echoed these sentiments when he addressed shareholders attending his company's annual general meeting in October 1973.

> Last year at this time, we believed that regardless of the federal election outcome, the exploration industry in Australia would continue to have the support of any elected government, and we were even hopeful that certain aspects of that support would be improved after the politically sensitive pre-election period was passed. Quite obviously we were wrong, as the support that was there for exploration companies before the election has now been completely eliminated.
>
> Having admitted that error in our judgement, I now presume to predict that under present conditions here in Australia exploration for new discoveries will virtually cease, except in those basins where significant discoveries have already been made.
>
> . . . The consumer may be well pleased with government price fixing of crude, as this appears to be saving money today. However, with crude price fixed in Australia and rapidly increasing throughout the rest of the world at a rate far exceeding global inflation, you can then begin to realize that exploration risk dollars are going to be directed to those countries where the prices are more attractive.
>
> What is the logic that would entice any company to spend exploration dollars in Australia, looking for oil to sell at $2 per barrel, when it can just as easily spend the same money in another country looking for oil that will bring $4 or $5 or more a barrel.[7]

Webb also referred to the proposed PMA.

> Economists and politicians become very wise about who should own oil and mineral deposits after discovery is made,

> but these same people would not risk their own money on original exploration where the odds are so high against them. This is the field of investment where professionally directed private capital properly supported by government incentives can and should be utilized.

The industry's advice, threats and then pleas all fell on deaf ears. Minister Connor pressed forward with his plans including a ban on foreign companies farming in to offshore permits; the abolition of petroleum search subsidies from July 1974; the removal of tax deductions for Australian resident shareholders' contributions to oil exploration companies' capital; the exclusive use of Australian flag offshore drilling rigs in Australian waters, and the refusal to allow foreign flag vessels into the country until orders had been placed with local shipyards for the building of their eventual replacements; and the setting up of the PMA and a similar body called the Pipeline Authority. By 1974 the country's exploration activity was heading for its lowest level in more than two decades as foreign companies, other than those with production interests, withdrew. Local companies fended for themselves in various ways.

Some simply cut back on exploration programmes and prepared to wait until more favourable policies were introduced. A few, like the Planet group, tried to keep up their work programmes, but soon found fund-raising virtually impossible and fell into the hands of liquidators. An exception was the unusual, one-off circumstance of small Brisbane-based explorer Exoil, which sold its name to the Exxon Corporation for $575 000. Exxon had decided to buy or register all similar names to keep them out of circulation, thus preventing others from using them. Exoil chairman, William (Bill) Siller told shareholders he had no qualms about the sale because it had no effect on the company's assets and yet it created a substantial increase in cash resources at a time when funds for exploration were not easy to obtain. Exoil became Oilmin NL from that point. A number of other Australian companies decided to try their luck in exploration programmes overseas. Even the local drilling contractors began to shift their best rigs and equipment to foreign theatres of operation. However politics was not the only motive for the exodus, as indicated by geologist Eric Webb, the (then) chairman of Melbourne-based Endeavour Resources.

> Like everyone else at the time I was concerned about the harm Labor was doing to the industry. But I also felt there was a lack of prospectivity for oil in much of the available acreage in Australia. I subscribed to Lewis Weeks's theories that Tertiary age rocks were the most likely to contain oil reservoirs and, as the Bass Strait region was taken, I decided to look at South-East Asia. If the geology was right I reckoned I could put up with the politics.

> Endeavour tried to strike a deal with Pertamina, the State-owned oil company in Indonesia, without success. From there I went to Malaysia where I knew the Surveyor General, and the company was offered an area off the west coast. But the seismic was uninspiring so I did not exercise the option. The Gulf of Thailand was next, but Endeavour was really too small to tackle this expensive area so I journeyed to the Philippines where we were the first foreign oil company to arrive requesting a permit deal. We didn't make a discovery, but nor did we spend much of Endeavour's money because other later farm-inees carried our costs.[8]

In a number of cases geologists and drilling personnel left the country as individuals, most of them made redundant by the cutback in operations. At the time, exploration round the world was booming in the wake of high world oil prices resulting from the OPEC rebellion of 1973, so most of the displaced Australians had little trouble in finding work. The slump of 1974–75 was seen as very much a local problem and, unfortunately, this exodus of exploration talent left a yawning gap in the indigenous industry at a time when all effort should have been made to stimulate the search.

Some geologists joined the Petroleum and Minerals Authority, although their stay was destined to be brief. Rex Connor's PMA Bill was passed in August 1974 and one of the body's first steps was to acquire 50 percent of Delhi International Oil Corporation's interests in the commercial oil and gas reserves in the Cooper Basin, and 25 percent of its interests in exploration areas outside those reserves. The deal was a substitute for a previously announced agreement, between Delhi and French company Societe Nationale des Petroles d'Aquitaine, which had been subject to authorization by the Reserve Bank of Australia. Connor informed Delhi that authorization had been withheld under the Federal Government's Banking (Foreign Exchange) Regulations and that the PMA would assume Aquitaine's role. Despite the Minister's high-handed approach, the agreement needed the consent of both the South Australian and Queensland Governments. Queensland chose to withhold its consent and so the deal was modified to cover only those Delhi interests within South Australia.

However, alerted to other parts of the PMA charter which directly clashed with State sovereignty — particularly the Federal body's right to enter, occupy, build upon and remove items from any land with a warrant made legal merely when signed by a Justice of the Peace — Queensland and Western Australia led the way in a constitutional challenge. After much wrangling the High Court found in favour of the States and the PMA was declared illegal. Connor was caught flatfooted in the midst of the Delhi deal when the decision was brought down but, as it happened, the result played neatly into the hands of the South

Australian Government, which had been wrestling with a problem of its own. Lee Parkin, a former Director of Mines in South Australia, recalled the unfolding of events.

> In order to carry out long-term planning the South Australians needed to build up at least 20 years of proven gas reserves, but the government was unable to persuade companies to fund exploration programmes where financial returns were so far in the future. So, when the High Court outlawed the PMA, the South Australian Government bought the Federal Government's Delhi interest and set it up as a separate company called the South Australian Oil and Gas Corporation.
>
> As a member of the Delhi/SANTOS consortium within the South Australian Cooper Basin, SAOG had the right to explore the acreage, including the right to sole risk — that is, finance the drilling of prospects on its own if none of the other permit holders wanted to join the programme. The company also initiated a downhole fracture programme within impermeable reservoirs to stimulate hydrocarbon flow. In other words, the South Australian Government became involved to accelerate exploration in the State.
>
> In the beginning Delhi, as operators, could not understand the government nosing around and the company was even prepared to build fences round the rigs to keep unwanted people out. But after a short time SAOG's role was accepted.[9]

Unfortunately there was no such acceptance between the oil explorers and the Federal Government — more specifically, Rex Connor. Relations between the two parties went from bad to worse in 1975 with a succession of speakers at the March APEA conference in Surfers Paradise that year pouring out vitriolic descriptions of the Energy Minister and his policies. Even managing director of Australian Oil & Gas Corporation, David McGarry, a man noted for his moderate approach and recalled to the APEA chairmanship as troubleshooter in difficult times, publicly called the Minister 'foolish'. McGarry went on to say during an interview several months later that he saw a bleak future for the oil industry while Rex Connor held the energy portfolio.

> This government has clearly indicated that it is anti-business. Let there be no doubt that the Minerals and Energy Minister is out for one thing... nationalization of the oil industry. As far as I am concerned the very future of Australia is at stake here. The fact is that we are not spending enough in the search for oil, and with a population of only 13 million we cannot ourselves handle the huge sums required for this high risk exploration.

> We've got to induce private industry, doctors, dentists, farmers, citizens who've got spare dollars, to invest in our search. More than that we've got to milk every buck that is available internationally.[10]

The drop in exploration activity had indeed been dramatic during the first half of the 1970s. BMR records showed that with the exception of 1971 the number of wells drilled in each year between 1968 and 1972 had been over 100. In 1973, however, the total had fallen to 69; in 1974 it was 55 and predictions for 1975 suggested the figure could be as low as 31, with only 13 companies indicating they would undertake any drilling programmes at all. An interesting sidelight to the BMR statistics which emerged from the Surfers Paradise meeting was the apparent consensus that, geologically speaking, Australia was rated about number five as prospective territory for finding oil. Delegates did not doubt there would be new finds, but it was generally thought that the probability of discovering major oil fields was lessening. As the *Australian Financial Review*'s reporter put it, the 1975 assessment lay somewhere between the heady optimism of the late 1960s when the major Bass Strait fields were discovered, and the older view that Australia was barren of oil. Much of the continental shelf, which looked so promising for oil discovery a few years previously, had been downgraded or discarded from calculations, with many of the favoured areas proving to be 'gas prone'. Gas was still valuable, but only a fraction of what oil would be worth.

Poor prospectivity certainly added to the economic risk of exploration, but there is little doubt that many of the gloomy geological predictions of the time stemmed from the artificially low Australian oil price. Yet there was some hope on the horizon because the five-year fixed price agreement set by John Gorton's Liberal–Country Party Government was due for review in September 1975. Rex Connor would be forced to consider a closer match with the spiralling world prices. It transpired that the Minister was somewhat pre-occupied at the time with the so-called 'loans affair' — an abortive attempt by him to fund his grandiose government-run oil and mineral schemes using money promised from shadowy sources in the Middle East. Nevertheless he insisted that he wanted a price set for existing fields based on the cost of production.

Consequently, when Prime Minister Whitlam announced the new pricing structure, the Bass Strait fields received a mere 23 cents a barrel increase. That figure had been assessed as the amount needed by Esso/BHP to make the 1960s discoveries of Tuna and Mackerel viable development projects. But the pleasant surprise for explorers elsewhere was that any new discoveries from that date (14 September 1975) would be priced at parity with the prevailing world values. This policy took a significant step forward in company eyes because it finally gave some encouragement to the beleaguered exploration scene in Australia.

The announcement gave no such joy to Rex Connor, who felt sure the big companies would suddenly trot out a discovery they had known about for years and claim the higher price. However by that time The Strangler was in no position to argue, having been forced to resign in disgrace over the loans affair. Within a few more months the whole Whitlam Labor Government was rocked by a series of ministerial resignations, leaving the electoral door wide open in December 1975 for the return of a Liberal–Country Party coalition under the prime ministership of Malcolm Fraser. This did not result in the immediate return of foreign explorers, or an instantaneous exploration revival round the country, but dialogue between the industry and government was at least re-established. Fraser created a new department named Trade and National Resources, which included the oil industry, and he appointed Country Party leader and Deputy Prime Minister Doug Anthony to the portfolio. Anthony retained Whitlam's promise of parity pricing for new discoveries and gradually a mood of cautious optimism began to creep back into explorers' ranks, bringing with it a more favourable perception of the country's oil prospectivity.

However as companies began to cast their eyes round Australia anew, they were conscious that one region had been declared strictly out of bounds to exploration programmes, and it seemed quite possible that more would follow. The region in question was the Great Barrier Reef off eastern Queensland. The reason for its exclusion from the permit map was centered on the rising tide of concern for the environment which had begun in that area during the late 1960s and culminated with a Royal Commission and a ban on drilling on or near the reef during the mid-1970s. Apart from giving technical evidence before the Royal Commissioners, the oil explorers took little part in the debate, although concern at what they saw as exaggerated claims from some members of the conservation movement did find occasional public voice. Delivering the concluding address to the 1970 APEA conference in his capacity as vice-chairman, Charles Easley, managing director of Delhi Petroleum, was one of the first to broach the subject.

> [Today] marks the beginning of a new era for the oil industry in a world that has suddenly grown hostile to those who seek to develop the earth's natural resources. Development of our natural resources is not something we have a choice to do or not to do. Our society is dependent on their development. We cannot return to the Stone Age even if some of us might want to.
>
> Conservation has become a fashionable cause and... almost overnight [it] has exploded into a worldwide issue... But amongst all this conservation publicity, a flood of irresponsible statements concerning the oil industry has come forth from people without even the vaguest notion of what occurs in oil drilling, particularly offshore oil drilling.

> There have been, of course, some reasoned and rational comments. However, reports about the effect of oil spills on the natural environment, more often than not, have been grossly exaggerated. We have been inflicted with an emotional campaign involving press and television statements and petitions to government, signed, I suspect for the most part, by unknowing persons calling for such things as licence cancellations and drilling bans for at least 10 years. As with other controversies, the fringe elements have been making some pretty wild statements which, as expected, have received a great deal of publicity. It almost seems that the most outrageous claims receive the widest coverage.
>
> Up till now we have been acting like technicians, totally absorbed in the important national task of finding oil... and in this posture perhaps we have been too busy to be bothered with the irresponsible statements of a vocal minority. The result... is that now the public looks on the conservation issue as something of a grudge match between the oil industry and the conservationists... as a conflict with no holds barred between two totally opposed points of view.
>
> Nothing could be further from the truth. The oil industry is vitally concerned with the problem of conservation... and will back the genuine conservationist.[11]

Easley went on to say that offshore oil rigs were fitted with complex and effective blowout preventers and that, if only for economic reasons, explorers would do anything in their power to prevent accidents. He also pointed out that fishing and the onslaught of the Crown of Thorns starfish had already done a great deal of harm to the Barrier Reef — much of it unpublicized. However his comments went virtually unnoticed in the continued blare of publicity surrounding overseas incidents like the *Torrey Canyon* tanker accident off the English coast in 1967, the Santa Barbara platform blowout off the Californian coast in 1969 and, closer to home, the Petrel gas blowout in the Bonaparte Gulf in late 1969 and the sinking of the tanker *Oceanic Grandeur* in Torres Strait during 1970. Explorers' protestations of technical excellence and extremely low accident risk failed to halt the groundswell of opinion. Fearing a political backlash the Federal and Queensland Governments decided to hold a joint Royal Commission into the subject and a moratorium was placed on Barrier Reef exploration permits.

The first sittings in the Commission took place in May 1970 and, from that date until the last in July 1972, a total of 88 expert witnesses presented 465 exhibits and filled 18 200 pages of transcript. A final report was released in 1974 but it was far from being conclusive. Two commissioners said more research was needed, although limited drilling in carefully controlled areas was still appropriate in the meantime. However the commission chairman, former New South Wales Supreme Court Judge Sir Gordon Wallace, ruled that exploration should be postponed pending further study, and so the drilling bans remained. Sobered by this experience, the explorers began to look more carefully at all their operations, including the effects of search programmes on land. Geologist/biologist Reg Sprigg was one of the first to examine the industry's onshore record.

> Geological exploration leaves practically no permanent imprint on the environment, and regional geophysics little more. In areas of established [exploration] target interest, gridding for geophysics and finally drilling becomes more concentrated. In a discovery area it is relatively intense.
>
> Most land exploration in the interior is concentrated in the desert regions. Principal need is for temporary tracks across literally hundreds of miles of dune desert. This conjures up a picture of mangled sand dunes and new areas of drift. While it is too early to be dogmatic, the experience of the last decade or so in relation to Australia's worst deserts has been heartening.
>
> Several tens of thousands of line miles of exploration roads and tracks have now been cut through the most formidable of the country's dune sea deserts. These include the Simpson, Strzelecki, Officer, Victoria, Gibson and Canning Deserts. In practically every instance, while the tracks frequently are visible for many years from the air, on the ground they are commonly extensively overgrown and unrecognizable. Often they are recolonized by grasses of far greater food value than the original spinifex and sandhill canegrass.
>
> In the case of the Simpson Desert... on two occasions parties have set out to [locate] the deeply bulldozed French... track via Poeppels Corner, and failed completely to find it despite seeing permanent track markers in situ.[12]

Nevertheless Sprigg recognized that there were impending problems for the delicate environmental balance even in desert country, the more so as search programmes intensified.

> The exploration industry has special responsibilities. As the group most intensively concerned in 'frontier' activities, more of its energy must be directed to preserving the status quo in the environment. Unnecessary road construction must be curbed, interference with waterholes minimized, and all operational centres and camp sites must be conspicuous only by their tidiness and operational cleanliness. A single drilling site left in disorder, or uncovered dumps, must be expected

> to attract increasingly unfavourable public reaction, to the detriment of the industry generally.

Sprigg's interest, coupled with the work of naturalist Harry Butler and of APEA's technical manager, Dr Graham Maxwell (who was appointed to the position in 1974), led the explorers' association to draw up a Code Of Environmental Practice in 1977. It was a pioneering document for all Australian resource industries and pointed the way for a generally more enlightened, responsible approach to the issue in the 1980s. But the Barrier Reef remained out of bounds. In 1979 the Fraser Government placed an indefinite ban on drilling in the region and declared much of the region to be a National Marine Park. Although the explorers had never advocated drilling on the reef itself, they now saw little likelihood of being able to search off any part of the east Queensland coast and so turned their attention back to the country's more accessible basins.

Chapter 16

UNEXPECTED SUCCESS AND A DEEP-SEA GAMBLE

Although the changing electoral tide swept the Fraser Liberal-Country Party coalition into Canberra during December 1975, two more years elapsed before the oil explorers returned to the field in anything like full strength. Part of the delay was caused by the tardiness of the new government in ratifying its electoral promises concerning search incentives, including the new pricing regime and tax write-offs for investors in companies looking for oil. This meant that public funds, which had contributed to the major exploration efforts of the 1960s and early 1970s, were withheld and hence the local explorers were restricted in the extent of their programmes. Provided a company did not need much cash it could mount a low-key and relatively inexpensive campaign in regions such as the Surat Basin, and expect a better than average chance of oil or gas success. However it would do so without any expectation that there would be finds of Bass Strait proportions. Funding and incentives were lacking for explorers to venture into the real wildcat areas — particularly offshore — where the possibility of large finds was still present.

Another reason for the delay stemmed from the degree of decimation inflicted on the Australian industry by the Labor years. Technical people had found good jobs overseas and were in no hurry to return. Similarly, field hands, drilling and seismic crews had also dispersed — many into more secure vocations. In addition, the foreign companies, especially the majors, had found other parts of the world, like the North Sea, to be very worthwhile exploration areas during the mid-1970s. Therefore they had little time or budget left for Australian programmes. As the vice-chairman of APEA, A.H. Brackenridge, remarked when summing up the Association's 1976 conference, about 85 percent of the exploration expenditure in the previous 15 years had been made by overseas groups. Australian investors, he said, had gained almost 50 percent of the petroleum reserves discovered as a result of this expenditure. During 1976 and 1977, however, the local industry struggled on with a much-reduced foreign presence.[1] By early 1978 the (then) chairman of APEA, Ric Charlton, felt the need to admonish the government's approach.

I believe the time has come when we must question, very seriously, the level of government control, and the way in which this control is exercised. I am not suggesting that government should withdraw completely — that would be both unrealistic and impractical. It is obvious that only governments have the power to establish the sort of economic and political environment in which the exploration for, and

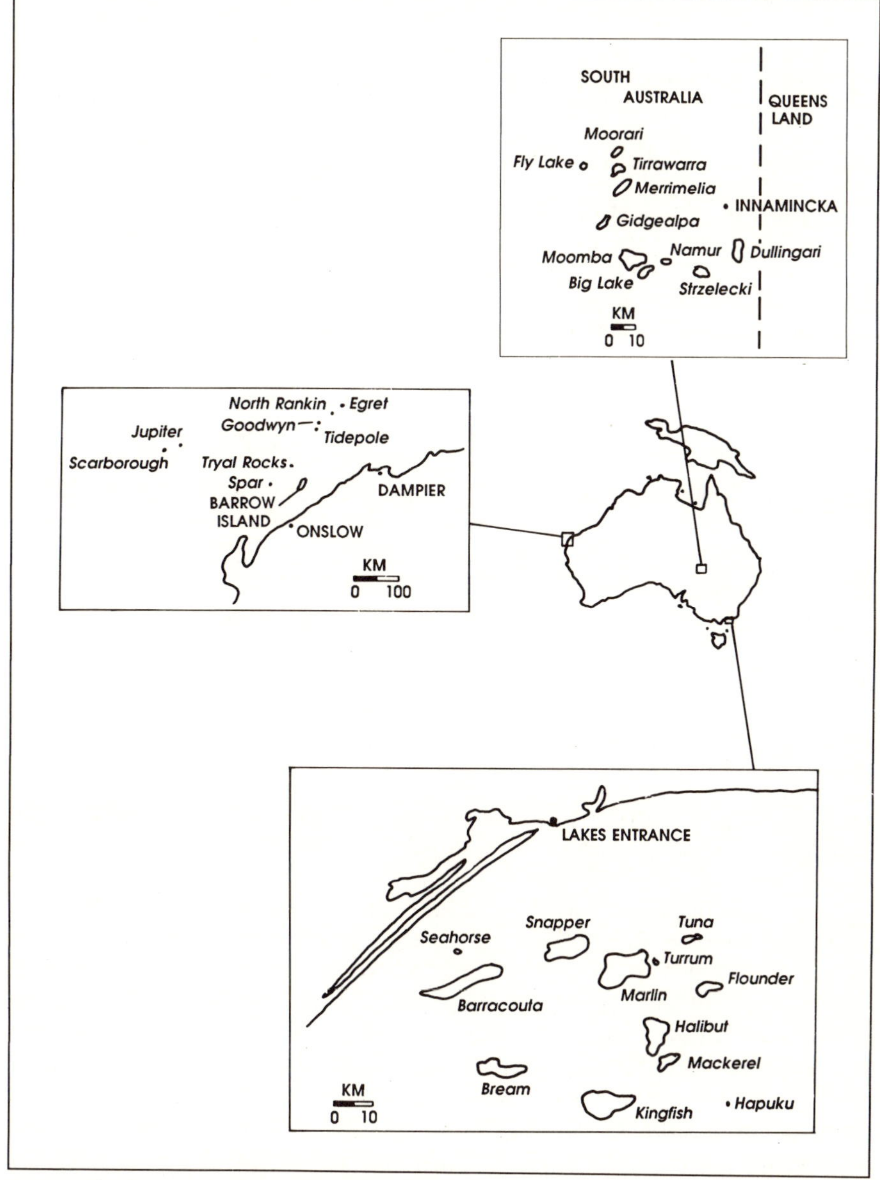

Location of key places mentioned in text.

> production of, oil and gas can be prosecuted with maximum effort and efficiency.
>
> It should, by now, be equally obvious that governments also have the unique capacity to create an environment in which investors are unwilling to venture their capital in an activity where the risks are high, the bureaucratic burden grows heavier each day, and there remains the threat of discriminatory taxation.
>
> The role of government, to my mind, is quite clear and simple. It is to provide the framework and establish the broad rules under which the petroleum exploration industry can get on with the vital task of exploration and production.
>
> ... The present Federal Government [of Prime Minister Fraser] was elected on a platform of a belief in, and an acceptance of, the private enterprise, free-market system. Those of us who are in the private sector are entitled to ask the government to release the petroleum and mining industry from the excessive control which has been progressively tightened over the years.[2]

Charlton's words were prompted by concerns that the government was considering a secondary profits-related tax over and above the complicated new oil pricing system imposed on production from existing discoveries introduced in the 1977 Federal Budget. Although complex, and destined to become more so, that policy did allow producers to receive more for their crude than under previous arrangements. But the concept of an additional profits tax was viewed as a government giving with one hand and taking away with the other. After considerable debate the plan for a secondary tax was shelved. More importantly for explorers, the concept of levy-free, full import parity (world-related prices) for new oil discoveries, belatedly introduced by the Whitlam Government in 1975, was allowed to stand. In addition, the Fraser Government decided to reintroduce a 30 percent rebate for expenditure on both onshore and offshore exploration and development. So, towards the end of 1978, the companies began to regain the confidence lost during the first half of the decade and activity began to increase once more.

In the end, however, the exploration revival of the late 1970s was sustained by two unexpected successes — one offshore and the other onshore. By 1976 almost two-thirds of Australia's indigenous oil production came from Bass Strait fields and, although exploration in the region had slowed during the Connor era, there was work done by reservoir engineers behind the scenes to evaluate the potential of previous finds. The price rise of 23 cents a barrel announced by Gough Whitlam in 1975, small though it was, prompted the decision to develop Mackerel and Tuna discoveries. Appraisal drilling raised the reserves in Cobia to a commercial level and further work on the west flank of the Kingfish field indicated that a third platform would be feasible there. However the overall drilling programme was certainly nothing like the continuous success story it had been during the 1960s. Between the Cobia No.1 discovery in August 1972 and the Cobia No.2 appraisal in May 1977, Esso/BHP drilled 27 exploration/appraisal wells. Twenty-five of them were dry and the other two — Turrum No.2 (gas) and Hapuku No.1 (oil) — were noncommercial.

Ironically, Whitlam's pronouncement of excise-free status for oil found after 15 September 1975 came just too late for Hapuku, which was discovered only 15 days earlier. But, to be fair, the extra fortnight would not have bridged the gap between viability and noncommercial status because the find was made in deep water and the accumulation itself was small. Geologically, however, it was of interest because it indicated that oil had probably been generated in sediments much deeper than those found in the inshore areas of the Gippsland Basin. It was food for future thought. The first 'new' oil discovery was a well called Seahorse No.1 drilled much closer to the coast not far from the Barracouta field in 1978. But it too was small (a few million barrels) and, although completed for potential production at a later date, was classified as noncommercial at the time. Described at best as encouraging, these results were soon overshadowed by a much more momentous discovery, the outcome of which was as unexpected as it was controversial. The Institute of Petroleum's quarterly magazine *Petroleum Gazette* referred to the sequence of events as 'The Fortescue Saga'.

> ... the saga began in September 1978 when a well to test the western extension of the big Halibut field found oil, but also hinted that the discovery was physically separate from the main Halibut reservoir. The West Halibut extension was hastily renamed Fortescue, and three subsequent wells drilled late in 1978 and early 1979 confirmed a good oil field. Through measurements of depths of oil/water contacts and pressure differentials, reservoir engineers postulated that an impervious siltstone layer cut the structure between the main Halibut field and the newly drilled extension.
>
> Esso/BHP... set out a geological proof that Fortescue was a separate field from Halibut, and added that if this was the case, it qualified as 'new' oil under the Federal Government's pricing system. In other words, Fortescue oil would receive full import parity price, free of levy — obviously an important point both for development of existing fields and as an incentive for further exploration in the area.[3]

The appraisal wells indicated that Fortescue contained 300 million barrels of recoverable oil, which made it the best discovery since Mackerel a decade before. Obviously its status as a new field was important to the explorers but, because the proof hinged on purely technical grounds, there was a lot of scepticism in political circles about the authenticity. To put it bluntly, some felt that Esso/BHP was trying to 'put one over' the government to ensure a much higher price than would be the case if the oil was classified simply as an extension of the previously delineated Halibut reservoir. However the joint venture geologists, people like Bernard (Bunny) Brown of BHP, had no such doubts.

> The definition of a field is always something that is difficult to pigeonhole but, because Fortescue was slightly unusual and had large reserves, it became a major focus of contention. The fact was that West Halibut No.1 was originally drilled looking for undrained oil within the Halibut structure itself. Instead it discovered a deeper oil/water contact than the one already mapped in the Halibut field — not just a small discrepancy, but 15 m deeper. It was obvious to the technical people that the new well had found a section of untapped oil and that there must be a seal between Halibut and West Halibut.
>
> Looking back at the records we could see the same thing had been noted with the West Kingfish extension (Kingfish No.7), but in that case the discrepancy was much smaller and it was rationalized as a well data-gathering problem. Opinions changed after West Halibut came in and the concept did cause a new wave of thinking, with drilling carried out on a number of westerly extensions of other discoveries like Marlin and Barracouta during the next few years. Unfortunately there was no further significant success.[4]

West Halibut was renamed Fortescue partly to avoid confusion and partly to try to minimize the controversy, taking the new name from a dry hole drilled in a nearby structure three months earlier. The boards of Esso and BHP were particularly sensitive to the politics of the issue and it took the technical teams some time to convince them of the accuracy of the claim. However, this done, the geological theory was submitted to the Bureau of Mineral Resources, acting as the government's adviser, and then to the Department of National Development before it was passed on to the Minister (Kevin Newman) and finally, the Cabinet. In October 1979 Fortescue was declared to be 'new' oil, confirming it would attract a price of $12.59 a barrel as opposed to $2.33 a barrel had the submission been rejected.

With the pressure obviously on the Federal Government to uphold its 'new oil' incentive, sceptics saw this decision as more of a political gesture than a genuine conviction in the correctness of the complex geological argument. But, political or not, when combined with the sudden new rise in world oil prices caused by the panic surrounding the Iranian revolution, it had the desired effect of kick-starting the country's stalled oil exploration effort. Esso and BHP were delighted and immediately announced plans to develop Fortescue and the more problematic 'old' oil field of Flounder discovered back in the 1960s. But, perhaps more importantly, the joint venture also announced a revival of exploration drilling in Bass Strait with plans to spend $250 million on new seismic surveys and wells during the following three years. By a stroke of what can be described as 'oil field luck', this offshore enthusiasm coincided with a series of events in the Cooper Basin of South Australia which ensured the onshore sector would be part of the industry's rejuvenation.

By 1978 the Cooper region had been firmly established as a gas province. Between the discovery of Gidgealpa field back in December 1963 and the beginning of 1978, the explorers, led by Delhi and SANTOS, had found 27 gas fields — 22 of them in South Australia and five more just over the border in southwest Queensland. Although many of the finds were rich in condensate, there had been very little oil discovered with the exception of that associated with gas at Tirrawarra, Moorari and Fly Lake to the north of Moomba. For virtually the whole of this 15-year period the companies had been concentrating on the known reservoirs of Permian age and little attention was paid to formations higher in the sedimentary sequence. Rod Hollingsworth, who joined Delhi as chief geophysicist in 1975, had just been promoted to exploration manager of the company when this geological oversight was remedied with startling results.

> It's probably not really true to call it an oversight because the early 1960s drilling in the region tested everything that looked as though it had reservoir potential. The presence of Mesozoic-age sediments above the Permian reservoirs were known well enough because they provided the main aquifer in what was broadly called the Great Artesian Basin. In a number of wells there were hydrocarbon indications at this level, but when tested they always flowed strong volumes of water. This happened so often that the geologists decided to save the money and drill on through the Mesozoic rocks without stopping. It got to the stage where a geologist was not even sent to the well site until the drill bit was nearing the Permian targets. In many cases the gas detector was not connected for the early part of the well.
>
> The first exception to this came in 1976 when a well called Namur No.1 drilled through a gas sand of Jurassic age (a part of the Mesozoic sequence) and established economic reserves. Even so the discovery was thought to be out of the ordinary and it was explained by assuming that gas had migrated up from an underlying Permian reservoir via nearby faults. But then the following year a small amount

> of noncommercial oil was found in the Jurassic section of a well called Poolawanna No.1 and some of us began to wonder if we might be missing something by targeting only the Permian section.
>
> But the real eye-opener took place in 1978 with a well called Strzelecki No.3. The first Strzelecki well had been drilled some years previously and found gas, but the follow-up appraisal had been dry so the prospect had been left for the time being. In 1978, with better seismic control, we decided to try again and sited the No.3 well right on the crest of the structure. It flowed oil at 2400 barrels a day from a Jurassic reservoir called the Hutton Sandstone which in fact is one of the main aquifers in the Great Artesian Basin.
>
> About the same time someone noticed that one of the gas wells in the Dullingari field nearby was leaking oil to surface round the casing. It was decided to drill a new well — Dullingari North No.1 — to check the Jurassic reservoir. When that found a thin oil sand in the Murta Member at the top of the Jurassic sequence we retested several other wells in Dullingari with the same result. It sent us scurrying back to relook at maps of the other fields in the region.[5]

By this time too the New South Wales gas market had been tied into the Cooper Basin gas supply and hence the production of Permian oil, condensate and other liquids associated with the gas stream had become a viable proposition. Sensing the future oil potential of the region might be even greater in the light of the new Jurassic finds, Delhi immediately went back to the drawing board and initiated a new drilling programme over the existing gas fields. Success was rapid. A total of 11 oil pools were discovered in the Jurassic sequence between 1978 and late 1981 including Strzelecki, Dullingari, Moorari, Merrimelia, Gidgealpa and Big Lake. By then geologists had discarded the name Great Artesian Basin and, for the first time, Australian investors caught up in the excitement heard the new discoveries referred to as Eromanga Basin oil. What is more, the mood of optimism manifested itself in a rash of new programmes right round Australia, particularly into parts of the country covered by Jurassic sediments originally mapped as being within the Great Artesian Basin boundaries. Even the oil majors became interested as indicated by the Galilee Basin project in central Queensland where Esso geologist Chris Curnow was in charge. It was soon seen that the big company approach differed markedly from that of the smaller Australian groups.

> The Galilee move stemmed from our basin studies group in Sydney and it began as a surreptitious thing. We felt that the region had not been properly evaluated in the past and

> that there were two, possibly three, potential target zones to look at. They included the Permian and Jurassic, but also the possibility of even deeper Devonian-age reef sequences. However to do the programme justice we decided that before any drilling could be attempted we would need to mount a preliminary seismic programme on a similar scale to those the company had been using in Bass Strait. To do that we needed a lot of acreage.
>
> That's where the secrecy came in. Obviously if it became known that Esso was interested in the region, others would try to move in too. That would put the price up and maybe spoil our efforts to secure the ground we wanted. So we put together a bid package in Sydney and used a go-between to approach the companies which held the permits in question. One of the permit holders was a company called Viking Exploration run by Chuck Einarson, an American with a Nordic background — hence the company name.
>
> He was based in Denver, Colorado, but I didn't meet him until the Exxon board in New York had approved our Galilee strategy. Then I went to Denver and revealed our hand. Einarson was an oil entrepreneur of the old type and at first he couldn't believe that it was Esso behind the approaches he'd had. It tickled him and we got on really well. He took me to his place in the nearby Rocky Mountains and taught me fly-fishing while we worked out the final details of the farm-in.[6]

A similar deal was negotiated with the other permit holder, Earth Energy of Calgary, Alberta, and the area involved in the $13 million programme which followed totalled 140 000 km^2, stretching from Winton in the north to Longreach in the south. The overall commitment for Esso was an amazing 4 500 km of seismic survey and 12 exploration wells. Because of the shortage of seismic crews in Australia, the company imported two US crews and began work with a fanfare of publicity which included a series of television advertisements pointing to the effort Esso was making to find new oil reserves in the country. Based in Longreach, the exploration teams quickly spread out over the permits. For the next two years, vibroseis trucks, followed by drilling rigs and a fleet of ancillary vehicles, raised the central Queensland dust as well as the anticipation of the programme overseers in Sydney. Unfortunately, after such a huge build-up, the results were an anticlimax. Although promising Jurassic prospects were delineated and then drilled during the early 1980s, the Galilee programme ended in virtual obscurity without repeating the Delhi group's rash of oil discoveries in similar formations further to the southwest.

There is no doubt that the onshore Queensland project had been a sizeable gamble, even for a company of Esso's financial and technical strength. But it was nothing like the dimensions of the challenge occurring simultaneously some 300 km off the central Western Australian coast. Fired up by the twin incentives of parity price for new oil discoveries and the possibility that the Iranian crisis would push that price to much greater heights, four exploration groups gathered at the frontier of drilling technology and prepared to tackle a region known as the Exmouth Plateau. Many of the big names were there — Esso, BP, Mobil, Phillips, Hudsons Bay Oil & Gas, along with local companies BHP, Mount Isa Mines, Australian Oil & Gas Corporation and Woodside Petroleum — arranged within five permits located on the outer edge of the North West Shelf.

The challenge was that water depths in the region ranged from 1 000 m to about 3 000 m and there were only seven drilling vessels in the world capable of working in those conditions. The gamble was that although large structures were known to be present, they would have to be full of oil to make any commercial development possible. A small oil find, or even a medium-sized one, would not be viable in such deep water so far from the coast. A gas discovery would not be commercial no matter what its size. Although the appearance of this region as a vital new exploration play seemed rather sudden and caused a great deal of media attention during 1979, the path to Exmouth Plateau had actually begun 10 years earlier during a BMR programme known as the Continental Margin Survey. Peter Cameron graduated from Professor Sam Carey's geology tutelage at the University of Tasmania in 1970 and immediately joined BMR's marine section, overseeing the survey.

> The work was done under contract to geophysical company CGG using a converted North Sea fishing trawler. We began in Torres Strait and surveyed down the Queensland coast right round southern Australia and back up to the North West Shelf. The whole data gathering 'voyage' took about three years.
>
> Alternating with the other BMR people in the team I'd spend a month at sea at a time running seismic, gravity, magnetometer and bathymetry surveys out from the coast to the 4 000 m water depth contour on grid lines about 40 km apart. Some of the bathymetry measurements were literally the first taken since Captain Cook's voyages in the eighteenth century.
>
> Then I'd come ashore for a month to plot up the data. The resultant maps and sections were of varying quality and, although we did gain a good overview of Australia's continental margins, a lot of the work was not particularly relevant to the oil explorers because it was in very deep water outside the permit boundaries. The exception was off the mid-West Australian coast where we recognized a number of huge structures and these later became the basis of interest for the Exmouth Plateau programmes.[7]

In fact the world's oil majors like Exxon, Shell and Gulf Oil were sufficiently interested in these results during the mid-1970s to run seismic programmes of their own over the region, and they liked what they saw. The structural framework of the Plateau seemed to be similar to that of the North West Shelf, and the thickness of sediments appeared to be sufficient for the generation of hydrocarbons. The final ingredient required to prompt a drilling programme clicked into place with the rise in world oil prices following the 1973 OPEC crisis, and when the five Exmouth Plateau permits were gazetted in 1977, the response was the most enthusiastic ever seen to newly opened Australian territory. Twenty-eight companies arranged in eight groups made application. Four consortia were successful and they outlined programmes involving a total of 34 wells plus another 3 400 km of seismic survey — all at an estimated cost of $300 million. The *Petroleum Gazette* for June 1978 recorded the unbridled enthusiasm of the time.

> The hounds of the hunt — geologists, seismologists, drillers — have been let loose to pelt ahead of the production engineers. If the hounds make a kill, the expectation is that the engineers will catch up, without causing undue delays in production schedules. In other words, an oil strike itself would be the major spur to accelerate development work and motivate the successful explorers to 'bite the bullet' and be the first to put the new hardware to the test.[8]

By 1979, when the first drillship appeared on the horizon to begin work, the Iranian crisis had made the potential prize even greater by pushing oil prices to an all-time high. Suddenly the Exmouth Plateau was front page news across the country, billed as Australia's last chance to find oil fields big enough to replace Bass Strait. Costing $10 000 per day each to operate, the three vessels — *Sedco 445, Sedco 472* and *Sedco 471* — embodied the most sophisticated pieces of equipment then possessed by the drilling industry. Conventional anchoring was not possible in such deep water, so the rigs were kept on station using submarine electronics whereby signals sent from transponders placed on the sea bed notified on-board computers of any movement off location. Thruster propellers in the hull automatically corrected the drift. The vessels were also capable of carrying enough supplies without replenishment for up to three months at a time, a factor which made the logistics of the operation a lot easier. On-board machine and fabrication workshops also gave offshore drilling a greater degree of flexibility and independence than had been previously experienced in the country.

Nevertheless the rigs were not immune to problems, as the Phillips Petroleum-led group found to its cost with a series of mishaps on *Sedco 471* in an early well, Jupiter No.1, during 1979. First the blowout preventer assembly — a vital piece of equipment in any drilling operation, but particularly in a new and unknown region — was dropped to the ocean floor, posing enormous difficulties in retrieval in 1 000 m of water. Then there were difficulties with the marine riser pipe connections and, finally, an electrical fault caused a fire in the vessel's vital computer controls. The fire was quickly extinguished, but not before it had damaged the station-keeping equipment and the rig had to be taken to shallower waters and anchored conventionally so that repairs could be made. The inevitable delays caused by all these accidents stretched the time for drilling Jupiter No.1 to five months and pushed the all-up well costs to around $30 million. On top of that came the disappointment of it being a dry hole.

Unfortunately for all concerned the Jupiter No.1 result was symptomatic of the whole Exmouth Plateau programme. The sense of anticlimax was more acute than it had been in the Galilee Basin and certainly the cost was far greater. Esso/BHP, working in two of the four permits, spent close to $130 million in drilling costs for eight wells, seven of which were dry. The joint venture did find and retain rights to a large gas field with the eighth (Scarborough) but, as the explorers knew from the outset, gas would not be viable as a development prospect for the foreseeable future. Phillips drilled a second dry hole in 1980, while the Woodside group also drilled two disappointing wells during the same period. Exploration programmes which had begun with such enthusiasm in 1979 petered out to nothing by the end of 1980. The final axe fell in 1982 when Esso/BHP became the last group to relinquish its Plateau acreage with the exception of a small licence kept round the Scarborough gas find. This remains as a rather forlorn reminder of Australia's first, and so far only, deep-sea gamble.

Following the success at Woodside's North Rankin and Goodwyn in 1971, exploration had continued along the North West Shelf during the second half of the 1970s (mainly through the Woodside, WAPET and Arco groups) in the hope that structures there might contain the hoped-for large oil prize that eluded explorers elsewhere. In the event some oil was found, but it was in small noncommercial pools like Tidepole, Egret, Caswell, Swan, and even next to Goodwyn itself. Frustratingly, the bigger finds — like Tryal Rocks, Spar and Brecknock — were all gas/condensate and, as such, stood in the shadow of development plans for Woodside's two earlier discoveries.

The latter half of the decade had brought a similar feeling of frustration for explorers in Papua and New Guinea. The original APC group watched Texaco discontinue its farm-in programme in 1974 after the failure of the Cecilia highlands wildcat, and then saw the 16-company Japanese Oceania Petroleum consortium walk away from the country in 1976 after its farm-in programme around the shores of the Papuan Gulf had registered three dry holes. A brighter note was sounded in 1978 when the US major Gulf Oil became interested in the highlands potential and decided to take over where Texaco had left off. Nevertheless, each step in building up the geological picture in New Guinea was still a huge physical and financial effort, particularly for the small Australian explorer Oil Search.

Despite the advent of modern data-gathering apparatus and the use of helicopters and better communications, the jungle was still king. Certainly it was easier to move equipment from one place to another, and the air lifting of food supplies to the field camps obviated the need for the long lines of carrier bois. But geologists still wanted to see and measure the rock outcrop on the ground and nothing could substitute the walked traverse line. Seismic surveys still required line cutters, so a single party might consist of up to 200 employees. In addition, the method of using a fixed area base camp and a series of canvas and stretcher field camps remained unchanged. Yet the second half of the 1970s did bring change, subtly at first, then more noticeably as the decade came to a close. In 1975 Papua New Guinea gained Independence. BP geologist Paul St John watched the early transition from his vantage point as an oil explorer within the APC group.

> The most obvious alteration was to the patrol officer system and its organization of the local communities. Before 1975 the patrol officers (kiaps) — mostly Australian expats — cooperated with the geologists on their traverses through the country and often accompanied them or sent police bois along. There were few territorial disputes and any trouble was dealt with quickly, usually on the spot.
>
> But after 1975 the patrol officer positions were given to local Papuans and unfortunately the system lost its discipline. The local officers somehow did not command the respect former kiaps had and, consequently, any educated Papuan did not want the job. Patrols through the bush visiting the villages became a thing of the past because the new officers liked to stay at their desks.
>
> This made things difficult for the oil explorers because jurisdiction within the various oil permits reverted to clan boundaries and it put restrictions on geological work. Instead of a patrol officer being able to help organize carriers and line cutters for a whole survey as they had in the old system, the oil companies were compelled to train local people in the work, only to lose them at each clan boundary and then start again by hiring people from the next group. So a lot of delay to programmes was caused by visiting local groups, talking and training.
>
> The locals became pretty sharp at bargaining and positioning themselves with their claims, but the field geologists and surveyors understood this and a pretty fair work system did develop. Sometimes though, the company top managements would undo a lot of hard bargaining by flying in for

> a day and 'magnanimously' granting some concession or other. It only made working relationships with the local people more difficult.[9]

Quite a number of expatriates stayed in the country after Independence and some set up businesses which specialized in organizing supplies, equipment and labour for the oil exploration teams. Many soon became local characters, like John Senior, the 'King of Kikori', an ex-APC man who married a local girl and set up a trade store at Kikori on the Papuan Gulf. Another was Bert Counsel, who also married a local girl and set himself up further round the coast at Vailala. He had a sawmill, bulldozers for hire, a hotel and trade store. Neil Ryan, a former kiap, chose the highlands capital of Mt Hagen to found his logistics company specializing in presurvey work such as liaison with local villagers, selection and construction of base camps, clearing helipads and building temporary bridges.

Gradually the explorers absorbed the change. Despite the drilling disappointments of the mid-1970s, the APC group hung on to its acreage. Survey work continued slowly in all its permits, but the accent slowly moved from the coastal regions in the south to the highlands making up the country's spine, as geologists began to realize the significance and potential of the fold belt structures. Gulf Oil's decision to farm-in to the two permits in this region in 1978 was a major boost, not only to APC, but to the country as a whole. For the newly independent New Guinea the agreement was an historic occasion because it was the first petroleum concession granted under the recently formulated laws regulating the nation's oil exploration and, hopefully, eventual production. For APC it was the start of an association destined to lead to the oil reservoirs which the group had sought for so long.

In both Australia and Papua New Guinea the 1970s had been a time of substantial change for oil explorers. For one thing the 'buccaneering' days of the individual geologists and entrepreneurs had given way to a much more ordered team approach to field work. The decade had also introduced the industry to a much greater political scrutiny in its daily workings than it had experienced before, and made it far more aware of the growing concern about environmental matters. But perhaps the most important advances of the 1970s were made in the realm of technology — the tools of the trade. By the time the decade had come to a close, Australia's oil explorers had become skilled in the use of new search techniques which ranged from LANDSAT imagery to the anchorless station-holding capability of drilling vessels over deep water targets. The decade had made the industry very much a part of the computer age.

Chapter 17
THE CHANGING TOOLS OF TRADE

As the membership of the Australian Petroleum Exploration Association grew in response to the amazing run of oil and gas discoveries during the late 1960s, there was a corresponding growth in awareness of the technical aspects of the industry. Many geologists and geophysicists even felt that the APEA Professional Division, formed in 1967 to disseminate technical information, had slipped too far into the shadow of the Association's political lobbying functions. The Federal Government upheavals of the early 1970s, and the industry's responses to them, greatly accentuated the pressure for separation, so much so that in 1974 it was decided to hive the professional division off as a separate, self-funding body. The new Petroleum Exploration Society of Australia (PESA) — was swiftly organized into a national executive overseeing branches in each State and drawing its membership from individual employees of oil industry companies. From that point, membership of APEA itself was confined to the corporate sphere.

Nevertheless the parent body was not oblivious to the need to take account of the increasingly sophisticated methods of exploration and production in its submissions to government and, in July of the same year, APEA appointed a full-time technical manager. Dr Graham Maxwell came to the post highly qualified. A petroleum geologist with the Shell group in the Netherlands and the West Indies early in his career, he had also been Associate Professor of Geology at the University of Sydney during the latter half of the 1960s. His appointment underlined APEA's decision to put its approach to government on a more detailed and formal footing. Quietly spoken and diligent in his research, Maxwell quickly assembled the background data for APEA submissions calling for subsidy replacements and for higher crude prices in Australia. He was also prominent in preparing the industry's case for the Barrier Reef Royal Commission as well as APEA's pamphlet entitled 'Petroleum In Perspective' — the first of a number of Association publications seeking to explain industry-related topics to government and the general public.

Within the industry itself, however, new search methods and technical breakthroughs were telegraphed via regular PESA meetings and through the annual APEA conference — the latter by that time an established event on the world as well as the Australian oil calendar. The tools of the oil explorer's trade were many and varied even by the close of the 1960s but, as computer technology and space research began to spin off into related commercial sectors, new methods and equipment literally exploded onto the scene in the 1970s. Much of this technology centred on survey systems, a good illustration being the advent of a satellite programme called LANDSAT early in the decade. Exploration vice-president of Exxon Production Research Company in the US, J.B. Coffman, reviewed the capabilities of this system for APEA delegates attending the 1979 conference in Perth.

> The LANDSAT satellite orbits the earth (via a near polar route) at an altitude of about 500 nautical miles [800 km]. It carries a multispectral scanner that covers the green to low infrared part of the spectrum, and the data are digitally recorded on magnetic tape with a five-channel system. As the satellite orbits the earth, it scans a scene which is 115 miles [185 km] square and has a nominal resolution of approximately 260ft [80 m]. The digital data are accessed by several earth stations and standardly processed images are made available to the public. Because of the potential use of satellite data, many companies have their own proprietary processing centres.
>
> Faults, lineaments, [rock] outcrops and formation contacts not easily mapped on the surface are readily apparent on the processed satellite photographs.
>
> [They] are also helpful in differentiating surface geological features because each rock type and mineral deposit has a unique spectral signature. By mixing the spectral bands with the computer, we can create false colours which better identify these deposits. If we know the location of one mineral deposit on the ground and determine its reflectance characteristics from the satellite, we can try several colour combinations until we find the spectral image that best identifies its location. Then we have the computer identify all other deposits with the same colour. Industry is now investigating this same application to hydrocarbon deposits.[1]

Perhaps the most dramatic improvements in oil search technology during the 1970s were the acquisition and processing of seismic survey data. While the methods of imparting a shock wave to the ground or through water did not change greatly, the advent of digital means to record returning wave data revolutionized the science. The recordings could be 'massaged' and 'manipulated' by computer to produce pictorial traces of extraordinary clarity. When this was combined with a survey grid of close-line spacing (as little as 50 m apart in some cases), a very accurate three-dimensional picture of the subsurface structure could be built up by the interpretation geophysicists. Because of the expense however, the 3-D grid was generally reserved for accurately siting delineation wells on a development prospect. It was only used to guide exploration drilling when the geology was known to be complex or the potential structures were very small.

Nevertheless the new found clarity, even on regional surveys, was a boon to all explorers because it introduced an interpretive tool known as seismic stratigraphy. Geophysicists learnt that different rock types (sandstones, shales, etc.) each gave characteristic traces on the seismic profile and, with experience, they could not only draw up an interpretation of the structure, but also fill in much of the lithology and compute a model for the depositional history of the region. Naturally enough when dealing with earth parameters, not all surveys were perfect and not all interpretations proved to be correct when ultimately put to the test. This was well illustrated during the early 1970s with a mooted breakthrough in what geologists called 'the impossible dream' — the detection of hydrocarbons without the drill. As the *Petroleum Gazette* of September 1974 remarked, no one had dared claim that ability since the days when water diviners hawked their services to gullible wildcatters.

> ... speaking in very qualified terms, the impossible dream has become a reality. Oil companies are successfully detecting the presence of hydrocarbons, mostly natural gas, from seismic charts in certain geological settings. Success is generally restricted to 'young' [Tertiary-age] rocks less than 65 million years old and to depths less than 2400 m.
>
> The new technique has not eliminated the risk factor in wildcat drilling, but it has certainly shortened the odds for gas exploration at least — shortened them five-fold in the opinion of one industry commentator. Previously the odds were about 25:1 worldwide against bringing in a wildcat well that would yield enough oil or gas to pay the exploration costs.[2]

The technique was called DHI, short for 'direct hydrocarbon indicator', but it soon became more popularly known as the 'bright spot' method. *Petroleum Gazette* went on to explain the phenomenon.

> ... when travelling through a medium, sound waves (or seismic shock waves) continue undisturbed until they reach some sort of change in material. Then some of the sound goes into the new material and some is reflected back... In travelling through ground, sound waves find that one kind of rock is much like another, so that geophysicists have to rely on only one or two percent of the sound being reflected back from a typical change in rock.

> But when the rock is porous and filled with water, the percentage of sound reflected increases markedly. And the porous rocks are even better reflectors when they are filled with gas or oil... The presence of hydrocarbons is indicated to the expert eye by an anomaly on the seismic chart. The 'bright spot' is, in fact, a dark spot — a deepening of the black on the black-and-white mosaic (caused by the greater intensity of reflections sent back to the surface from that point). Even so the bright spot on the chart is only an indicator — something to be assessed and argued about.

By the mid-1970s, however, it was realized that this phenomenon was of limited occurrence, and a more widely detectable indication of hydrocarbons was found to be the reflection of seismic waves from a fluid interface — such as gas over water, gas over oil, or oil over water — within a reservoir. These reflections were characterized on the seismic chart by a horizontal surface. The most likely explanation of the so-called 'flat spot' was gas over water, as the change in wave velocity between those two media was the most marked. But a few more years experience showed that this technology was not perfect either because a 'flat spot' could also be produced by thin coal beds and other material buried within denser rocks. Where the indicator did correctly diagnose gas, there was no guarantee that it would be in viable quantities, nor that oil would also be present. So although geophysicists had made substantial inroads with seismic technology, in commercial terms the impossible dream remained essentially intact. Nevertheless, explorers' horizons were very much broadened by the new computer techniques.

Another direct hydrocarbon detection technique tried in the late 1960s and early 1970s was the geochemical survey. The concept centred round the belief that gaseous hydrocarbons from a subsurface reservoir diffuse to the surface under the influence of pressure and temperature in a process known as microseepage. The geochemical survey was an attempt to detect this gas by taking samples of the soil in a regular grid pattern over a likely area. It sounded straightforward but there was potential for false readings. Jaap Poll, a geologist with Woodside-Burmah Oil in the mid-1970s, pointed out its problems to delegates at the 1975 APEA conference.

> In order to minimize the quantity of (misleading) syngenetic gas in the sample [i.e. gas formed from compacted recent organic material in the topsoil] care is taken that each one is collected from beneath the topsoil layer... Minute quantities of gas are always present on a regional scale and it is necessary to distinguish between regional 'background' anomalies and the local gas anomaly (which might indicate the presence of a hydrocarbon reservoir at depth).

> However gas anomalies are often due to other factors which cannot be readily recognized. This has caused controversy and nonacceptance of the geochemical method.[3]

About 14 geochemical surveys were carried out in eastern Australian basins between 1967 and 1974, but there was no evidence that any of them led directly to a subsequent oil or gas discovery and the technique fell into disrepute. Geologist Jack Mulready, who worked with Alliance Oil Development and then Delhi during the 1970s, noted its fall from grace.

> After an initial tryout during the early part of the decade, geochemistry as a surveying method in the oil industry took a back seat for a number of years. The major oil companies seemed to be more interested in developing its potential than the local Australian explorers, particularly from the early 1970s on. Despite being a relatively cheap preliminary method of screening potential hydrocarbon-bearing areas, it was seen as an evolving science and treated with caution because of its simplistic theories.
>
> There were more variables which could affect or influence results than many realized or understood at first, and that caused some disenchantment. If a geochemical anomaly was found, it tended to be treated as a qualified positive factor in the exploration programme. But if the survey found no anomaly, the results were never taken as a definite negative for the potential of the area.[4]

While the geochemical techniques were directed at onshore programmes, a more visual set of technological advances reaching Australia during the 1970s were directly related to the offshore sector where explorers continued to push at the industry's frontiers. In Bass Strait, for instance, divers were required to work in water depths of up to 76 m for long periods, thus diving bells and decompression chambers became vital additions to the equipment inventory. Drawing upon experience gleaned from work in various parts of the oil world during the 1960s, including the North Sea, the Gulf of Mexico and Bass Strait, companies had become more familiar with the hazards of deep-sea diving. It was known that each 10 m of water depth added another atmosphere of pressure, and that complications could occur in the human body as pressure increased. The potential problems a diver faced during an oil operation were laid out in simple terms in the June 1970 edition of *Petroleum Gazette*.

> [With increasing pressure] more and more oxygen and nitrogen is carried in the bloodstream, impregnating the body tissues. Too much oxygen can lead to oxygen poisoning. Decrease

> the oxygen content and the danger looms of nitrogen narcosis — a drunk-like form of irresponsibility. The answer to the nitrogen problem (particularly for dives below 55 m) is to substitute an alternative, inert gas — helium. This too is dissolved in the bloodstream and carried to the body tissues, but it has no harmful effect.
>
> There remains the problem of decompression. The degree of compression, or saturation of the body tissues by inert gas, increases proportionately to depth and time spent at that depth. Decompression, the reverse process of exhaling excess inert gas via the bloodstream and lungs, takes time. The environmental pressure has to be gradually returned to normal, otherwise the inert gas will emerge as bubbles in the bloodstream — the cause of divers 'bends'.[5]

The standard diving technique at, for example, 80 m depths allowed divers a bottom time of about 50 minutes. But then each man needed several hours in a decompression chamber afterwards. So, for jobs requiring a longer bottom time, it made more sense to keep the divers in compression for the duration of the job, rest periods as well as work periods — a process known as 'saturation' diving. But even that technique became impractical when the exploration rigs moved out to water depths of 300 m and more on the North West Shelf. A dive to 300 m, for instance, would require a day for compressing the diver and 11 days to decompress him. Clearly a more flexible, time-efficient solution was needed. It came during the mid-1970s when Australia's North West Shelf explorers were introduced to an apparatus the divers had nicknamed JIM. Once again *Petroleum Gazette* recorded the event with colourful prose.

> He lurks in the depths of the sea off the northwest coast of Australia: a strange, ponderous, robot-like creature. Arms and legs he has (of sorts) but no hands. In their place are devices that swivel and clamp and hold objects like a vice. Atop the trunk, or rather, growing out of it, neckless, is a monstrous, bulbous head with four great eyes, like portholes.
>
> Oilmen know and respect him, even address him by the familiar and disarming name JIM. Alternatively they describe him as the ADS, which stands for Atmospheric Diving Suit.
>
> ... At one stroke the suit eliminates the problems of saturation. In it, the diver operates at normal atmospheric pressure — like a submariner. He can attend or retire from a work site just as quickly as it takes a winch to lower and raise him.
>
> The suit is made of manganese and steel and weighs about 410 kg out of the water. It has articulated arms and legs which are neutrally buoyant so that the diver only needs

> to overcome the minimal friction of the joints. Understandably, a man in an ADS, working with manipulators instead of hands, performs his tasks more slowly than a man in a wet suit. But JIM offsets this against the great penalty of saturation diving — the hundreds of man-hours lost during decompression.[6]

The suit was the forerunner of a series of increasingly sophisticated deep sea 'vehicles' — manned and remote controlled — introduced to the oil industry during the 1970s for mechanical and inspection tasks. Many were equipped with TV cameras so that engineers on the surface could guide the manipulators, or at least give instructions and advice to the divers below. Although most of these devices were developed in places like the North Sea where diver conditions were extremely harsh, a number did find a place in Australian waters. By the end of the decade the country had also benefited from industry-wide improvements in drilling technology and combined many of these features in its own offshore exploration rig construction business.

Australia had built its first exploration rig — the semisubmersible called *Ocean Digger* — at BHP's Whyalla shipyards in South Australia during the 1960s, and it had successfully operated in Bass Strait as well as on the North West Shelf since its launch in 1967. Partly due to this success, and partly due to the Whitlam Government's preference for Australian-built vessels for use off the country's coasts, plans were drawn up for the construction of three more rigs carrying the national flag. The first of these was a drillship named *Regional Endeavour,* owned by a Sydney-based company called Drillships Ltd which had previously owned a supply boat of the same name. Signing an agreement with the big US drilling contractor Atwood Oceanics to operate the vessel, Drillships planned to convert the hull of an existing vessel — a former BHP ore carrier — into a self-propelled exploration rig with its drilling derrick and moonpool centred amidships. The conversion work was carried out in Newcastle on the New South Wales coast during 1973 and 1974.

Regional Endeavour was launched in September 1974, and immediately sent to continue Esso/BHP's long-running exploration/appraisal programme in Bass Strait, replacing the drillship *Glomar Conception* which had dominated the work during the previous two years. The new vessel was 30 m long and had a displacement of 14 224 tonnes compared with 9 144 tonnes for the Glomar rig. However one of its most important features was its capability to drill in water depths up to 350 m when most conventional drillships of that period had a limit of 200 m. Although the new rig was unlikely to need this capacity in Bass Strait, Australian explorers by that time were eyeing the potential of prospects on the outer continental shelf where such depths were the norm.

The two other rigs featured in Australia's mid-1970s construction programme — *Ocean Endeavour* and *Southern Cross* — were of semi-submersible design similar to the *Ocean Digger* but much more sophisticated in concept. Built in diagonally opposite points of the country, both employed innovative construction and launch techniques. *Ocean Endeavour* was half owned and fully operated by US drilling contractor ODECO, while local explorer Ampol Petroleum and the Australian Government-backed Australian Industry Development Corporation shared the remaining 50 percent ownership. Self-propelled to allow ease of manoeuvrability and carrying a construction price tag of $25 million, *Ocean Endeavour* was one of the most modern rigs in the world during the 1970s. But there was as much focus on the unusual launch method as there was on the rig itself. Because Australian drydocks were in short supply in 1972 when the construction contract was signed, it was decided to use a beach site for the job. As *Petroleum Gazette* remarked at the time, normally a vessel is launched by taking it to the water. In this case the water was brought to the vessel.

> At Woodman Point near Kwinana [just south of Perth in Western Australia] the waters of the Indian Ocean are already lapping the 12 000-tonne drilling vessel... as it sits securely in a beach pond behind great walls of sand. After moving millions of tonnes of sand from the beach and pumping in millions of tonnes of water, Transfield Pty Ltd [the rig builder] is close to proving that it is possible to complete a major shipbuilding job in Australia without an orthodox shipyard.
>
> Alongside the construction site, a deep pool had been dredged within the pond to a depth of about eight metres below low-water mark. The first task was to winch the vessel across the pond to the pool. With the rig safely in that position, the water could be lowered to sea level and float-out to sea would be possible [after breaching the retaining sea wall].[7]

Although there were temporary problems with faulty pumps and leaks in the plastic-lined pond sides, the operation was successfully completed early in 1976. By that time, over the other side of the country in Queensland, *Southern Cross* had reached the half-way mark in its construction schedule. Owned 51 percent by the South Seas Drilling Company (an associate of US contractor Santa Fe), it had an Australian content made up of the AMP insurance company and local rig operator ODE sharing the remaining 49 percent. The rig was built in two halves on the Brisbane River within a stone's throw of Storey Bridge leading into Queensland's capital city. Once the twin pontoon hulls were completed and joined, this lower section was ballasted down allowing the deck section to be floated over the top, lined up using laser technology and welded into position. Although smaller than *Ocean Endeavour* and without self-propulsion, *Southern Cross* possessed state-of-the-art chain and wire mooring cable which gave it a superior anchoring system to most of its contemporaries.

A downturn in exploratory work off Australian coasts soon after the new rigs became available, plus the politically imposed limitation on their ability to take work overseas, made it difficult for the three vessels to secure long-term contracts and fully utilize their potential. The main victim was *Ocean Endeavour*, which waited more than 18 months for its first well, and that was in Bass Strait where there was no real need for its size, manoeuvrability or water depth capabilities. Then, in 1978, when there was a revival in offshore activity, the accent shifted to the really deep waters of the Exmouth Plateau and the outer edges of the North West Shelf where conventional anchoring was out of the question. Hence the oil companies once again bypassed the Australian-built rigs and contracted the highly sophisticated, dynamically positioned drillships for the work.

One of the first vessels to break the deep-sea barrier was the research ship *Glomar Challenger*. Launched back in 1968, it was built specifically as a scientific rig to study the sediments of the Pacific and Atlantic Oceans. As such it was designed to operate in 7 000 m of water and continue working in severe sea states and adverse weather conditions. At the time it was described as a dramatic extension to the needs of offshore petroleum explorers but, as the industry entered the 1970s, it was realized that such technology had become a commercial requirement. Several globe-trotting, deep-sea drillships were in operation by the middle of the decade and more were under construction. Not only had these new vessels incorporated the *Glomar Challenger*'s unique design characteristics, but they had also improved them to a high degree. The most important advance — that of dynamic positioning — was explained to APEA conference delegates in 1979 by J.M. McKinney of Exxon's Production Research Company.

> The system uses acoustic signals from beacons on the ocean floor to monitor vessel position. The signals are received by hydrophones affixed to the vessel's lower hull and are then processed by an on-board computer. The computer processes the signals together with environmental data to determine what force must be applied through the thrusters [propellers located fore and aft on the sides of the hull] in a given direction to maintain vessel station above the wellhead.
>
> The computer then provides signals directly to the thrusters and main screws to control their output. The sequence may be likened to a humming bird imbibing lunch on the wing, continuously consuming fuel but never remaining absolutely still.[8]

Although these major advances in offshore work tended to grab the technology headlines, there were a number of other less spectacular, but nonetheless important, advances in the oil drilling industry. Many of them made life easier and safer for the onshore explorer as well as his colleagues off the Australian coasts. They included the introduction of lighter and stronger steels for drill pipe, casing and longer lasting bits; bigger and more powerful mud pumps; more competent and hence safer blowout and choke monitoring equipment; and more sophisticated and efficient downhole logging and testing tools. As with seismic surveys, computers and digital recording techniques revolutionized logging and well testing programmes. Logging techniques, in particular, were rapidly improved to a point where the raw downhole electrical, nuclear and thermal measurements could be converted by computer into direct readings of rock type, porosity, permeability and the identification of hydrocarbons in place. The programmes eliminated much of the previous hand computation work and analyses of the rock formations could be carried out at the well site. Data could also be instantly transmitted via telephone or radio link to major processing centres in town.

The content of the mud systems used during drilling grew much more diverse too during the 1970s, as engineers paid more attention to the chemistry of the formations through which they were drilling. It was found that using solutions of compounds like potassium chloride, instead of a water base mud, inhibited the tendency of some rock formations to swell or break down, thus allowing more accurate subsequent test runs to be carried out. Significant advances had also been made in the test procedures themselves, including improvements in the manufacture of valves and well seals and the development of multiple test chambers. The latter samplers allowed for a number of flow and pressure tests to be run without the need for bringing tools to the surface between each one.

At the time such advances seemed minor but, looking back over the decade in 1979, it became apparent that the exploration industry had accomplished a great deal on the technical front. When compared with the 1960s, the overall nature of the business had changed to the point where every field programme required computer-age technology and every operation involved a team effort. There was also a feeling among geologists, rightly or wrongly, that the surface work had been done in the most prospective basins and there was little to be gained from additional mapping at close quarters. Satellite imagery plus sensitive seismic recording and processing were thought far more likely to yield positive results. Geological debate during the 1970s centred instead on the two broad questions of permit size and the quantitative assessment of the country's remaining oil reserves. Consultant John Blumer outlined the permit arguments for delegates attending the 1982 APEA conference.

> The practical implications of granting very large permits are considerable and arguments can be made both for and against the practice, on such grounds as efficiency and maximization of exploration activity. It has often been stated that the availability of large permits has been the major inducement for

> overseas companies to become involved in Australia and that the total level of exploration over the last 20 years would not have been as great otherwise.
>
> Another argument in favour of large permits is that in relatively unexplored areas, permits covering entire basins or sub-basins enable the acquisition and compilation of geological and geophysical data on a regional basis. The development of coherent geological models which can act as the basis for further exploration is thus facilitated. Evidence supporting this argument can be seen in most of [Australia's and Papua New Guinea's] major onshore and offshore basins... and it is doubtful whether the geological models [which ultimately resulted in discovery] would have been developed to the same degree of sophistication in the same time if initial exploration of these basins had been carried out by a number of different companies operating in relative isolation.
>
> Against this it can be argued that regional exploration by a single operator involves the danger of wrong geological concepts becoming entrenched. It is all too easy for an entire basin to be effectively 'written off' because of lack of success by a single operator exploring with an inadequate geological model. Once the basic stratigraphic and structural framework of a basin has been established the argument for granting very large permits loses its force. At this stage it is important that a number of different operators be given the opportunity to test a variety of exploration concepts within the area.[9]

Blumer added that it was important to have a reasonable turnover of permits and that the most effective and equitable way of doing so was through regular relinquishment of a proportion of each permit under lease. The provisions for such relinquishment existed in the Petroleum Acts of each State, but there were wide variations in the details of permit term and proportion to be relinquished, as well as in the enforcement of the regulations. Similar debate ensued over the difficulties in assessment of Australia's oil reserves, and this subject was picked up by C.S. Robertson of the BMR during his wide-ranging talk about the country's petroleum prospects delivered to APEA delegates in 1988.

> Until the 1970s all published assessments of Australia's petroleum prospects were qualitative rather than quantitative. A number of attempts were made to classify basins according to perceived prospectivity as 'good', 'moderate', 'slight', etc. and/or to rank basins in order of prospectivity, but no estimates were given of the amounts of petroleum likely to be contained in the nation's sedimentary sequences.
>
> Quantitative assessment of an unknown quantity, such

> as the amount of undiscovered petroleum in a given area, is fraught with difficulties... One approach is have an individual or group of experts who are familiar with the area in question make an educated guess... This method, which reaches the answer required in one step, is relatively quick, but is... highly subjective.
>
> The other approach is to consider all factors which contribute to the occurrence, or nonoccurrence, of petroleum separately, and then combine the effects which all these factors have on the final outcome using a mathematical model, with the aid of a computer programme... this method requires many unknown components to be estimated separately... [and] involves much more thought, expertise and knowledge than the 'single guess' method.
>
> [A third] method... relies on the extrapolation of historical data relating to petroleum discoveries into the future.[10]

Robertson went on to record several quantitative assessments made during the 1970s. The first, by M.C. Konecki in 1972, suggested Australia's continental shelf could contain 120 000 million barrels of oil equivalent (including gas reserves) and optimistically assumed that the prospectivity of the country's continental shelf is representative of continental shelves worldwide. Then, in 1976, Frank Jeffries of Esso Australia published a detailed account which suggested that Australia had an 80 percent chance of finding more than 1.8 billion barrels of oil, but only a 20 percent chance of finding more than 5 billion barrels. Esso's average value for all possible outcomes was 3.6 billion barrels.

A slightly different tack was taken by Robert (Bob) Smith of the BMR in 1977. Without attempting to estimate the quantities of undiscovered oil, he made quantitative estimates of the chance of occurrence of petroleum accumulations in what he considered to be the most important nonproducing basins in the country. Smith's assessments were interesting given the benefit of hindsight. The Adavale Basin was rated as having a 1 in 8 chance of finding petroleum, the Otway 1 in 10, the Galilee 1 in 16, the Canning 1 in 25, the Bonaparte also 1 in 25, and the Eromanga 1 in 33.

The fact that new exploration in the Eromanga Basin the following year began a series of oil discoveries in the region, while a similar programme in the Galilee had no commercial success, bore no reflection on Smith himself because he was only using the conventional wisdom of the day. But it did indicate that the pitfalls of relying too much on such geological predictions had not lessened over the years. Perhaps more to the point, the attempts at quantitative analysis during the largely unsuccessful 1970s reflected a growing concern that the 'bonanza' days of the 1960s had truly passed and might not be repeated. As the 1980s approached, a new slant on the problem was suddenly provided by high oil prices in the aftermath of the Iranian revolution and the rush, albeit short-lived, for alternative energy sources.

PART FIVE

THE 1980s
PROGRESS AT A PRICE

Chapter 18
IN SEARCH OF A SAVIOUR

APEA stated the exploration industry's position at the outset by naming its 1980 annual conference 'New Decade — New Challenges'. The Association's chairman and managing director of local company Alliance Oil Development, E.H.C. (Ted) Garland, came directly to the point during his opening address.

> ... the seventies were responsible for shocking us into the stark realizations that oil is a finite resource; that what oil we have must be used more efficiently; that we now have a responsibility to step up the oil search for our short, medium and perhaps long-term needs; that we must develop, particularly for the long term, alternative fuel sources.
>
> Australia is currently 71 percent self-sufficient in its supply of crude oil, but without additional discoveries our rate of self-sufficiency will fall towards the end of the decade... The search for petroleum — onshore and offshore — must be continued and expanded vigorously.[1]

Garland went on to make the point that except for a few mature areas, such as the Gippsland Basin, Australia was still rank wildcat territory. There had been only 2 200 wells drilled in an exploration effort which, by that time, spanned more than 100 years. The APEA chairman conceded that improving technology was continually upgrading the worth of each wildcat, but he added that explorers were struggling to keep up with the increasing costs of progress. In 1979 Australia's oil exploration expenditure had totalled $220 million, which was double the previous peak year of 1972. Costs of the continuing search during 1980 were again expected to top the $200 million mark, but Garland wondered if these expenses were understood by people outside the industry.

> Figures of millions of dollars roll quite smoothly off the tongue when we are talking among ourselves... The cost of drilling

> an offshore well in shallow waters ranges from \$35 000 to \$55 000 per day. In the deeper waters such as the Exmouth Plateau, [costs are] in excess of \$100 000 per day. Onshore drilling costs range from \$7 000 to \$15 000 per day depending on the location.
>
> Mobilization costs for an onshore well could be as much as \$90 per metre drilled, that is, \$300 000 just to get a rig on site preparatory to spud and [then] rig down after the well is completed or abandoned. [All] this is without taking into account the prerequisite of seismic work, which ranges from \$700 to \$800 per kilometre offshore and \$2 000 to \$3 000 plus per kilometre onshore.

Yet in 1980–81 these relative costs did not register with politicians or the public at large. After all, exploration had suddenly boomed again right round the world. The Iranian crisis had seen to that by creating a fear of oil supply shortages and pushing prices past the \$A25 per barrel mark. Australia's explorers were soon out in force, new companies joining old, all hoping to cash in on the Federal Government's parity pricing incentive for a discovery. In many ways it was reminiscent of the heady

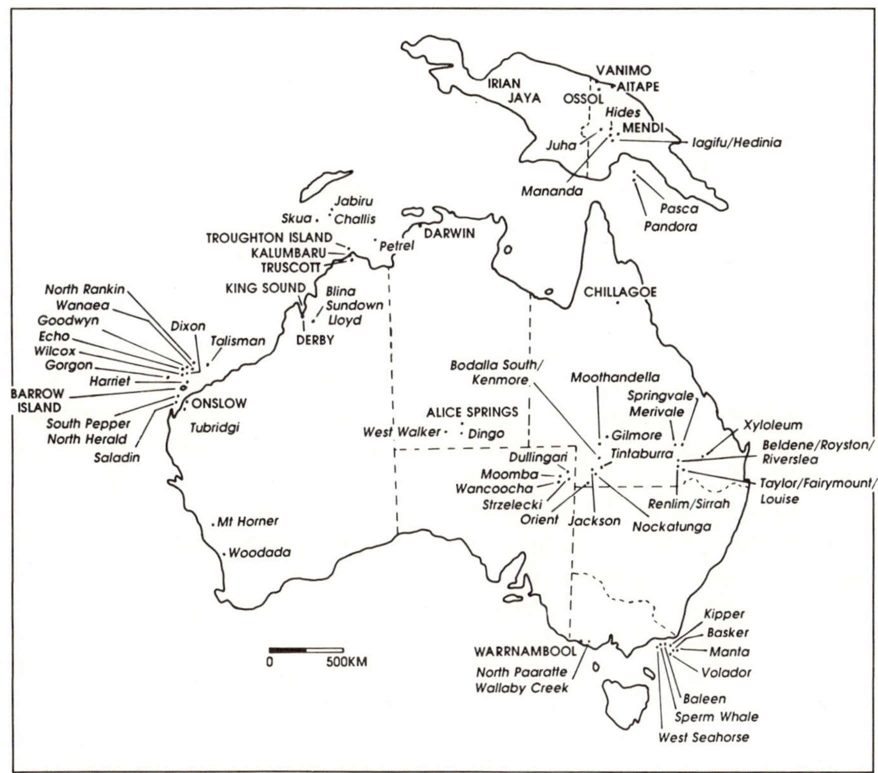

Location of key places mentioned in text.

days of the mining 'rush' a decade earlier. Exploration permits were at a premium, particularly if they had the right 'address', next to known discoveries. Farm-ins commanded high, often exorbitant, prices and the saying of the day was 'You haven't had a good day till you've done a deal'. A number of colourful oil entrepreneurs leapt to prominence during this time. For the most part they were men who found excitement in being where the action was, but preferred not to become fixed stars and often moved on before the companies they created became involved with development projects. Some accused these people of fostering instability in the exploration industry. Another school of thought acknowledged, however, that they put in a lot of hard work and that they were needed to keep the industry from stagnation.

Perth-based Alan Burns, an engineer/surveyor with some geological knowledge as well, had his appetite for resources whetted when he was involved in the discovery of mineral sands deposits at Eneabba, south of Geraldton on the Western Australia coast, during the late 1960s. He also had some peat mining licences round Perth and in the State's southwest, but at the time WAPET still had a large percentage of Western Australia's basins under licence so it was difficult to break into the oil industry proper. However, during the early 1970s when the difficulties of the Whitlam era began to cause a waning of interest in the Australian oil scene, Burns saw his chance to gather some exploration acreage. About that time he had acquired a nickel-boom company called Magnet Metals as part settlement of a commercial transaction and, very much an ideas man, he proceeded to build it up into a successful oil explorer. He was joined by Derek Gascoine, a skilful Perth lawyer with a reputation for being able to draw up and execute business deals when time was of the essence. Within a short time Magnet had amassed interests in 52 oil permits throughout Australia, more than any other explorer in the country. The two men overcame their relative inexperience in oil exploration by recruiting geologist Chris Curnow from Esso, and the group maintained the operatorship in 13 of the areas through three subsidiaries: Stirling Petroleum, Lennard Oil and Monarch Petroleum. Although a number of the permits were acknowledged to be quality areas with some good discoveries made during the high oil priced days of the late 1970s and early 1980s, the strain of servicing and financing such a large portfolio soon began to show. Further hampered by the tightening money supply in the mid-1980s as the boom came to an end, Burns and Gascoine sold the group and went on to other ventures. Fortuitously for them the sale went through in late 1985, just ahead of the oil price plunge in 1986 followed by the 1987 world stock market crash which cut a swath through the Australian resources industry and saw a number of other oil entrepreneurs fall by the wayside. While some of their new interests were outside the oil patch, both men remained in touch with the industry through private participation in exploration programmes. Alan Burns in particular believes that oil, and also gas, will rise to prominence in the 1990s as an investment opportunity and is repositioning himself with this in mind.

Geoff Albers was also inexperienced in oil exploration, but he too knew an opportunity when he saw one. Such was the case when he found a company called Oil and Minerals Quest 'on the shelf' during the late 1970s. After re-establishing it he also resuscitated companies called Zanex, South Eastern Resources and Cue Minerals, at the same time floating a new company he named Bass Strait Oil and Gas. Basing the group in Melbourne, Albers met geologist Colin Glazebrook, who had spent most of the 1970s in gold exploration, but also had oil experience in Australia and the North Sea dating back to the 1960s. Concentrating mostly on the local Victorian scene, the pair took the bit between their teeth and plunged into offshore permits in the Otway, Gippsland and Bass Basins, a region still considered to be the domain of the large explorers. Nevertheless, by the early 1980s, the Oil and Minerals Quest group (OMQ) had built up one of the largest offshore acreage portfolios of any small company in the country. At one point the Albers companies had interests in 32 areas on and offshore eastern Australia, including the first of three so-called 'total energy package' permits in the Clarence-Moreton Basin of northeastern New South Wales. This involved a licence which allowed companies to explore for oil, oil shale and coal in the one area. Unfortunately there was no commercial success with any of the group's exploration work and, like Burns and Gascoine, Albers stepped out of active oil dealings during the mid-1980s.

An explorer of quite a different character was Melbourne-born Ian Sykes. Vowing at age 13 to make a pound for every person living in Australia, he spent 10 years at Melbourne University in the 1950s graduating in economics/commerce, accountancy, geology and metallurgy. His first job was with the Decimal Research Organization before moving to stockbroker Walter P. Ham & Co. During the latter period he bought the company Jervois Sulphates and became chairman — a position he was to hold for 23 years. Sykes's first contact with the oil industry came when working as assistant to Eric Avery at the Associated Australian Oilfields Group during the 1960s. Meticulous, intense and hard-working, he formed the cut-price petrol retailing company XL Petroleum in 1966, tackling the major companies head-on. He even found time to write and publish a small book satirizing what he saw as their merciless domination of Australian and world oil markets. Then, in 1970, he registered an exploration company named XLX NL. Quickly accumulating interests in the Surat, Perth and Eromanga Basins, his next move was to buy his own exploration drilling rig at auction during the slump in Australian activity in 1975. By that time well on the way to achieving his boyhood aim, Sykes embarked on a course that singled him out as the country's only US-style wildcatter.

> I bid about $A108 000 for the rig [the auction was actually in US dollars] and then went on to build a camp and purchase drill pipe. Although the industry was pretty low at the time in Australia, I could see there was a good chance of the oil price following the rises seen overseas in the mid-1970s. Sure enough it happened about 1977 and in the final years of the decade it was hard to get rigs or equipment for jobs in Australia because they'd all gone overseas. Consequently my rig was in demand. I went over to the US myself in the late 1970s on a trip to see how they worked an operation there. The low cost was impressive and I felt it was pretty poor that the same jobs in Australia were so expensive.[2]

Sykes quickly built up a reputation for a no-frills approach to exploration. Many of his permits had been leased during the slump when few other companies were willing to take the field. Governments, however, were anxious to encourage any participating company and so the work obligations were not made too onerous. Sykes therefore had time to carry out drilling programmes for other operators during the subsequent revival. In 1980 however, he turned exclusively to his own areas, spending his own money and carrying out many of the rig tasks himself, including the geology and often the log interpretation and downhole cementing programmes.

> My crews were mostly made up of farm and station people in Queensland who'd taken up drilling in the off-season. Sometimes I borrowed off-duty people from other oil operators in the region so that they would come and work for me on their two weeks break. After a couple of programmes a lot of them would seek me out when I came up to spud a new well. In the end I found it best to just do the 12-hour daylight shift. It meant I only needed one crew of about seven men.
>
> Around the Surat acreage we had rig moves down to a fine art, leaving the drilling mast in one unit rather than dismantling it each time. I reckoned on an average $250 000 to drill a well. That was for basic fuel, wages and materials. Seismic expenditure had to be added to that to cost out the whole exploration programme. I had a good cook who did his own catering while his wife attended to the laundry and cleaning in the camp.
>
> I'd never use planes to get to locations, preferring to travel by bus or car. I built up a rig library and would run the operation and the company from the drilling site. There was one well in the Eromanga Basin of western Queensland where I didn't leave the rig for six months. There were no communications set up at the site so I had to use the pay

phone in the main street of town. It was sweltering in the booth and I had to prop the door open to let in a draft.

Few of Sykes's wells loomed large in published drilling activity during the 1980s, although Moothandella in the Queensland Eromanga found noncommercial oil and Xyloleum in the Surat was a noncommercial gas discovery. His most successful effort was a part-interest in an oil find at Mt Horner in the north Perth Basin, which he later sold in order to concentrate on other projects in Queensland. Yet 16 of his wells were drilled with his own finances — something no one else has attempted in Australia. Then, after watching stock market trends for over a decade, Sykes suddenly decided in 1986 that there would be a devastating and long-lasting stock market crash and he began 'shorting' the market. His timing was astray and he lost about $32 million in the exercise. Unrepentant, but unwilling to finance further exploration work, he put his remaining oil assets including the permits, rig, and camp up for sale early in 1989. Despite the 1987 stock market fall, he believed a much more horrendous financial crash was on the way and he did not want to be in business when it happened.

Another of the more colourful early 1980s entrepreneurs was Leon Le Grand, investment manager with the real estate company Hooker Corporation in Melbourne, where he made a fortune during the 1960s and 1970s. A solidly built man who had a penchant for after-lunch cognacs, drove a Rolls Royce and sported a big moustache, he was both flamboyant and generous. He also characterized one of those people willing to take a chance with their own money by setting up a company called Le Grand Minerals during the 1970s, prospecting for copper, antimony and gold in the Chillagoe district of north Queensland. In 1979 Le Grand anticipated the coming oil boom and had the foresight to reconstruct the company and begin taking up petroleum exploration permits round Australia, changing the company name to Metals and Energy Minerals in the process. The areas included interests in the Bowen/Surat, Clarence-Moreton and Drummond/Galilee Basins, plus a 10 percent farm-in to what proved to be a key region of the early 1980s — the Naccowlah block of Queensland Eromanga Basin permit ATP 259, operated by Delhi Petroleum. Having secured a number of vendor shares, Le Grand became deputy chairman and chief executive for a time. But the company 'rebirth' was a difficult process and there were frequent clashes within the management circle. Le Grand resigned in late 1981 soon after the company name was changed again (to Claremont Petroleum) and steered a new course, first becoming a successful novelist, then finding religion and conducting tours to the scene of a reported miracle in Yugoslavia.

Le Grand's move from Claremont came just after the company had participated in the Jackson discovery — one of the most significant oil finds onshore Australia since the 1960s. Once the Delhi/SANTOS groups had been alerted to Eromanga Basin oil by the finds in Strzelecki and Dullingari in South Australia during the late 1970s, the companies began to look more closely at the other parts of their vast acreage. These included the Naccowlah block in southwest Queensland contained within the permit known as Authority To Prospect (ATP) 259. In 1981 Claremont and two other Australian explorers — Ampol Exploration and Oil Company of Australia — were contemplating the best way to complete their farm-in obligations to earn a percentage interest in the Naccowlah area from original partners Delhi, SANTOS and Vamgas. Geologist Colin Glazebrook had not long left the Albers group and joined Claremont as exploration manager when the crucial joint venture meeting was held.

> By the time of the meeting we at Claremont had already earned 5 percent interest of our planned 10 percent in the block. Ampol and OCA had also earned 5 percent between them, which was half their projected 10 percent total. There were two prospects put up for drilling — Naccowlah South and Jackson — but SANTOS and Delhi were not keen on either. They were worried that there was not enough seismic confirmation of the structures to regard them as mature for drilling.
>
> At the time the farm-in partners were very keen to move to their full earning position which, by agreement, could only be done by drilling. More seismic expenditure might not have been a valid substitute. Anyway we insisted on drilling and we picked Jackson because it looked to be the bigger of the two prospects.[3]

The discovery of oil in three zones in the Jackson No.1 wildcat caused a sensation in geological as well as in stock market circles. The multi-reservoir find was quickly confirmed in 1982 and the group received an added bonus with another discovery in a smaller nearby feature known as Jackson South. Total recoverable reserves were put at between 40 and 50 million barrels — easily the biggest field in the Eromanga Basin — and plans were made to construct a pipeline running east across the bottom of Queensland to link up with the Moonie field line feeding into Brisbane. But perhaps more importantly for the oil explorers, the discoveries triggered an entirely new phase of exploration as geologists began to look more closely at the eastern section of the Eromanga Basin, particularly where it was underlain by the deeper Cooper Basin sediments. Suddenly all the permits in southwestern Queensland became 'hot property'.

Attracted by what it saw as similar geology in a trend stretching up to the northwest of Jackson, the Delhi/SANTOS group in the Naccowlah region quickly went on to record oil finds in prospects like Gunna, Chookoo, Naccowlah South, Wilson and Tinpilla during 1983 and 1984. Success also came in neighbouring blocks within the ATP 259 permit and to the Pancontinental Petroleum-led group at a discovery called Nockatunga in ATP 267 immediately to the east. Although the new fields were smaller than Jackson, the presence of a pipeline to Brisbane gave them the commercial status needed for development. However the full potential of the Eromanga 'play' did not become apparent until Christmas–New Year 1983, when a large group led by Hartogen Energy drilled a well called Tintaburra No.1 in permit ATP 299A. The location lay another 100 km east of the Nockatunga discovery so, not surprisingly, the lead-up to this well, like the original Jackson decision, was a contentious affair. Again Colin Glazebrook was involved in the crucial meeting, having left Claremont and joined US company Texas Gas with a mandate to set up that company's Australian operations.

> Part of the Texas Gas portfolio was obtained from the Magnet Group which was cutting back its operations at that time. We obtained 11 permits which included the Bonaparte Gulf Petrel field, some areas in Western Australia and also permit ATP 299A in the Eromanga Basin of Queensland. In the latter area a well proposal on the Tintaburra prospect was introduced at a joint venture meeting soon after we became involved.
>
> It was the only structure within a very large area over which any seismic work had been carried out. The only drilling in the permit had been Lou Smart's Orient wells in the south, way back during the 1960s.
>
> The joint venture was a big one and the majority of members did not support the drilling proposal. They wanted to run more seismic first. But Magnet (which still had an interest), Texas Gas and Hartogen wanted to drill straight away and eventually our plan prevailed.

From the moment of discovery it was realized that Tintaburra oil field was not as large as Jackson and that formation water could be a production problem. But it was commercial thanks to the proximity of the Jackson–Brisbane pipeline. Like Jackson, however, its major importance was in extending the potential of the Eromanga reservoirs in a great leap to the east. This lead was quickly followed up and confirmed during the next two years with further discoveries like Talgaberry in the same permit and Lasmo Energy's Bodalla South and Kenmore finds in the adjacent block to the north. Suddenly Eromanga acreage as far to the northeast as Longreach in central Queensland was in big demand. Even the old charts of the 1960s Gilmore gas discovery were exhumed and several wells drilled to investigate what was newly interpreted to be a possible oil leg in sediments above the gas reservoir. These were unsuccessful, but similar moves to the southwest parts of the Cooper/Eromanga region in South Australia continued to reveal oil reservoirs in structures where earlier drilling had failed. The

Wancoocha No.2 oil find southwest of Moomba during 1984 was a prime example.

Despite this drilling explosion during the first half of the 1980s, the physical conditions in the field camps had not changed a great deal since the 1970s. It was true the rig crews enjoyed a few more home niceties during their off-duty hours; for instance, they could watch videos in air-conditioned comfort instead of the old way of projecting a movie on to the white outside wall of one of the huts while the watchers sat round in the sand. But the work itself remained hard and long, and life could be lonely for those not suited to the isolation. Canadian-born drilling supervisor Ray Ell came to Australia in 1981 after answering a SANTOS job advertisement in an Alberta newspaper. Settling his family in Adelaide, he found the rig routine and even the scenery of the Cooper Basin not unlike the life he had left on the Canadian prairies.

> Maybe the winter cold was worse in Canada, but the heat is much the same; and you look out over a similar bald-headed plain. But really there's no time for gazing round. We run two 12-hour shifts a day seven days a week. The rosters vary a bit depending on your position. The drilling hands do two-week shifts and then have a week off. The catering staff usually do three weeks on and one off, and we drilling supers generally do two on and two off, although we are on call at any time day or night during our tour at the rig.
>
> Most of the guys come from farms in Queensland, ringers, station hands and the like. We don't get many city kids — they're generally not used to the life out here. It gets pretty hot and dusty. Mid-summer temperatures can reach 50 degrees Celsius. The roughnecks are usually in their mid-20s, pretty strong physically, and if the crew get along well together they stay 12 months or so. The routine doesn't vary much — work, eat, videos, sleep. Some read, but there's no grog. Normally we are too far out from either Moomba or Jackson bases to travel in to the amenities the production guys enjoy.
>
> Money keeps most of them at it. Drillers can earn maybe \$40 000 to \$45 000 a year while the roughnecks get maybe \$30 000. Then of course they have accommodation and food — and the food is good. Thing is, they don't spend anything while they're here so they can save quickly providing they don't blow it in town on their time off.
>
> I enjoy it out here myself. With a good crew we can roll things along. Mostly we know what to expect downhole and for a rig move we can dismantle the unit, take it 100 km and have it drilling again in four days. Not too bad with 45 to 50 truck loads of equipment.[4]

The early 1980s boom in drilling activity restored the industry to good humour, so much so that an unidentified wag in Richter Drilling's Brisbane head office sent off a memo to all the company's personnel round Australia concerning the previously touchy subject of staff retrenchments. Supervisor Bob McCulloch kept a copy.

> Attention all staff:
>
> Due to the economic situation, the management has decided to reduce the current work force, and has devised a reduction of employees programme.
>
> Under the plan older employees will be placed in early retirement permitting the retention of employees who represent the future of the department.
>
> A programme to phase out the older personnel (over 40) by the end of the financial year will be put into effect immediately. This programme will be known as RAPE (Retire Aged Personnel Early). Employees who are raped will be given the opportunity to seek other jobs within the department provided that while they are raped they request a review of their employment status before actual retirement takes place.
>
> This programme will be known as SCREW (Survey of Capabilities of Retired Early Workers). All employees who have been raped and screwed, may apply for final review.
>
> This phase will be known as STUFFED (Study of Termination of Use For Further Educations and Developments).
>
> Programme policy dictates that employees may be raped once, screwed twice, but can get stuffed as many times as the management sees fit![5]

New onshore discoveries during this period were not confined to the Eromanga Basin. The Bowen/Surat region of southeastern Queensland hosted a number of discoveries, notably: oil at Sandy Creek, Waratah, North and South Kincora (for Hartogen); at Yellowbank Creek and Taylor (for Bridge Oil); Riverslea (for BHP); and Fairymount and Louise (for Minora/Auspet); plus gas at Royston, Beldene, Newington, Avondale, Newstead, Yambugle and Yarrabend (for Hartogen); at Renlim, Sirrah and Roswin (for Bridge Oil); and at Merivale, Springvale, Punchbowl Gully and Murtleville (all in the north Bowen region known as the Denison Trough, for CSR). Beach Petroleum broke the long drought in the Otway Basin of western Victoria with commercial gas finds at North Paaratte and Wallaby Creek. The Amadeus exploration group in central Australia, led by Pancontinental Petroleum, found gas at Dingo and West Walker, although these could not be regarded as immediately commercial given their isolated location. In Western Australia, the Hughes & Hughes/Strata Oil combine found gas at Woodada in the north Perth Basin, while the Pan Pacific group found a similar-sized field at Tubridgi just south of Onslow in the Carnarvon Basin. But there was one onshore find in the 'West' which, for a short time, stood out from all the others and even jostled with the Jackson discovery for a place in the headlines of the period. The spring 1984 edition of the *Petroleum Gazette* looked back at the excitement generated by a group led by Home Oil.

> Drillers working in the shadow of the spectacular King Leopold Ranges in the Kimberley district, about 160 km east of Derby, startled oilmen around the nation in mid-1981 with the first commercial oil flow from the Canning Basin. Blina No.1 (taking its name from the cattle station surrounding it), drilled into a buried Devonian-age limestone reef akin to the craggy outcrops further north, culminated more than 50 years of efforts in the region. Previous explorers had found signs of oil and gas in a number of wells, but never a flow to surface, nor anything that looked even remotely commercial. Consequently the news sent wild speculation through the industry and through the stock market.[6]

Much of the excitement was generated because reef deposits had proved to be lucrative oil producers overseas, particularly in Canada. The fact that Home Oil Australia's parent was Canadian added to the stock market confidence. A few publications noted that the presence of Devonian reefs had been recognized in the 1920s by the Freney Kimberley Oil Company, but that company had preferred the standard anticlinal targets elsewhere in the basin. The *Petroleum Gazette* even shed a tear for WAPET, the company which had held the Blina lease for 25 years — right up till the late 1970s.

> Poor WAPET! WAPET started drilling about 30 years ago, giving some attention to the reefs. But its luck with interpretation was bad. It had some small shows of oil, but nothing worth recovering. As late as 1974 its geologists wanted to drill what would have been Erskine No.1 well. The shareholders said 'no'. The company gave up its last Canning Basin permit in 1978. This was the Meda block. And of course the Blina find is in the Meda block.[7]

Poor WAPET indeed, particularly since the Blina discovery was actually made in the rejected Erskine prospect. But oil exploration round the world was full of such tales and, in any case, the pain eased during the next few years as further exploration indicated the oil reservoir was relatively small and, despite a rush of new drilling in the northern Canning region, no further reef discoveries were made. Home did find three other small oil accumulations within its permit during the mid-1980s (Sundown,

West Terrace and Lloyd) which were added to the Blina development plan. These also triggered a burst of exploration activity in nonreefal plays in the the region, including permits to the east and to the south. Although oil shows were miserably few and far between during the Canning work outside the Blina permit, the explorers met and eventually overcame some intriguing logistics problems in this remote basin. One of them cropped up in the Esso-held permit to the west of Blina and centred round the need to gather seismic information from beneath the tidal mudflats of Point Torment, the aptly named peninsula jutting out northwest of Derby and forming the western shore of King Sound. Vibroseis surveys were carried out without problem over hard ground, but it was soon found that work on the mudflats themselves would not be so easy. Ron Angove from Esso Australia took part in the operation.

> Deeply incised tidal channels meander through the area and the mudflats are bordered by thick impenetrable bands of mangroves often several kilometres wide. Salt-water crocodiles infest this environment. Elsewhere the surface of the mudflats consists of a hard crust up to 80 mm thick which overlies a fairly consistent grey mud. Past experience had shown that where vehicles disrupted the surface crust, considerable efforts were required to extricate them.
>
> The crust... develops as the tides recede or as the area dries following heavy rain. Accordingly, the survey had to be conducted during the dry winter months... Operations [also] needed to take place during a period of low maximum tides to ensure that only the minimum area was flooded periodically while work was in progress
>
> To avoid disruption of the surface crust... Honda tricycles equipped with large balloon flotation tyres and similarly equipped eight-wheeled eight-passenger Argo all-terrain vehicles were used extensively. These units towed not only [shot hole] drilling rigs, but also specially fabricated trailers to transport survey equipment, geophones and the like.[8]

Esso used helicopter support in the mangrove areas, but the survey lines had to be cut by hand using three teams of six men each, equipped with chainsaws and machetes. Ten-metre tides flooded the lower mangroves and crews had to move to other uneffected parts of the line until the water receded. Tide movements also dictated when the crews could ford the creeks, although it was a moot point whether most attention was paid to the water ebb and flow or to the likely presence of crocodiles during these crossings. An average line clearing rate of 600 m per crew day was considered satisfactory. Seismic recording was achieved using a radio telemetry system which had the advantage of not requiring cables to transmit the seismic signal from the geophones to the recording cab.

The telemetry units in tidal channels were fitted with flotation collars and recording ranges of up to 20 km were possible between the central station and the seismic line. The recording crew averaged 3.36 km per day over the 200 km survey, and on several days exceeded six kilometres.

Unfortunately for Esso, the follow-up exploratory drilling was not successful and the group withdrew. Some teasing oil indications were found in several wells drilled in the central part of the basin — notably by Western Mining and BHP — but there was no further commercial success. By the second half of the decade the Canning had lapsed to its former enigmatic status of much oil promise, but no big discovery. In fact, despite the number of discoveries made onshore Australia during the first years of the 1980s, in volumetric terms there had not really been a lot of oil found. Many of the new fields were gas and of those that were oil, most contained only a few million barrels. It was only a drop in the bucket compared to the Bass Strait fields. The point came home even more forcibly when it was realized that six Jacksons were needed to equal the 300 million barrels in Fortescue, the most recent find off the Victorian coast.

With these sorts of comparisons aired openly in industry circles, the conventional wisdom of the time was that Australia's only real hope of big new discoveries was offshore. What was more, when the Exmouth Plateau programmes proved to be a disappointment, it seemed natural that explorers should look once again at the only basin where there had been major oil discoveries. Was it possible, they wondered, for Bass Strait to have more undivulged riches? In 1980–81, reading this mood well, the Victorian and Federal Governments regazetted four areas that Esso/BHP had been obliged to relinquish from its exploration permits.

Although one block was virtually ignored, response for the three main areas was overwhelming. A total of 154 companies within 28 consortia made applications. Local Australian companies were prominent in the lists, although most groups were led by one or two international companies. Even 'unlikely' explorers such as media company News Corporation and transport company TNT became involved. But perhaps the most earnest newcomer in the licensing round was a small explorer called Sydney Oil, whose style and enthusiasm very much reflected the outlook of its two principals — brothers John and Alan Conolly. Alan, a lawyer keen to establish a successful company by building from the 'grass roots' exploration stage in Australia, was chairman. However it was John, a geologist with a flair for creative exploration and a willingness to put his ideas to the test, who provided the technical input. John Doran, a Welsh-born geologist with experience in the North Sea during the 1970s, migrated to Australia in 1980 to join the Sydney Oil team.

> I arrived to find the company offices based in an old church in the heart of Sydney, a fairly basic, no-frills — at one stage no telex — operation. But it said something about the style.

Having worked for Conoco I was used to the plush, big company approach, but as John Conolly had told me, he could shut Sydney Oil in the morning.

I'd first met him in London and liked his laid-back approach. He was a charismatic character, a graduate from Sydney University who had done post-grad work in the US petroleum industry and, for the most part since then, had taken an academic role. He had been president of the New York Geological Society and had also joined BP in the US for the important Baltimore Canyon licensing round off the US east coast.

During the exploration revival in Australia in the late 1970s he had gone back to Sydney and set up a company called Era South Pacific with brother Alan who became chairman. But it wasn't long before John wanted to get back into oil so, in 1978, Era acquired the mining company Minerol and changed its name to Sydney Oil. It began with a few Surat Basin permits, but also found a corporate friend in Canadian company Northern Michigan Exploration (NOMECO) and the two became allies in a Surat joint venture in the late 1970s.

Always on the lookout to try new geological concepts, John was excited by the licensing round in Bass Strait and he put together a study group of Australian companies to evaluate the data and make an application. We lacked a major operator until Diamond Shamrock joined and then our efforts became known by some wags as 'the Little People's bid'. We worked about 500 man-days on that application. Apart from no telex, the Sydney Oil office had no Xerox machine either and messages to and fro staff and study group partners were all hand written by John Conolly.

The finished bid comprised nine volumes and we struggled down to Melbourne with two suitcases full of documentation and data to make our presentation to the Victorian Department of Energy.[9]

In the end the group did not receive a permit, but the exercise had shown that Sydney Oil was a serious contender and that it could be ranked with the major explorers in terms of effort and knowledge. During 1981 and 1982 the company went on to develop the study group concept to a fine art, particularly looking at areas in the Surat, Canning and Eromanga Basins, and also the shallow-water offshore Carnarvon Basin area south of Barrow Island.

The Bass Strait permit awards were announced in early 1981 and within a few months the three successful applicants — led by international companies Shell, Phillips Petroleum and Aquitaine — had begun work. Based on open-file data, industry opinion suggested that the Shell group in the far east part of the basin had the most prospective permit. During the following few years that opinion proved to be correct inasmuch as Shell was the only consortium of the three to make any discoveries. The finds, named Volador, Basker and Manta, all contained gas and some oil. Unfortunately none were large enough to be declared commercial. Another prospect named Kipper, drilled later via an Esso/BHP farm-in to Shell's permit, had a 300 m gross hydrocarbon column — the longest ever found in Bass Strait. But it was gas. If it had been full of oil it would have contained millions of barrels.

During the same period a group led by Hudbay (later to become Lasmo), which had been given a permit closer to the east Gippsland coast in an earlier licence round, found noncommercial gas accumulations at Sperm Whale and Baleen and a small oil field at West Seahorse. So it seemed, after all, that Esso/BHP had combed the region thoroughly during the previous 20 years and the large, obvious prospects had been drilled. Any new finds were likely to be small and marginal. As with the Exmouth Plateau effort, Gippsland Basin programmes which had begun with enthusiastic hopes faded away as the explorers cast around for new potential in Australia's other offshore regions — in particular the country's far northwest.

Despite the nagging realization that all the major finds made in the offshore Carnarvon Basin during the period leading up to the 1980s had been gas or gas/condensate, there had been sufficient small (albeit noncommercial) oil finds in the region to keep alive hopes of finding a big oil field. And there was Barrow Island. Surely that could not be the only oil accumulation in the North West Shelf province. Armed with these thoughts, plus the benefit of improved technology, explorers tackled the area again. Success came fairly rapidly, with oil at South Pepper and North Herald (for Mesa), at Harriet (Occidental), Talisman (Marathon) and Saladin (WAPET). Yet the accumulations were still relatively small — in the 15 million- to 50 million-barrel range. As before, the only big new finds — Gorgon, Wilcox and Dixon in the areas to the west and northwest of Barrow Island — were gas. Apart from their being noncommercial because of the dominance of the nearby North Rankin and Goodwyn gas developments and the lack of any new markets, they were no help at all in replacing the declining oil reserves in Bass Strait.

Explorers were also conscious of the mounting cost of operations. Despite the advantage of modern drilling equipment, problems like bad weather, unstable seabed and substrata conditions, plus unexpected shallow gas pockets all combined to make the work slow and hazardous. However the worst mishap did not occur in drilling mode at all, but to a rig under tow along the West Australian coast. In August 1983, the jack-up *Key Biscayne* was moving slowly down to Cockburn Sound near Fremantle after completing a drilling programme for Esso off Darwin when it was caught by rising seas and high winds. The vessel's three legs, towering above its deck in towing mode, were like giant steel sails as they caught the brunt of the 65-knot gale. A crew of 52 was still on board when a towing cable to one of the two tugs snapped and the rig began to drag the remaining tug backwards with the storm. When repeated attempts to re-establish the line failed and the rig began to list, it was decided to evacuate all personnel. Fortunately this was accomplished without injury by helicopters from the RAAF and the contract company Okanagan due to the skill of the pilots in landing on the rig's pitching helideck and avoiding the wildly swinging legs. The rescue operation spanned nine hours during the early morning and afternoon of 1 September. Several hours later the *Key Biscayne* lost contact with the remaining tug and sank.

The most publicized drilling failure during this period was the West Barrow No.1 wildcat drilled by the Offshore Oil group, which cost around $30 million and had to be aborted after repeated unsuccessful attempts to free the drill bit stuck in the well. Another $30 million dry hole, drilled in 1982 by a group led by US independent Cities Service Company (Citco), was located further to the north in the Timor Sea on a prospect named Vulcan. Using the veteran Australian-built semi-submersible *Ocean Digger*, Citco ran into difficulties right from the start when the upper part of the hole caved in and forced a respud. The problem occurred a second and a third time and it was not until the fourth try that *Ocean Digger* was able to drill through to the target formation. When the effort went unrewarded, Citco decided it had had enough and began to seek farm-in partners to the Vulcan permit (designated NT/P2) and its other holding in a lesser-known permit further to the northeast (designated NT/P26). Little did anyone realize that Citco's decision would soon lead to an oil discovery that would prompt the most amazing wave of share market hysteria the country had seen since the mid-1960s. Weeks Australia was part of the Citco group and Peter Cameron, then Weeks's exploration manager, was in a good position to see the events unfold.

> We'd become involved with the Timor Sea through Tom Patrick, who was exploration manager of the Weeks international company, but who had also worked with Arco in the 1960s and 1970s when the Puffin oil discovery was made. He was enthusiastic about the prospects, particularly of the NT/P2 block which contained the Puffin find and also some oil shows in a structure called Swan. The Vulcan dry hole was a disappointment, but not a disaster in Patrick's eyes.
>
> The other permit, NT/P26, had only been looked at with seismic at the time and there were a couple of semidefined prospects which had not even been named when Citco decided to farm-out. In fact, that block was a bit of a liability because there was a well commitment due and no firm place to drill

it. Several companies walked away because the drilling obligation was imminent.

Then BHP Petroleum came along. That company was beginning to come out from under Esso's shadow and operate for itself. It had been in the Surat Basin for a while and it had just put together a pretty good bid package for the South China Sea licensing round. BHPP was not put off by the NT/P26 commitment and the farm-in deal was done with Citco in 1983. The first job as new operator was to drill the necessary well, so they looked at the map and chose the structure with the most seismic control and where Citco had already marked a possible well location. Then they named it Jabiru.[10]

The group soon had another taste of the Vulcan problems and within a short time Jabiru No.1 had been abandoned and Jabiru 1A spudded a short distance away. Drilling supervisor Mike Imbert experienced the early frustrations, but he was also on duty when Jabiru oil came in.

The Timor Sea made for difficult drilling in top hole because the sediments were always unconsolidated and the sea currents worse than expected. Often the trouble would be at 400 m subsea, and then again at 800 m where caving could occur. There was always a danger of lost circulation, hole washouts and difficulty with cementing, which made the temporary guide base unstable and the subsea assembly hard to position securely. Thus the careful design of a well programme became very important. In our first attempt, the 20-inch [50 cm] casing was programmed too deep, but it was too late to change and so we lost the well.

Anyway we got over the early problems the second time and I was on the rig one night when a drilling break [a change of rock type downhole usually signalled by swifter drilling progress] was reported just below the Cretaceous and near the potential target formation. We decided to run a core but, when it took just 16 minutes to drill the next 10 m, I was sure we had gone through some sort of washout and wouldn't have any core in the barrel.

When we brought it up I was amazed to see the barrel was full of friable sandstone. I'd done a bit of mudlogging in my time and unofficially estimated it had 20 percent porosity and excellent permeabilities. We ran the barrel down again, kept up mud circulation and recorded fluorescence, and then a light oil which dripped right through the core.[11]

Explorers and sharemarket investors were stunned by the news — momentarily. Then came an excitement that quickly rose to fever pitch as brokers, journalists and 'industry spokespeople' scrambled over each other to predict the discovery of a new Bass Strait. Jabiru was the name on everyone's lips in late 1983 — even those who normally did not follow the run of oil discoveries. The only question asked was 'How big?', and so each prediction outdid the next. The figure of 800 million barrels became a common assessment and even a billion barrels was heard in some quarters. Some brokers painted extravagant pictures of the Timor Sea's rosy future.

> This sparsely explored area (one well per 14 000 km^2) has undertaken a quantum leap in hydrocarbon prospectivity following the intersection of 57 m of highly porous and permeable oil reservoir. Sediments to a depth of 15 000 m exist regionally and they are known to contain the necessary source, reservoir and trapping rock sequences. Until Jabiru, the region was believed to be 'gas prone'; accordingly, exploration became next to nonexistent. Now exploration consortia will be scrambling to tie up drill rigs to prospect their permits. The depth of sediment and areal coverage suggest the region may well contain over 100 prospects similar to Jabiru.[12]

The BHPP group itself was more cautious, believing the find might have 200 million barrels. The companies were confident enough to put a development design on a 'fast track' and, by the time the Jabiru No.2 well was spudded in 1984, about $20 million had been spent on hiring engineering staff and putting the plans into action. A report in the *Petroleum Gazette* of spring 1984 analysed the Jabiru phenomenon with the benefit of hindsight.

> There has never been quite the hysteria greeting a new oil find in Australia as that of Jabiru in the Timor Sea far into the country's northwest corner. Even the early Bass Strait discoveries, although attracting a good deal of attention, did not warrant such a stream of often quite reckless commentary and share market speculation. Perhaps because of the realization that in 1983 Australia's long-term oil self-sufficiency was not secure, the Jabiru discovery was seized upon as the saviour — the new oil province to replace a fading Bass Strait.
>
> And indeed the first indications from the BHP group's well were very encouraging, having strong test flows over a 60 m pay interval. But in many quarters outside the oil industry, the cardinal rule of never judging a discovery on the first well — particularly within a little-explored province — was either forgotten or disregarded.
>
> That made the disappointment, when it came in the first half of 1984 in the form of two dry appraisal wells, all the

> more profound. Now, in the third-quarter of 1984, BHP, although still contemplating a commercial prospect at Jabiru, may only have a recoverable reserve of 25–30 million barrels instead of the 800 million that some outside analysts were predicting at the end of 1983.[13]

The disappointments turned to dismay as Jabiru No.2 then Jabiru No.3 and Jabiru No.4 all failed to live up to the promise of the discovery well. The group had realized from the outset that the geology was complex, but the full implications of the fault control on the structure and the complications within the reservoir had not been appreciated. Yet, while disappointing, the three appraisals did give geologists a much better idea of what they were dealing with and the development plan was honed down to accommodate the reduced economics of the field. Jabiru No.5 was a better well and the work went ahead on the assumption that a minimum of 13 million barrels of oil could be recovered. Then, in October 1984, the BHPP group took another step out of the gloom when its Challis No.1 wildcat 20 km south of Jabiru in the same NT/P26 permit came in with a new discovery. Not surprisingly, the chastened share market response to this news was less than enthusiastic, although three appraisals confirmed the find as another commercial development project. A third discovery, at Skua further to the southwest, was more problematic, but it helped to revive the BHPP group's original hopes of the Timor Sea becoming an important new oil province, if not a full-blown replacement for Bass Strait. The company assembled and led several more groups in the region, setting up a major exploration base in Darwin and a front-line helicopter base on Troughton Island just 20 km off the northwest tip of Western Australia.

At one point during 1988, BHPP was operating four rigs in the Timor Sea and spending over $700 000 a day in exploration costs in the region. To reach the work place, crews assembled in Darwin (some after already spending five hours on a plane up from the southern capitals) and took a BHPP-chartered fixed-wing flight in a small Bandeirante or Cassa aircraft to Troughton base, two hours and 540 km further west. They then boarded the helicopters for another hour's flight to the rigs. Occasionally crews would stay on the island overnight if circumstances dictated, such as bad weather grounding the choppers, but mostly the schedules were arranged so that the passengers went straight on to the rigs and the drillers coming off duty boarded the planes back to Darwin. Accommodation on Troughton was principally for the helicopter pilots, the mechanics and a catering staff.

The island itself is a low, treeless and windswept piece of land only four metres above sea level at its highest point. The 850 m long airstrip takes up most of its length and the aircraft hanger is the largest building. Fringed by white beaches alternating with mangroves and surrounded by shoals and reefs teaming with fish, the pilots jocularly dubbed the place 'Club Troughton'. Previously used as a manned beacon station for BHP's iron ore ships sailing from Dampier to Japan, it was seen as an ideal forward base for the company's oil exploration programmes. Supervision of the day-to-day operations on the island was carried out by aircraft contractor Lloyd Aviation. Scotsman Allan Mackenzie, an ex-Royal Navy pilot who later settled in Western Australia, was one of those in charge.

> We have a basic population of 13 and most of us are on a two-week on, two-week off roster. Everyone has their own room and we have satellite TV, videos, plus a library of sorts. There's a swimming pool which we officially call the 'emergency fire water tank'. For those inclined there's a makeshift gym and a jogging track round the island perimeter. We do a bit of fishing, although the seas can be deceptively treacherous for anyone in a small boat. Plenty of tiger sharks round too, and we've even had the odd crocodile swim over from the mainland. It's not a lot different from life on the rigs really, except we can move about a bit more.
>
> We strictly forbid alcohol and drugs and anyone violating that is sent straight out on the next plane. We make our own fun and there are lighter moments like the time the Dalgety Flyer came through in his ultra-light plane on the way from London to Sydney for the Bicentenary. Troughton was his first landfall in Australia after a hairy flight from Indonesia, and he thought his compass must have gone wrong when he was met by a mad Scotsman in a kilt (me) proferring a plastic bag of fruit and a bottle of orange juice!
>
> But it's not a holiday here. Jabiru field is 220 km offshore from the island and the exploration rigs are scattered in an arc of about the same radius. The choppers are the main personnel transport links and, apart from routine ferrying of rig crews, we have to be ready for any emergency. That includes night flying if necessary and we do search and rescue practice-runs at regular intervals. We have one machine on standby at the base at all times, and everything except the major chopper maintenance is done right here on the island.
>
> The closest communications base is Kununurra in Western Australia and that's also our evacuation point if there is a cyclone alert. Every flight is different and we are in the air two and sometimes three times a day depending on the traffic. The pilots keep in touch with Troughton and the rigs by VHF radio when they're in the air. There's always a half-way call, and if that is not received a search is mounted immediately.[14]

The rescue drill was put to the test during a real emergency on 28 March 1988 when one of the helicopters crash-landed in the sea between Troughton and the drillship *Energy Searcher*. The pilots had sent a mayday signal and put the aircraft into autorotational glide to land upright on the sea, but its flotation bags were not adequate in the two-metre high swell and the machine quickly turned over and sank. The thirteen rig crew and two pilots spent two hours clinging to life rafts after surviving the crash and managing to swim clear of the stricken helicopter. Signals from the life rafts were picked up by an RAAF Hercules travelling up to the Butterworth airbase in Malaysia and their position fixed. The Troughton standby chopper then located the group and winched them to safety.

Life on the Timor Sea exploration rigs was a cosmopolitan affair, with people coming together from all walks of life and a range of countries. The routine was strict and, in some eyes, fairly monotonous. Like their counterparts onshore, the drilling crew worked 12-hour shifts seven days a week for a two-week roster, before coming ashore for two weeks. John MacErlich, a New Zealander who was one of the toolpushers on the semisubmersible *Zaparta Arctic,* knew the value of a keeping a good crew together.

> There's often a bit of turnover when moving to a new country, or even a new region within one country. Most operators like to use the local labour as much as possible, so for Timor Sea work a number of the blokes come from the Darwin area. Some don't like the life and leave again after their first tour, or earlier. But most make it OK. With modern communications we can phone anywhere in the world from the rig, and that knowledge helps a lot in some of the personal problems people have in being away from home for long periods.
>
> There's little conflict on board. The old days when a toolpusher would yell at his drill floor crew and kick butts have gone. Now there's communication and preshift briefings so that by the time everyone gets to work they are all thinking as a team. Sure it's a buzz if we find oil, but there's also a pride in just making good, trouble-free time with the drilling. We have regular awards as incentive for the individual crews.
>
> Drilling contractors work hand in hand with oil operators round the world and really it is a pretty small industry. Word soon gets around if someone has been dismissed for any wrongdoing.[15]

The offshore drillers also worked alongside marine crews responsible for location moves, anchoring and keeping the vessels on station. Every rig had one or two 'cards' in its complement with jokes and good-natured 'ribbing' commonplace — almost obligatory — in such a close-living community. But there was respect for skill and knowledge of the job too, no matter what the person's background, age or sex. That was something BHPP geologist Cathy Elliot appreciated when she began regular visits to the Timor Sea on well-site work. Having fought hard to convince the company to allow it, she became the first woman in the Australian industry to live and work on the offshore exploration rigs.

> The first time out was an awful feeling. Walking into the mess on the first day everyone put down their knives and forks and just looked at me. I was sort of prepared for an initial prejudice and I knew I would automatically stand out, being the only woman in a man's world. I realized I would have to speak out and be forceful. It was almost a case of having to do better than the males to be equal.
>
> But I loved the outdoor life and looked forward to each two-week trip out there. After a bit all the guys started to become used to the idea. Then one of them said that he couldn't tell me from the rest of the men on the rig. I suppose he meant it as a compliment, but it spurred me to go and buy a pair of lilac and a pair of pink overalls on my next two weeks ashore.
>
> The swearing was toned down a bit in my presence, but I didn't mind it too much and would soon let them know if it went over the top. I was conscious of not appearing closed-minded and the only time I really felt different was when they all watched porno movies, even though they closed the door in deference to me.
>
> One big disadvantage I had was never being able to show my true feelings. If I'd been up for 24 hours or more during an important period in the drilling, I couldn't be sharp with anyone even though I felt tired and irritable, whereas a bloke probably could get away with it.
>
> The other thing I could not afford to do was sit at the same table or with the same group each meal time. I made sure I moved round and talked to everyone. I used to get so many 'Dear Doris' questions though — fellas just wanting to talk about home. Quite a number were divorced, some for the second and third time. It was a bit sad really. Most saw rig life as a long-term job, although some came back for each tour only because they had to — they'd drunk or gambled their wages when ashore. That probably sounds depressing, but on the whole they were a fun bunch with some outrageous characters amongst them. The cooks let me use the galley sometimes and on one tour I made personalized gingerbread men for everyone on board. They all thought that was hilarious.

> I shared quarters with BHPP people on the rig. It was a sort of staff cabin and there were a few women visitors. Mostly I shared with a bloke, but that didn't cause any problems except sometimes with the bathroom. Often I'd use the one in the hospital, and there was one time when the medic and a patient came into the adjoining surgery when I was in the shower. I didn't want to burst in on them, but they didn't seem to be leaving, so I waited. Eventually I yelled out for them to hurry up because I wanted to go to my cabin. It turned out they'd been waiting for me too. We all had a laugh about that.
>
> Workwise it was a big responsibility being the BHPP geology rep on board. I was on call the whole time. It was a thinking job and it meant there were some important decisions to be made. There was also diplomacy required. The point was to make the decisions quickly and stand by them. They had to be 'yes' or 'no' answers. There was no room for 'maybe' or any hesitation.
>
> Probably the hardest thing to cope with though was the transit from rig life back to domestic life on my off-duty shore time. It was a common thing with the men too. I found it hard to jump straight back into the home routine and yet I couldn't talk about work for several days either.[16]

Despite BHPP's concentrated effort, along with that of several other operators in the Timor Sea, there was no further success in the region during the 1980s. Although the effort had lost its initial shine and the extravagant predictions had been toned down, many geologists still believed the area to have significant potential for more oil discoveries, and further programmes were being planned for the 1990s. In particular, a group led by SANTOS set up a forward helicopter base near the Kalumbaru Mission on the northwest tip of the Western Australian mainland. Using the airstrip of the old World War II Truscott bomber base as a focal point, the company set up huts and hangers in the area in much the same configuration as BHPP's Troughton Island camp just offshore. However, while this preparation was still in progress, the most intriguing, although expensive, oil discovery of the late 1980s was made back on the North West Shelf only a stone's throw from the North Rankin and Goodwyn gas finds of the early 1970s.

The Woodside Petroleum group had returned to exploration drilling in 1988 after several years absence during which the company had carried out a thorough review of its acreage surrounding the existing discoveries. During the study, which included reprocessing of old seismic data and the running of new surveys, two prospects (Echo and Wanaea) were delineated and matured for drilling. Contracting the semisubmersible rig *Margie*, Woodside spudded Echo in mid-1988 and several months later recorded a strong, but noncommercial, gas/condensate discovery. Moving on to the second location, Wanaea No.1 wildcat was spudded in on 8 November 1988 but, just over seven-incident packed months and an estimated $30 million later, the well was finally suspended as an oil producer on 16 June 1989. The early drilling had gone according to plan and excitement rose when a thick hydrocarbon interval was encountered in the main target section. But when continuing to a lower secondary target, Woodside began to encounter some downhole mechanical difficulties and these came to a head when the drilling bit was inadvertently cemented into the well during April. The troubles were compounded when a very late season cyclone bore down on the rig before the bit could be freed. The explorers were forced to slice through the drill pipe at the wellhead to disconnect the riser and evacuate personnel before the storm hit.

Cyclone Orson swept right over the *Margie* with winds of 250 km per hour, snapping three 450 000 kg breaking-strain anchor lines and dragging the rig about two kilometres off station. Although other damage was only superficial, it was two weeks before the vessel was back over the Wanaea location. Even then the Woodside group's troubles were not over because the original drill bit still could not be retrieved. It was decided to drill a sidetrack hole round the obstruction into the reservoir zone so that tests could be run on the hydrocarbon column seen earlier in the programme. More mechanical problems were encountered during this process and not all the planned tests could be run. However, much to the relief of all concerned, the main section of interest did flow a light oil at strong rates. Wanaea was obviously a good discovery although Woodside refrained from declaring it commercial until a full appraisal programme could be mounted in 1990 to confirm the results of the first well.

PAPUA NEW GUINEA

To participate in equal or perhaps even more frustrating intrigue during the 1980s, oil explorers had no need to venture further than the continuing saga of Papua New Guinea — both on and offshore. In 1984, the strange sight of a roaring jet of flame rising skyward from an apparently boiling sea in the middle of the Papuan Gulf was symbolic of the country's hydrocarbon potential. And yet, at the same time, it represented the difficulties in harnessing the resources when they were found. The flame in the sea, which burnt for most of that year, was the result of a spectacular gas blowout in an exploration well called Pasca No.2 drilled by the Australian Superior Oil group. After pulling the rig away, the escaping gas was deliberately lit to burn it off and minimize the danger, in much the same fashion as the treatment of the Petrel blowout in the Bonaparte Gulf 14 years earlier. However, unlike Petrel, the Pasca well was left to flow wild until it collapsed in on itself and died of 'natural causes'.

Some explorers felt that the unmeasured amount of gas which had escaped from the buried limestone reef structure may have depleted the reservoir. Others thought that there might still be as much as 30 million barrels of condensate and gas liquids trapped in the prospect, giving rise to speculation about a possible development project. But Superior chose not to return to the area, and no further work was done in the region until the second half of the decade; even then it was brief.

At the instigation of geologist Ed Durkee, an entrepreneurial Swiss-Canadian company called International Petroleum Corporation took up a permit immediately south of the Pasca block. Durkee was an American who had worked in and around Australia during the 1960s with the Amoseas group, and had more recently been a consultant to the PNG Government, advising on the country's 1984 licensing round covering the onshore and offshore sectors. Visualizing the potential for oil in some of the shallower buried reef plays on the fringes of the Papuan Basin, he supervised a new seismic survey in the area. In 1987, he and IPC were amazed when the interpretation showed what appeared to be a huge reef structure clearly defined on the section profile.

> It is just begging to be drilled. In all professional seriousness it is the largest, best defined, and the most magnificent reef prospect with which I have ever had the chance to be associated. We are all familiar with the old adage 'that many a mine has been spoiled by sinking a shaft', and so it is that this is a marvellous structure.
>
> If there is an 'elephant' to be found in 1988, then this is a candidate. We will not know the answer till it is drilled. But we believe we are in a relatively low-risk area considering the good track-record of the Miocene reefs in the Papuan Gulf for containing hydrocarbons.[17]

In an almost mischievous reference to Greek mythology, Durkee called the prospect Pandora and set about finding farm-in partners to help defray the drilling costs. Although the body of industry opinion suggested the structure would, if anything, contain gas and perhaps condensate rather than oil, Australian companies Ampol Exploration, Oil Search, Claremont Petroleum, Pacific Arc Exploration, plus two international companies shared the belief that such a large feature should not be ignored. In June 1988 the drillship *Chancellorsville* sailed into the permit to begin work in 130 m of water. An old rig which had been without work for some time, it experienced a number of mechanical problems before reaching the target and finding that the pre-well suppositions had been correct. Pandora was a large gas discovery. Tests in the basal zones revealed no hoped-for oil leg and, in fact, the gas itself contained no condensate. IPC entertained the idea of a gas development in the form of export LNG and even a possible pipeline south to supply mining projects and communities in northeastern Australia, but these seemed unlikely for the short term in the face of strong competition from other sources during the late 1980s. In September 1989 IPC was also granted a licence over the Pasca area 70 km to the north and the company continued to pursue gas markets based on the reserves of both Pasca and Pandora discoveries. By that time though, exploration in PNG had refocused on the onshore sector.

The Australasian Petroleum Company was still the main explorer during the first part of the decade although there were some internal changes. In 1981 Mobil withdrew and BHP Petroleum was invited to join BP and Oil Search. A short time later another Australian company, Pioneer Concrete, also came into the group. Later, Ampol Exploration and Bond Petroleum were added and finally APC itself was dissolved allowing the individual companies in the former combine to hold permit interests in their own rights. (More recently still Bond sold its shareholding to a group led by Japanese company Mitsubishi.)

Virtually all work in the field had been transferred to three permits in the central highlands, two of which included 1970s farminee Niugini Gulf Oil. It was in the westernmost of these during 1984 that the explorers made their first significant discovery at a prospect named Juha. Located close to the headwaters of the Fly River, the test results indicated a large gas accumulation which also contained a high percentage of condensate. This liquid content was confirmed with two appraisal wells and, for a while, the joint venturers toyed with the possibility of a development based on the condensate alone. But then, in 1985, the Gulf-operated group (Gulf was taken over by Chevron soon afterwards) switched from the previous 'slim hole' drilling method back to the use of a full-sized rig and made the breakthrough that explorers had craved for nearly 60 years. A well called Mananda No.3, in the permit immediately east of Juha, found oil shows in the important Jurassic-age Toro Sandstone reservoir for the first time.

In 1986, with pulses quickening, the group moved on to a nearby prospect called Iagifu. Originally named by APC geologist Keith Llewellyn in 1953, the giant anticline had remained undrilled until a well in the early 1980s slim-hole programme tried in vain to reach the Jurassic target zone. To the companies' collective joy this second try tapped a strong and unmistakable flow of oil. The discovery was confirmed several months later with Iagifu No.3, but dry holes at the No.4 and No.5 wells indicated that commercial development was not a foregone conclusion. The *Petroleum Gazette* of autumn 1987 pointed out that complex geology was just one of the difficulties of the operation.

> Iagifu is located in the tectonically active mountains of the Southern Highlands, still being built by the pushing and clashing of the great plates of the Earth's crust. Earthquakes are frequent. The terrain is more than rugged. This is some

> of the wildest country on Earth. Many of the mountains are around the 3 000 m mark. Jungle is dense. Thick clouds swirl in rapidly. The monsoons dump as much as 10 m of rain a year. Though the locale is almost on the equator, altitudes easily bring frosts.
>
> The Iagifu base camp is 15 km southeast of Mendi. All supplies and equipment have to be trucked in 500 km on the winding Highlands Highway from the port of Lae and then on another road to the old Poroma airstrip. From Poroma, heavy-freight jet turbine helicopters (twin-rotor Vertoles) take the crates, drill pipe and whatever to the drill site 1 300 m up in the mountain tops. The site is on the top of a treacherously steep ridge.[18]

Despite the difficult logistics, the explorers persisted and were rewarded during the late 1980s with new discoveries at Hides (gas), Hedinia and Agogo (oil and gas). Hides discovery, made by the BP/Oil Search joint venture, led to the award of the first petroleum production licence in Papua New Guinea in September 1990, which involves using the gas to generate power for the new Porgera gold mine in the central highlands. Continued appraisal work at the Chevron group's nearby Iagifu/Hedinia oil finds developed into what has become known as the Kutubu Project which, from mid-1992, will pipe the country's first commercial oil from the highland fields to the Gulf of Papua coast and thence to export. But not all the exploration in the country was confined to the central highlands. One of the most active programmes during the mid-1980s was carried out by BHP Petroleum in the Aitape Basin of the Sepik region, a territory which lies in PNG's northwest corner against the border with Irian Jaya. It was the company's initiation into jungle operatorship, and work began with a 180 km, three-month seismic survey early in 1984. At one stage the survey team consisted of over 600 people, 90 percent of them nationals from the local villages. The advance party, armed with bush knives and axes, hacked a rough line through the undergrowth to be followed by shot hole drillers who bored 10 m deep holes using hand augers. About 20 expatriates from the seismic contractor SSL handled the explosives and recording equipment.

Three drilling locations were chosen from the seismic profiles and the advance parties set out again to prepare the sites, first by cutting enough of a clearing to bring in a sawmill, and then a bulldozer and fuel. BHPP soon learned that no matter how rugged the terrain, the jungle canopy viewed from the air was always flat because trees rooted in the valleys grew taller than those on the ridges in their endless battle for light. This fact made assessment of ground topography from helicopters a trial-and-error affair, and as much as seven metres had to be sliced off the top of hills to provide enough flat ground for the rig, helipad and camp. All the cleared timber was cut up into planks for use on site, most of it for the rig substructure as a 'raft' to cover waist-deep mud produced by torrential rain on the disturbed soil.

BHPP set up its coastal base in the town of Vanimo, 48 km east of the Irian Jaya border, then leased and cleared two hectares of land for a jungle base camp near the village of Osol about 40 km inland. A track between the two places was widened and upgraded to facilitate transport of drilling materials and equipment to Osol camp, but from there to the three drilling sites everything had to go by air. Mike Imbert had been brought in from BHPP's Timor Sea operation to run the Sepik programme.

> The technicalities of drilling and seismic work were much the same as anywhere else, but getting the equipment onto location and maintaining supplies needed careful planning in a detail not often necessary in Australian conditions.
>
> It wasn't long before we saw that at least a third of the cost of the $50 000 a day drilling programme would be spent paying for helicopter time. The maximum chopper load in the humid conditions was 1 800 kg. Broken down to comply with these limits, a single D6 bulldozer took 18 loads to move. The entire rig and its camp and supplies worked out to be 750 loads, and each rig move took between two and three weeks depending on the weather conditions and whether two or three helicopters were used.
>
> We also had to put up with indifferent telecommunications because of the climatic conditions, the lack of oil field machine tools in the area to cope with repair and maintenance, unstable ground conditions which made landslips a constant danger at the rig sites, and just sheer isolation from any major town or hospital in case of emergency.
>
> We had to be careful of the local laws and customs which were not always spelt out or even apparent until we'd transgressed them. There were numerous attempts at intrigue which had to be dealt with on the spot. One 'con' tried was cutting a certain type of vine and putting it in the river to stun the fish, then suggesting oil company negligence to gain compensation. But there was no malice involved. It was more that the locals saw the whites as providers.
>
> We went to great lengths to refrain from calling the chopper fuel 'kerosene' because the locals would pinch it for their lamps. When they did realize what it was, we lost about a drum a day. They'd also think it natural to 'find' equipment and then innocently try to sell it back to us.
>
> The pay was set by the government at the equivalent of 50 Australian cents per boi and 75 cents for boss bois. Generally the boss boi was the oldest and he was given the

> most respect by the others. But he was not necessarily the best worker.
>
> At the rig most of the roustabouts were locals, but it was the devil's own job to get them to do night shift. We'd build a long cabin for them from the timber cut to clear the site and put on a corrugated iron roof. At the end of each well we always presented the hut to the elders of that particular area.[19]

But not all the locals were born in the Sepik. Throughout the programme BHPP called on the services of a Dutch woodcutter named Gim who had a portable sawmill. Based in Wewak where he'd married a local girl, he would trek out into the bush and work virtually nonstop to clear the drill sites before the camp and rig arrived. Another local identity was Brother Jim, an Australian who ran the Ossima Mission not far from BHPP's jungle base camp.

> Brother Jim was a real character who'd have a beer out of the fridge to offer you before you'd reached the mission verandah. He had the place in pristine neatness and was pretty self-reliant. There were sweeping lawns up to the buildings and he ran cattle and pigs and chickens. He'd use the manure to make methane gas to fuel various things including the electricity generating plant. Jim was also the butcher and he built an abattoir. He was always into a new building project.
>
> There was one occasion during our drilling programme that he was scavenging about the mission for steel to make something or other. Just then one of the choppers flew overhead and it was obviously struggling to lift its load of drill pipe. The hydraulics had failed and the pilot was in danger of crash-landing in the jungle so he jettisoned the bundle of pipe. As Jim watched, the load of steel landed right on the edge of the mission airstrip. He told me later he had just raised his eyes skyward and said softly, 'Thanks Lord'.[20]

BHPP failed to find any hydrocarbons during its three-well programme and withdrew from the Sepik permit. However, spurred by further appraisal successes in the Iagifu/Hedinia region and the possibility of a development project involving a pipeline to the Gulf coast, exploration interest blossomed throughout the whole of the Papuan Basin in the southern half of the country. A number of small Australian companies took up permits surrounding the highland successes, but they also spread out into the foreland areas where the PNG search had originated 60 years earlier. Late in the decade, when drilling commitments in these areas were beginning to fall due, a number of the world's oil majors began negotiating to pay seismic and drilling costs. Their aim was to take over the operatorships and earn substantial interests in what they judged to be one of the few places left on earth where big oil discoveries could be made.

Australian activity during the last years of the 1980s, however, was on a narrower path and had contracted to the basins where finds of the past had been most prolific. Offshore, that meant Bass Strait and the North West Shelf/Timor Sea; onshore, it was the Cooper/Eromanga and Surat/Bowen Basins, with some scattered interest also in the Otway region. Few companies were prepared to explore in the more isolated parts of the country away from existing infrastructure, when the risk of doing so did not match the perceived reward for any discovery. That yardstick itself was not new. It governed exploration throughout the world. Certainly the price of oil was always a large contributory factor, but in Australia during the mid to late 1980s three other factors had begun to discourage explorers: the first was geology — the fact that there was a distinct lack of significant discoveries made in new areas like the Canning Basin during the early part of the decade at a time when the price of oil could have sustained an isolated development project; the second was taxation — the perception that the system would heavily penalize a good discovery but not compensate for the costs incurred in failed exploration attempts elsewhere; the third was land access — the growing concern that preferential treatment was being given to the environmental and Aboriginal land rights lobbies, thus excluding large areas of the country's sedimentary basins from the exploration map. All three had a strong influence on exploration decisions by the time Australia stood on the threshold of the 1990s. What is more, despite some reasonably good discoveries during the decade, there was still no real saviour in sight.

Chapter 19
A MATTER OF INCENTIVES

> Considering its size, Australia is a very tough continent in which to find oil. Nevertheless, by a quirk of geological uniqueness, it also possesses the Gippsland Basin. For the last 10 years, this unique area has provided the nation with a security of petroleum supply bettered by only a handful of industrialized countries. At the same time, its discovery and development lulled this nation into a sense of false security. It all appeared so easy and it all appeared so cheap. The hangover has, unfortunately, been hard to cure.[1]

With these words, during the concluding address to delegates at the 1980 APEA conference in Surfers Paradise, Ken Richards, the Association vice-chairman and exploration director at Esso Australia, summed up the country's position at the beginning of the new decade. He believed that everyone should recognize that Gippsland was not the norm; in fact, it was a rarity and, in production terms, it was entering its twilight years. To maintain oil production for the future, explorers had to make a concerted effort to look for new reserves elsewhere round the country, but clearly the success rate was unlikely to be as great as the discoveries in Bass Strait had first indicated. Thus, in the decade ahead when companies weighed the risks against the possible rewards, the matter of incentives would have a vital influence on their activities in the field. Richards had levies and taxes specifically in mind. In terms of government input, he said that policies which worked in 1970 would not work in 1980.

> Exploration companies certainly take note of assurances on what will happen if they are lucky enough to make a new discovery, but base their hard-headed expenditure decisions on what actually happens to existing producers. The industry's willingness to continue exploring at present levels will bear a direct relationship to how the government actually implements its levy policy over the next few years.

Richards added that APEA endorsed the Liberal Government's parity pricing policy as being essential to the long-term economic strength of the nation. However he treated with suspicion the Opposition's scheme to introduce a resource rent tax on the industry if and when the Australian Labor Party was returned to power.

> We [APEA] deduce... that the Opposition's resource rent tax scheme will be designed to raise more revenue than the existing levy, that it will apply to 'new' oil as well as 'old' oil, and that it will be based on a before-tax rate-of-return trigger... This will be a harsh and regressive regime. We hope the Opposition clearly understands that industry's current assessment of what we deduce from their resource tax scheme is quite negative and, in the event of their coming to power, an unacceptable basis on which to continue exploring.

Although Richards could not have forseen the extent of the long battle ahead between the explorers and the Hawke Labor administration in Canberra during the middle years of the decade, he was aware that the Labor machine — albeit in Opposition at the time — was a vastly different animal to the one that held government during the mid-1970s. Behind his words to APEA delegates there was a realization that, in such people as energy shadow minister Paul Keating, the industry was dealing with skilful politicians who could mix comfortably with senior resource managers and would not be beaten on rhetoric. There was also the fact that new people like John Howard were emerging within the the conservative side of politics, and the parallel ideals forged between the oil industry and senior Liberal/Country Party leaders during their opposition days in the Whitlam era were beginning to diverge. The public service had changed too and people like John Stone, head of Treasury with strong ideas on key issues such as resource taxation, had come to the fore.

Within the APEA Council, Richards and others including Peter Gester of WAPET, Joe Wilding of Shell and Des Wittwer of BHP had all reached positions of authority by the late 1970s and their major-company backgrounds had a strong influence on the way the Association sought to promote itself in the 1980s. In particular, they felt there was a need for pro-active public relations and issues management rather than the low-key, almost reticent approach to politicians and the media that had been the rule during the last years of the 1970s. A serious illness which forced the resignation of Graham Maxwell from the APEA executive director's job in 1980 focused thought on the Association's future to an even greater degree. There may have been an initial temptation to appoint someone from one of the member companies to fill the vacancy but, at the end of the day, the APEA Council decided to seek an appointee from the wider sphere of the Australian business community. This approach brought to the fore an advocate who, during the following 10 years and on into the 1990s, played a major role not only in presenting the oil explorers' case to federal and state politicians, but also in lifting the industry's profile in the eyes of the general public.

Keith Orchison assumed the post of executive director of APEA in October 1980. Born in South Africa where he was a journalist and at one point editor of *Drum* magazine, Orchison had come to Australia in 1970 and worked for the Petroleum Information Bureau in Melbourne before its amalgamation into the Australian Institute of Petroleum. He then went into public relations for the Associated Pulp and Paper Mills and Melbourne's La Trobe University before his appointment to head the APEA secretariat in Sydney. Energetic and resourceful, he also brought a high degree of political nous to the role — a quality that was important in the rapidly changing world and Australian oil scene.

> Even in the early days of the 1980s it was obvious there had been a great change since the previous decade. The world was different and so was Australia. APEA Council appreciated that an 'ad hoc' approach to issues affecting the oil explorers would not do, so it sought to ensure that the Association's policies were cohesive and consistent. The existing Association committee system, made up of members from many companies, contained a lot of expertise, but was not always being used to best advantage. So we began to tighten up and go into a lot more planning for the issues of the day. We also brought in good consultants for some of the more specialized topics.[2]

APEA Council set out to maintain and, if possible, improve the collective health and standing of the petroleum exploration and production member companies. Allied to this was the more lofty goal of maintaining the country's welfare through a continuing high level of indigenous production. The Association well understood the fact that incentive lay at the root of any Australian oil search, given the high-cost/high-risk nature of exploration on the continent. Such incentives included not only a high world oil price and optimistic perceptions of geological potential, but also the comparative level of Australian taxation (versus other investment opportunities in Australia and fiscal regimes in other parts of the world) and removal of barriers to access exploration areas. The latter point had been a major problem for APEA with regard to exploration on the northeast continental shelf (beyond the Great Barrier Reef off the Queensland coast), from where the industry had been barred in the 1970s after a long battle. Now the same problem reared its head rather unexpectedly on the other side of the continent. Orchison was still working out his notice at La Trobe University when he and the Association had to face a crisis that tested their issues management skills to the utmost. The incident itself, concerning the right of Aborigines on a cattle station in Western Australia to block an oil drilling programme, became known as the Noonkanbah Affair, but the whole question of land access had its roots back in the 1970s.

As early as 1969 the industry had noted a rising tide of environmental concern off the Queensland coast culminating in a large region surrounding the Great Barrier Reef being placed out of reach in a national marine park despite a concerted effort in arguing the case for oil exploration. During the same period there was also a growing desire among Aboriginal people for a return of their tribal homelands and the right to govern them as they saw fit. That could involve the exclusion of all resource industries in many parts of the inland sedimentary basins, or at least the attachment of onerous conditions for entry. Such restrictions had never been imposed on the early oil explorers like Roy Hopkins of Magellan Petroleum, who worked in the Amadeus Basin of central Australia.

> When Magellan first applied for the exploration leases in the 1960s the Haarst Bluff Reserve was the only restriction in the area. We had to obtain a permit from the Department of Aboriginal Affairs to enter, but that was all. The Department's main concern was that we did not take any diseases in and, after that clearance, we could then carry on the normal business of exploration. There was a royalty of 10 percent to be paid in the event of commercial production and that went to the Aboriginal Trust Fund. But similar royalties were applicable wherever explorers went in Australia.
>
> For the Palm Valley exploration work we obtained permission from the Hermannsburg Mission. The Aborigines there would be told where we wanted to run seismic lines and then drill wells. Occasionally they'd ask us to relocate a well or part of a survey line.
>
> Our exploration drilling gave them water wells and access tracks into the country. We also hired a few of them onto the crews from time to time and generally the relationship was a pretty good one.[3]

With the formation of land councils, access to legal counsel, and the Whitlam Labor Government's legislative Land Rights initiatives during the 1970s, Aboriginal groups began to make more demands on oil explorers and would-be producers. Plans by Magellan and its partners to develop the Palm Valley and Mereenie fields were delayed for several years, in part by an Aboriginal stance to gain higher royalties and a carried interest in the production consortium. However progress was made in negotiations during the late 1970s and the go-ahead which followed for both field developments was hailed as a step forward in relations between Aboriginal groups and oil explorer/producers. Nevertheless it was realized that a 'blanket' approach to all Aboriginal land access questions would be difficult to achieve. Each case would need to be treated separately. Unfortunately the ink was barely dry on the Magellan agreements when any hope of an early breakthrough in a national solution was blown away by events in Western Australia. In mid-1980 the Premier of that State, Sir Charles Court, ordered a heavy police guard to accompany a drilling rig to a well location on the Noonkanbah Aboriginal cattle station in the northern Canning Basin. This was the culmination of a long battle in which the oil explorers, led by Amax Petroleum, had become caught between State Government regulations that required the drilling of a well, and the forces backing Aboriginal resistance to it. Max Reynolds was head of Amax's Australian operations at the time.

> The Canning exploration permit EP 97 which took in Dalgety Australia's Noonkanbah Station area was granted to a company called Whitestone Petroleum in mid-September 1976. A few days later the pastoral lease was bought from Dalgety by the Noonkanbah Pastoral Company, the majority of shares in which were held by the Aboriginal Lands Trust. By the time Amax Petroleum farmed into the oil permit and became operator in December that year, Whitestone had already carried out a seismic survey with the full permission of the Aboriginal owners.
>
> In 1978 Amax received permission from all adjoining station owners, including Noonkanbah, to run a second seismic survey. About 83 km of the total 313 km was on Noonkanbah land. After completion and interpretation, a drill site was chosen and its location, which happened to be on Noonkanbah, was discussed with the Aboriginal community who were satisfied there were no sites of significance involved.
>
> A number of conditions were placed on our formal request for access to carry out the drilling, including the drilling of several water wells, construction of a new airstrip and a ban on alcohol and firearms. That was in early April 1979 and acceptable to us. But a few days later we had word that the Noonkanbah community had suddenly changed its mind and the Aborigines 'totally and unequivocally opposed exploration and mining in all forms and by all persons on its station property'. From that point things became pretty difficult.[4]

It soon became apparent that the matter had suddenly become one of land rights and not sacred sites. Noonkanbah had been transformed into something of a test case and, as such, a political football. Players included the State and Federal Governments, State and Federal Oppositions, the companies, the industry through APEA, the Aboriginal Legal

Service, the West Australian Museum, Aboriginal community advisors, the unions, church representatives, the media and the Aborigines themselves. As part of its permit obligations Amax was obliged to drill a well or relinquish the permit. The consortium chose to drill and, with the support of the State Government, it attempted to do so. When this met physical resistance on Noonkanbah Station in March 1980, the well was postponed and the company withdrew to allow more talks to take place. Soon afterwards, Premier Sir Charles Court pointed out that the Noonkanbah community were not lawfully entitled to exclude explorers from a normal pastoral lease, and that drilling would go ahead in the national interest. The Aboriginal Legal Service responded that the Aborigines would make a stand even if that meant confrontation. When further talks failed to resolve the impasse, the State Government decided to take the rig to Noonkanbah under police escort. More fruitless talks and union bans on drilling effectively tied the explorers' hands so that the government stepped in once more and took over the operation. The well, Fitzroy River No.1, was finally spudded on 29 August 1980. When there was little disruption from the Aborigines to the actual drilling of the well, union bans were lifted during October and Amax resumed operatorship from the government. Fitzroy River was plugged and abandoned at the end of November having failed to find its main target.

In the final analysis, the build-up to drilling had been by far the most acrimonious part of the affair and, although an Aboriginal delegation journeyed to Geneva to petition the United Nations subcommittee on Racial Discrimination, there was no tangible result and media attention to the Noonkanbah Affair dwindled. For Amax and its partners, however, there was little joy. In addition to drilling a dry hole, the drawn-out exploration of permit EP 97 had been a very costly exercise. It had also been a very visible and traumatic one, and there was no guarantee that the same thing would not happen again if they wanted to drill a second well. Even with legal rights on their side, the disincentive was strong and Amax withdrew soon afterwards. The importance of the affair for APEA, however, was far-reaching, as Keith Orchison explained.

> The Noonkanbah Affair occurred just as the announcement of my appointment to APEA became public. As I had to complete three months notice at La Trobe University, I had to cope with academic and student emotions over the issue and endure the frustration of a major problem effecting my new job without being in the seat directly dealing with it. The day-to-day management of the issue fell to Ted Garland, then chairman, and he and I used up a fair bit of telephone time as he coped with the Prime Minister, the Western Australian Premier and many others.
>
> The Association called on a wide range of skills in the industry and among consultants to obtain information and

> guidance in dealing with the Noonkanbah Affair. My first task on taking up the job as Director was to work with the Aboriginal Affairs Committee of Council to produce a code of conduct for member companies dealing with Aboriginal communities.
>
> This document was crucial in persuading the Fraser Government not to impose special legislation on the industry to cover dealings with Aboriginal land issues. The Deputy Prime Minister, Doug Anthony, played a central role in cabinet deciding to accept that APEA had produced a workable alternative to legislation.
>
> The issue had a number of important repercussions for APEA. Firstly, and perhaps most importantly, it brought home the need to identify potential problems as early as possible and put policies in place to cope with them, rather than trying to cope with crises like a fire brigade. Secondly, it underlined the need for APEA to work as hard as the mining industry in lobbying for government policies that treated Aboriginal land issues fairly, but did not unnecessarily restrict exploration access. In fact that battle is still going on. Thirdly, it brought the role of the more general Association code of environmental practice into sharper focus. The attention this point received in the early 1980s was to pay off in the latter days of the decade when 'green' politics really took off.[5]

Perhaps an equally significant point was that the Aboriginal exercise, coupled with a separate campaign aimed at deregulation of drilling rigs offshore, clearly illustrated that properly organized lobbying based on factual presentations meticulously researched and presented at the right pressure points in government could be successful. It also vindicated the Council's choice of a non-oilman as industry spearhead and set the tone for the Association's activities for the rest of the 1980s.

Importantly for industry's case, the Noonkanbah experience was not the norm. Elsewhere there were exploration programmes run in harmony with Aboriginal wishes, one of the most notable being Shell Australia's work in the Officer Basin of Western Australia across territory that was the tribal homeland of the Pitjantjatjara people. Solicitor Phillip Toyne, based in Alice Springs, was legal advocate for the Pitjantjatjara Land Council at the time.

> Generally described as desert country, it does in fact vary widely from areas of shifting sand dunes through to hilly ranges with spinifex and stunted trees. Although they are sometimes a great distance apart, the region does contain areas of significance and the Pitjantjatjara, like other Aboriginal groups in Australia, did not want them interfered with by mining and oil exploration companies.

> Shell was already drilling the Yowalga No.3 exploration well, which was not on Pitjantjatjara land, but I was concerned about the company's forward programme in the area and pointed out that there was an enormous risk of disturbing areas of significance to the Aborigines if discussions were not immediately held between the two parties. Shell, which did have plans for an extensive seismic programme on its recently granted permits, agreed and the first meeting was held near the Yowalga drill site in September 1980. At the time we were all aware that the Noonkanbah dispute in the northwest had reached flash point.
>
> The practice in negotiations elsewhere had been for the companies to ask Aborigines where the sensitive areas were located so they could be avoided. But we pointed out that it would be a difficult and slow process in the desert country and also that some of the areas were secret and the Aborigines did not want to reveal their whereabouts to anyone.
>
> So our approach was to turn that usual practice around so that the company was required to submit an exact plan of its survey programme to the council in advance of any work actually taking place on the ground. The survey lines and proposed campsites would then be traversed by Aborigines and a council-employed anthropologist, and this advance party would indicate whether or not the lines passed through or too close to sensitive areas. If any did, there was enough time and flexibility in the seismic planning to reroute the sensitive sections.
>
> In practice it worked very well and the one instance of sensitivity was easily avoided. There really could only be one expert opinion on the whereabouts of significant areas and that was the Aboriginal [one]. There was no point in disputing it, but the Shell negotiations had shown that confrontation could be avoided provided there was frank and honest dialogue from both sides.[6]

Whether the Officer Basin solution would work in areas of greater sensitivity, like Arnhem Land, was another question, and both Shell and Toyne pointed out that their agreement was not an overall solution to the relationship between the Aborigines and oil explorers. It was, however, a professional solution, similar to a dialogue between two companies. Unfortunately in the political sphere the way was not so smooth. The triumph of the Hawke-led Labor Party at the 1983 federal election brought a change in approach to the Aboriginal question, including a new look at the path to legislation. The resources industries immediately became concerned, and when Labor introduced its preferred model for national land rights in 1985, oil and mining explorers protested loudly that it would result in 25 percent of the country being available for Aboriginal claim. Aboriginal groups also protested, but their reasons were that the explorers and miners would be given too much access. They wanted the power of total veto, and consequently it seemed the two sides were being pushed further apart, not brought closer. In the face of opposition, not only from explorers and Aborigines but also from the individual States which felt their constitutional rights to govern such matters were being infringed, the Federal Cabinet finally abandoned its proposed national land rights legislation in March 1986. In terms of oil exploration this meant that individual companies would still deal directly with the Aboriginal groups in whose territory they had permits to explore.

Although APEA had taken no active part in the individual negotiations and agreements made by its member companies, executive director Keith Orchison saw the whole issue of land access and environment remaining one of major importance to explorers.

> Based on the Noonkanbah experience, the Association's approach to this and other issues had become one of combining the research and lobbying exercises to ensure that as far as possible the facts were not only available, but also understood by the groups influencing decisions. The Council was also at pains to ensure that APEA's antagonists were challenged publicly and effectively on claims that were false or misconceived. That meant a great deal of hard work behind the scenes in assembling information and gaining broad member acceptance of the various policies. It also meant meeting politicians, public servants and media commentators, not just to establish contact, but also to establish a credibility so that the Association's views would be taken seriously.[7]

As the decade progressed, the number of Acts and regulations governing designation and management of land grew swiftly and Orchison felt that APEA should strengthen its pro-active approach still further. In May 1988 the Association published a booklet outlining the oil industry's perspective and recommending that an economic/resource impact statement be undertaken at an early stage in the decision-making process.

> Decisions to designate land for conservation purposes only are taken without recourse to the economic consequences in most cases, and often without any stocktake (let alone an objective one) of an area's resource potential or other economic use.
>
> The lack of a consistent, open and objective process has made it very difficult for the petroleum exploration and production industry to contribute to decision making on individual areas in many instances. APEA believes that input on

Illustration showing the decline in the total area of the country under exploration permits during the 1960s J. Blumer, *APEA Journal*, 1982.

French geologists from the Institute Francaise du Petrole arrive in Australia to consult for the Federal Government. From left, M. Tissot, host John Casey of the BMR, Dr Daniel Trumpy and M. Rambeau.

David (Dave) McGarry was elected chairman of APEA in 1968, when managing director of Australian Oil & Gas Corporation, to bring unity to the local and foreign companies within the Association. Courtesy APEA, Sydney.

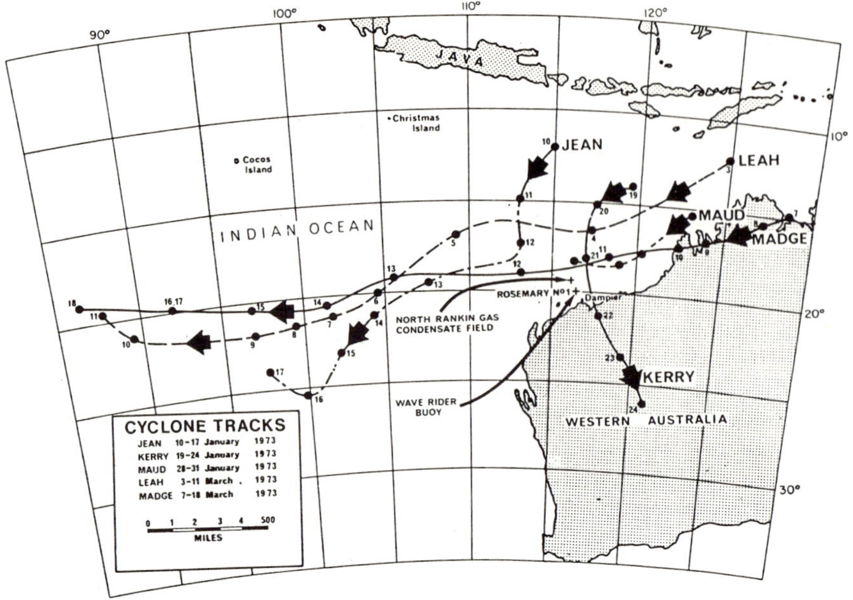

Cyclone tracks on the North West Shelf (cyclone alley) during 1973. Courtesy Woodside Petroleum, Perth.

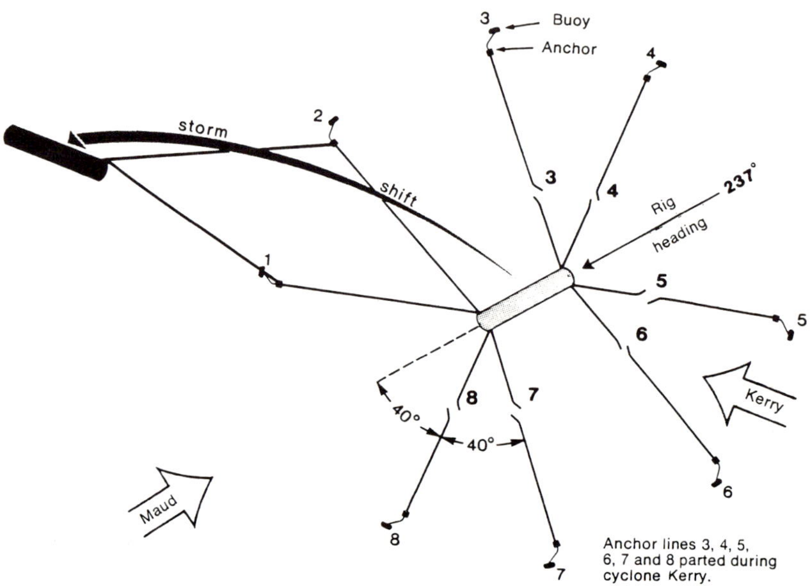

Diagrammatic representation of the anchor pattern of drilling barge *Big John* and the disruption caused by cyclones 'Kerry' and 'Maud' on the North West Shelf during January and February 1973. Courtesy Woodside Petroleum, Perth.

Drilling barge *Big John* drilling for Woodside on the North West Shelf during the 1970s. Courtesy AIP Library, Melbourne.

J.G. (Geoff) Donaldson, chairman of Woodside Petroleum from the 1950s through to the 1970s and their development during the first half of the 1980s.

Drillship *Glomar Tasman* drilling for Woodside during the early 1970s on the North West Shelf. Courtesy AIP Library, Melbourne.

Gas blowout at Big Lake No.2 well in the Cooper Basin of South Australia in 1972. Courtesy South Australian Department of Mines and Energy.

One of the fleet of helicopters based in Broome to service the Woodside group's exploration programme on the North West Shelf.

R.F.X. (Rex) Connor, the Whitlam Labor Government's Minister for Minerals and Energy from 1972 to 1975. Courtesy *Canberra Times.*

Sir Thomas Webb, chairman of Alliance Oil Development, was outspokenly critical of the Connor regime.

Des Wittwer, head of BHP's oil subsidiary Hematite Petroleum and chairman of APEA during the first year of the Whitlam Labor Government in 1973.

The APEA Council at a meeting in 1972 chaired by Eric Avery (centre far right). Ken Horler, the Association's executive director, is directly behind Avery (top far right) and chairman-elect, Desmond Wittwer, is on Horler's immediate right. Courtesy APEA, Sydney.

Ric Charlton, general manager of exploration for Shell Australia, then executive director of the North West Shelf project and chairman of APEA during the late 1970s. Courtesy Shell Australia, Melbourne.

Rod Hollingsworth, geophysicist and exploration manager of Delhi Petroleum during the discovery of Eromanga Basin oil in South Australia and Queensland.

APC geologist John Senior leading a survey party through a Papuan village during the 1950s. He later set up a trade store at Kikori on the Papuan Gulf and helped supply the survey parties during the 1970s. Courtesy BP Australia, Melbourne.

Dynamic positioning drillship *Sedco 472* on the Exmouth Plateau off Western Australia during the late 1970s. Courtesy Val Foreman Photography.

The size of semisubmersible drilling rig *Ocean Digger* contrasted against the Melbourne Cricket Ground, one of the largest sports arenas in the Southern Hemisphere.

The semisubmersible rig *Ocean Endeavour* under construction in its excavated beach site near Kwinana in Western Australia. Courtesy AIP Library, Melbourne.

Graham Maxwell — APEA technical manager in the early 1970s who was prominent in preparing the industry's case for the Barrier Reef Royal Commission. Appointed APEA executive director in the late 1970s. Courtesy APEA, Sydney.

The hull sections of semisubmersible rig *Southern Cross* under construction on the Brisbane River during 1976. Courtesy AIP Library, Melbourne.

The modern diving helmet, strong but lightweight, replaced the 1960s heavy version and allowed divers much more comfort and freedom of movement. Courtesy AIP Library, Melbourne.

The JIM atmospheric diving suit enabled divers to work in water depths of up to 500 m without the need for decompression. Courtesy AIP Library, Melbourne.

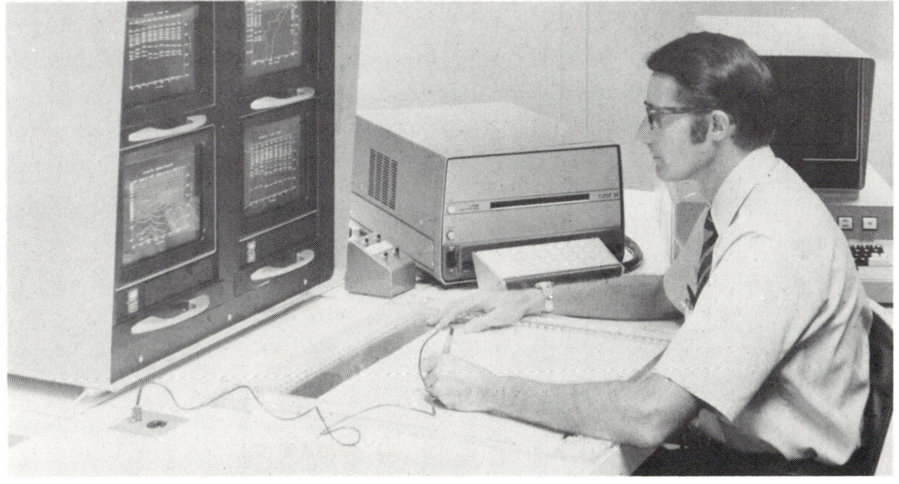

The 'massaging' of seismic data using computer techniques produced results and interpretations of exceptional clarity during the 1970s. Courtesy AIP Library, Melbourne.

Alan Burns, Perth-based entrepreneur who turned the Magnet Group into a successful oil explorer during the early 1980s.

Geoff Albers, Victorian-based entrepreneur who chaired the Oil & Minerals Quest group of companies including Zanex, Cue Minerals and South Eastern Resources.

E.H.C. (Ted) Garland, managing director of Alliance Oil Development and chairman of APEA at the beginning of the 1980s.

Ian Sykes, economist, geologist, metallurgist, entrepreneur, and co-founder of petrol retailer XL who branched out into oil exploration with his own rig and drilled at his own expense.

An aerial view of Troughton Island off the northwest tip of Australia, BHP Petroleum's forward helicopter base for its Timor Sea operations. Courtesy BHP Petroleum, Melbourne.

Brothers Alan (left) and John Conolly who built up the 'grass roots' exploration company Sydney Oil.

Oil workers go through a helicopter 'dunker' course in which they are submerged upside down in a simulated offshore helicopter ditching and taught to free themselves from the aircraft.

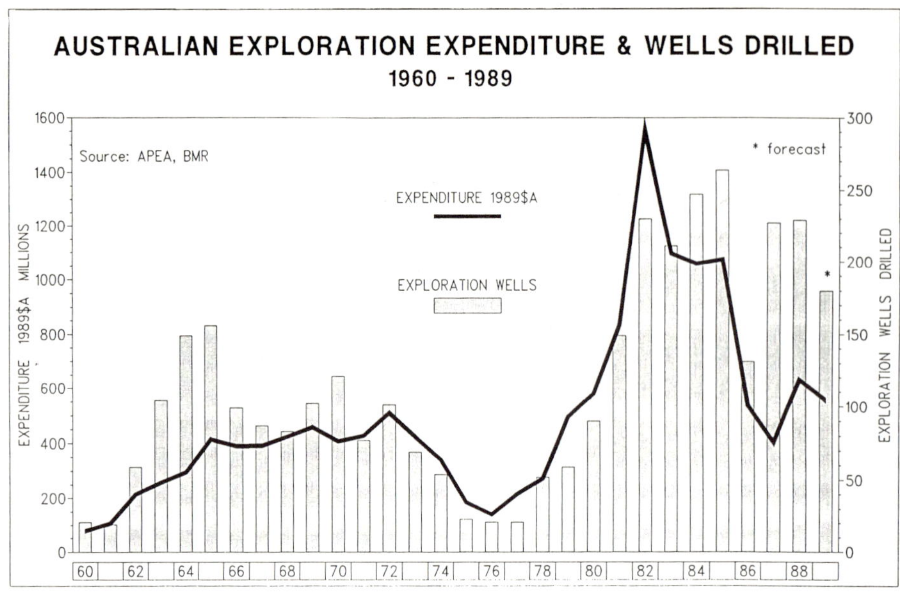

Graph showing the oil industry's exploration activity and expenditure in Australia during the 30 years since 1960.

The Highlands Highway winding between Watabung and Daulu Pass in central Papua New Guinea — the only road access to Chevron's Poroma base near its Iagifu and Hedinia discoveries. Courtesy AIP Library, Melbourne.

Keith Orchison, executive director of APEA during the 1980s and one of the main forces behind the Association's high profile opposition to secondary taxation on the industry and its work in the fields of conservation and Aboriginal land rights. Courtesy APEA, Sydney.

Dennis Benbow, director of Magellan Petroleum and chairman of APEA during the Association's campaign against resource rent tax in the mid-1980s.

Senator Peter Walsh, the Hawke Government's first Minister for Minerals and Energy, 1983-1984.

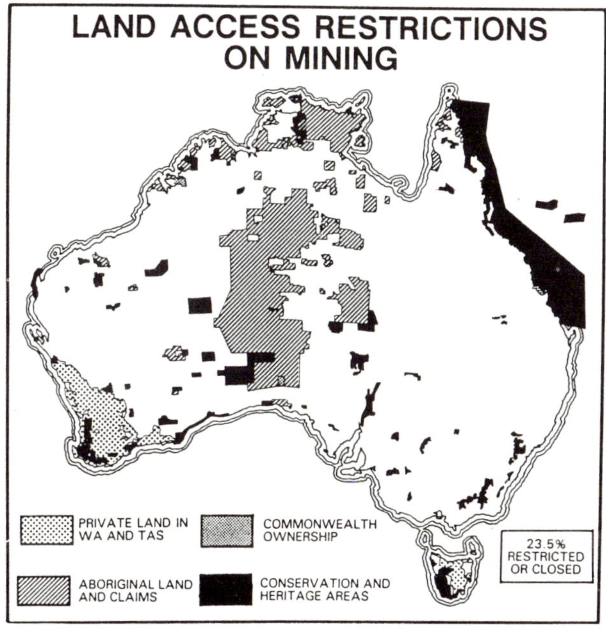

Restrictions on land access to mining and oil activity in Australia during the 1980s.

Senator Gareth Evans, the Hawke Government's second Minister for Minerals and Energy, 1984-1987.

The late Ken Richards, one of Australia's best modern geological thinkers and oil explorationists, a chairman of APEA and director of Esso Australia.

Colin Glazebrook, exploration manager for Oil & Minerals Quest and then Claremont Petroleum during the early 1980s. Now a geological consultant.

Graeme Foley, founder and director of Sydney-based Rimcorp Capital.

Steven Koroknay, executive director of Bridge Oil, Sydney.

John Armstrong, executive general manager of exploration for SANTOS in Adelaide.

Bevan Devine, general manager and director of SAGASCO Resources, Adelaide.

Eric Webb, consultant geologist, former exploration manager of Woodside (Lakes Entrance) Oil, former chairman of Endeavour Resources and former managing director of Drillsearch in Melbourne.

Jaap Poll, managing director of Petroz based in Perth.

David Maxwell, managing director of Arrow Petroleum (formerly Barrack Energy) in Perth.

John Doran, managing director of Command Petroleum in Sydney.

The late Sir Hermann Black who, as Chancellor of the University of Sydney, delivered a thoughtful principal address to delegates at the 1977 APEA conference.

Murray Johnstone, senior exploration geologist for Esso Australia in Sydney.

> resource potential, and on whether such resources can be sought and developed in a way compatible with conservation/ heritage (including land rights) objectives, is essential if government ministers are to make well-founded decisions in land use.[8]

The booklet covered a wide range of land access restrictions like land rights, national parks and world heritage listings, along with issues such as ecosystems, the fragility of coastal zones and wilderness areas. While recognizing the rights of Aboriginals to maintain aspects of their traditional lifestyle and protect sites of cultural significance, and noting the merit in conservation of the natural world, APEA's theme was that these things and petroleum exploration/development were not necessarily incompatible. The Association suggested that in many regions multiple land use was a valid option, and the booklet pointed to the success of just such a programme in the Innamincka region of South Australia. Located in the northeast of that State, the area contained important historic features, including Aboriginal engravings, and natural features like the Coongie Lakes. It was also prospective for oil exploration. After a number of combined field trips, the explorers, led by SANTOS and the South Australian Government, agreed it was possible to explore within areas of high environmental sensitivity and significance without harmful effects. SANTOS environmental officer Oleg Morozow described the proposal to APEA delegates attending the 1988 annual conference in Brisbane.

> The first test of the [South Australian] government's new policy arose when the Minister for Environment and Planning communicated his intention to proceed with the proclamation of some form of reserve status for the Innamincka Pastoral lease, essentially to protect the Cooper Creek, Coongie Lakes and associated wetlands areas. It was intended, however, to adopt multiple land use principles and therefore allow the continuation of the existing pastoral and petroleum exploration and production activities over most of the area while providing protection for sensitive areas as well as implementing some form of management control over tourist visitation.
>
> This would entail the declaration of some form of exclusion or control zone around the Coongie Lakes and the immediately adjacent areas. While potentially restrictive, this proposal was a reflection of SANTOS/Delhi's long-term insistence to government that smaller, specifically delineated areas around sites of environmental sensitivity or historical, scientific and cultural significance were a much more effective tool of management and protection than the broad declaration of reserve status of vast areas using existing boundaries.
>
> Thus, instead of declaring the entire 14 000 km² of the

> Innamincka Pastoral Lease as a national park in order to protect discrete areas of sensitivity within it, the government was indicating its willingness to adopt an imaginative and pragmatic new form of multiple land use management.[9]

A key factor in these moves from the explorers' standpoint was the perception that the Innamincka region was prospective for oil and gas. A favourable geological environment which translates to a good chance of finding commercial hydrocarbons has always been one of the most important stimuli for companies joining the oil search. An interesting observation the world over is that while the geology does not change, perceptions of oil potential in some areas do. Sometimes just one well is enough to change the conventional interpretation of a region, or even a whole basin, and this was clearly demonstrated in Australia during the late 1970s and into the 1980s. Esso/BHP's unexpected discovery of Fortescue field in the Gippsland Basin showed that there was potential for further traps of that type in Bass Strait and this perception stimulated a strong response to the gazetting of new permits during 1980–81. The same stimulus to renew exploration was seen following the discovery of Jurassic-age oil at Strzelecki in the Eromanga Basin, then again after the discovery of oil at Blina and Sundown in the Canning Basin, and yet again in the Timor Sea/North West Shelf region following the success of Jabiru and Wanaea. Going the opposite way, the Exmouth Plateau programmes began with gusto in response to excellent survey data in 1979, only to die out in the early 1980s after a series of disappointing dry holes. But while the geological incentive waxed and waned in the various basins round Australia during the decade, the overall perception of oil prospectivity of the country remained fairly low.

During the early to mid-1980s Australia's supposed poor prospectivity became a political football, and the notion of the country being an unlikely place to find much more oil was given unwitting credibility by APEA and some of the leading companies in their fight against secondary taxation. Keith Orchison admitted that the Association fell into the trap of overkill in its arguments to convince government of the difficulty and costliness in finding new reserves and of maintaining the country's oil sufficiency.

> What we succeeded in doing was conveying a message of NO prospectivity to some influential quarters in Canberra. There has had to be a concerted effort since 1988 to restructure the argument to ensure it is understood that what we have been saying is that oil is there to be found with the right incentives.[10]

Politics aside, the perceptions of how much oil remained to be found in Australia were also greatly influenced by the volatile exploration progress in the Timor Sea. Extravagant predictions of 'a new Bass Strait' which greeted the initial Jabiru discovery just as rapidly gave way to gloomy 'no substantial new oil' forecasts when three follow-up appraisal wells were dry. But, taking the middle road, two computer-generated quantitative assessments published during the mid-1980s — one by the Bureau of Mineral Resources and one by Esso Australia — were noted by C.S. Robertson of the BMR during his review of petroleum prospects for delegates attending the 1988 APEA conference. He found Esso's estimates of probable reserves still to be discovered were a little more optimistic than those of the BMR, but that the two organizations were suggesting numbers in the same order of magnitude.

> ... BMR's latest assessment [1986] is that there is a 50 percent chance of finding about as much additional oil as there is in Australia's remaining discovered reserves (1.8 billion barrels). Esso's 1984 assessment indicated we are likely to find slightly more than twice this amount, or about as much oil as has been discovered in Australia to date. Thus Australia's oil prospects are currently perceived to be only fair. It is considered likely that we will find sufficient additional oil to make a substantial contribution to the nation's oil requirements, but not enough to make Australia self-sufficient or a significant exporter of oil.[11]

A later (1987) estimate from Esso indicated that between one billion and five billion barrels of oil remained to be found in Australia, with the most likely volume being three billion barrels. The prospects for gas and gas liquids were more hopeful with large volumes of new discoveries expected in both categories. Esso suggested the gas liquids still to be found might equal that of the undiscovered crude oil potential. In relation to oil, the BMR's 1986 assessments also looked at the prospectivity of individual basins and found that the Bonaparte Basin (which included the Timor Sea region) was still the most favourable place to explore. With an estimated potential of just under a billion barrels of oil still to be found, it was a long way ahead of the Carnarvon Basin with about 500 million barrels. Northeast Queensland and the Otway Basin followed in the BMR's eyes with about 250 million barrels each, while the Gippsland, Browse, Canning and Eromanga Basins made up the other top order regions with between 150 million and 200 million barrels still likely to be found in each.

While these assessments and others like them (made by individual companies for internal use) did not put Australia high on the list of places to explore for oil in the 1980s, geologists out in the field continued to try new ideas in the hope of proving the theorists wrong. At the same time they began to realize that with the aid of increasingly sophisticated technology and engineering efficiency, smaller and smaller oil accumulations could be developed economically. In the explorers' realm, computerized logging techniques ushered in a new era both at the well site and the processing centre. Barry West of the BMR outlined some of the advances when presenting his history of well-site services in APEA's 1988 bicentennial publication *Petroleum In Australia.*

> The latest mud logging units are now leading the way towards downhole logging whilst drilling, giving real-time [instant] availability of bottom-hole information. This service may be linked by the latest satellite communications technology to large-scale office-based computer facilities... A slimhole [small diameter] measurement-while-drilling tool known as 'Skinny Minnie' has recently been developed by the company Exlog. It provides a low cost, easily transported and serviced tool and, after testing in the North Sea in 1987, will undoubtedly see applications in offshore Australian areas in the near future.
>
> Advanced tool combinations and some new tools have been developed [in wireline logging], such as the spectral gamma-ray tool [clay and shale volume and radioactive mineral identification], the litho-density tool [porosity and lithology determination], and the nuclear magnetism tool [estimates free porosity and producibility]. Some tools, such as the six-arm dipmeter, provide much improved resolution and data quality even under extremely adverse borehole conditions, allowing detailed sedimentological analysis of reservoirs. The most significant advance has taken place in the field of data analysis, interpretation and presentation.[12]

Developments in drilling technology itself also became more widely used during the 1980s — particularly the practice of steeply inclined and even horizontal drilling to penetrate targets difficult to reach economically via the standard vertical well. Several groups in Western Australia used this technique to drill from land into prospects located just offshore. In other cases the inclined well allowed explorers to drill into deep-water targets from a shallower water location such as on the edge of the steep continental slope in Bass Strait. It was also extensively used in Papua New Guinea so that two appraisal wells could be drilled from the same surface location, thus obviating the need for preparing two difficult jungle sites. In all cases the motive was to cut front-end costs and give any discovery a better chance of becoming a commercial proposition.

The dollar was always 'king' for oil explorers and a key barometric test for most programmes was the world oil price. During the late 1970s and the early part of the 1980s, events surrounding the Iranian Revolution shot the official OPEC crude price to a high of $US34 a barrel with spot prices in some instances going $US10, and even $US15, higher.

In these circumstances exploration in expensive, high-risk regions like the Exmouth Plateau became worth the gamble. But so too did research into the production of oil from alternative sources, and Australia was quickly seen to be rich in the raw materials to make 'syncrudes'. Schemes based on conversion of gas and coal to liquid hydrocarbons, and on the revival of extracting oil from oil shale deposits, were rife during this period — many of them unreservedly endorsed by the Federal and State Governments of the time. However, by 1983–84 it had been shown that there were a number of problems to be overcome, including the huge scale of mining needed and the thorny question marks hanging over pollution and environmental controls. In addition, the Saudi Arabians had forced world crude prices back into the $US20 range and so the gloss had faded from syncrude economics in Australia.

Then, in 1985, the world oil price began a slide that not only cancelled alternative oil projects altogether, but also cut the ground from under the feet of the country's explorers who had been actively rebuilding the industry during the first half of the decade. The lowest point was not seen until August 1986, when the world price dropped below $US10 a barrel making the Australian equivalent price $15.75. Fortunately that trough price was short-lived but, even so, most exploration companies were forced to drastically cut costs. That included the retrenchment of many employees, some with long experience in the Australian industry. The retrenchments read like a battle ground casualty list and APEA estimated there would be between 3 000 and 4 000 jobs lost by the end of 1986 if the low prices continued. The figures became more stark in the case of the small Australian exploration companies because the lay-offs represented between a half and two-thirds of their work forces. The tragedy of the downturn was that it extended far beyond Australia and, unlike the retrenchments of the mid-1970s, there was nowhere for geologists, geophysicists and oil technicians to go. There was a real danger that many would be lost from the industry completely.

The true story of a retrenched Australian geologist who went to Spain to take up management of a sausage factory smacked of black humour, but it underlined the predicament the country was facing. Other geoscientists went back to university studies, some bought businesses like coffee shops and some went into teaching; others became waiters and taxi drivers. A few, like Jack Mulready who was retrenched as exploration manager of The Moonie Oil Company, kept their hand in by offering geological consulting services. Mulready also suggested that the government could make some grants of $3 million or $4 million to pay the salaries for geologists and geophysicists to work on broad studies of the country's oil potential.

> I felt it was a time when a number of displaced geoscientists should be put to work on basin studies to assess the real, rather than the imagined, potential for future oil discoveries.

> I thought then Australia would be in diabolical trouble in the 1990s with the forecast decline in Bass Strait production. What we needed was an assembly and evaluation of all the data gathered in the last five hectic, but good, years leading up to a clear indication of where exploration should go in the future.
>
> I knew the BMR and some of the State mines departments did run programmes of that sort, but I felt they could be extended and put on a faster track so that the country would be in a better position to meet the Bass Strait decline.[13]

Helped by a recovery of oil prices to around $US18 a barrel, the industry bounced back surprisingly rapidly from what had been dubbed its 'blackest hour' and there was no time or inclination to follow up Mulready's scheme, or others like it. By July 1987, in addition to the exploration companies that had weathered the downturn, there were 23 new companies formed and contemplating the investment market to float as public listings. Clear reasons were not readily apparent for the resurgence of interest, but it was noticeable that behind many of the new organizations were young technical people dissatisfied with the approach former employers had taken to some regions. They wanted to reinvestigate the permits using their own theories and methods. The cyclic nature of the industry was also thought to have generated the new crop of explorers as replacement for those in the middle or top order who had left the oil industry of their own accord or, in the case of companies, had been taken over.

Whatever the reasons, the bloom of new companies did not last long. Only a few managed to fulfill their float and public listing aims before 19 October 1987 — the day the biggest stock exchange crash since the 1930s began. Share prices round the world tumbled to severe lows, not only in the volatile resource company stocks, but also in the usually more stable industrial companies. Fortunately the world oil price itself remained at around $US17–$US18 and most of the established Australian oil explorers were able to weather the difficult few months till the end of 1987–early 1988. In fact there were a number who saw the opportunity to begin, and in some cases expand, their overseas interests. The USA was a favourite place, but others also looked at Canada and South-East Asia.

The beginning of 1988 also saw the end of Australia's parity pricing scheme and a new, direct linking of oil prices in the country to the world free-market system. It meant that producers large and small no longer had the obligation of selling their crude to Australian refiners and, conversely, the refiners were not bound to buy it unless the producer's volume was too small to attract a realistic price on the world stage. The lead-up to the Federal Government's free-market proclamation had been a hotly debated affair with the majority of smaller Australian companies arguing that there was likely to be a fall in realized prices for indigenous oil of about $A1 a barrel. This, they said, would be just one more disincentive to exploration in the country at a time when a much greater activity in the field was desperately needed.

However, not surprisingly, the explorers' main cause for concern on the question of exploration disincentives during the 1980s was the one APEA vice-chairman Ken Richards alluded to when closing the first conference of the decade — taxation. When Robert Hawke led the Labor Party to a sweeping electoral victory over Malcolm Fraser's Liberal/ National coalition in March 1983, one day before APEA's annual conference in Melbourne, the Association's avowed intention to fight against resource rent taxation suddenly became a reality. Richards, by that time APEA chairman, opened the conference cautiously by pointing out that although the exploration industry did not agree with many points in the Labor Party platform, he hoped a dialogue could be swiftly set up and maintained with the new government.

> History has shown that misunderstandings between industry and government have hampered oil exploration in the past. We believe this unfortunate situation can and must be avoided in the future.[14]

Nevertheless, Richards did not let the opportunity go by to send a clear message to Labor.

> ... stability of government policy must continue. If ever there was a time for bipartisanship and consistency it is now, irrespective of the party in power. The policies which culminated in 1982's record achievement (230 exploration wells drilled and seismic programmes the highest level for 10 years) cannot be changed without seriously harming this industry... oil explorationists must remain overly sensitive to the price producers receive and the amount of taxes they pay. Any changes to the import parity pricing policy and 'new' oil policy, or any attempt to increase government revenues from 'old' oil, will guarantee reductions in exploration effort and production output.

The Labor Party had learnt from the Whitlam Government's mistake of trying to introduce reforms too quickly and the first Minister of Resources and Energy, Senator Peter Walsh, was in no hurry to change the oil taxation system. However he left no doubt that the Resource Rent Tax (RRT) proposal was on the agenda. This prompted APEA chief executive Keith Orchison to point out that the exploration year 1982 had completed a cycle of optimism and high activity which had been fuelled by (Liberal) government policies designed to encourage exploration. He added that exploration in Australia in 1983 was at a crossroad similar to the mid-1970s, when government policies then turned a cyclical downturn in exploration into a slump of much more severe proportions. However, no sooner had these sentiments been uttered than the Jabiru oil discovery unleashed a wave of euphoria and excitement about the potential of a new exploration province in the Timor Sea. The extravagant predictions of 'another Bass Strait' (from the BMR, industry commentators and some explorationists) acted as both a spur and a justification for the government's planned RRT and Senator Walsh released a discussion paper on the subject late in 1983, intending that the new profits-based tax would blanket the whole industry and replace the existing system of levies and excise. After much debate with APEA and individual explorer/producers, he conceded that it would be too difficult to apply the regime to Bass Strait and the North West Shelf gas project because historical costs from those areas would be virtually impossible to quantify accurately for deductions against the tax.

Walsh then decided to continue with the scheme, but confine it to what he termed 'greenfield' regions where there had been no previous production. Jabiru was the prime candidate but, by a twist of exploration fate, appraisal work at that discovery had shown it to be much smaller than first thought and most of the explorers concerned became more inclined to accept an RRT regime. They saw that profit, after deduction of development and operating costs, would not trigger the tax until late in the field's life. Taking this as an acceptance of the idea, albeit a grudging one, Walsh then gave the industry two RRT alternatives. One contained an exploration subsidy with a low threshold and higher tax rate; the other no subsidy, but a higher threshold and lower tax rate. Keith Orchison was immediately on the offensive, pointing out that Australia's oil explorers would slash future exploration commitments. During an interview with the *Australian* newspaper he said the inevitable result would be a significant decline in exploration throughout the country.

> It is quite clear from the reaction of APEA's member companies that offshore wildcat exploration just will not be worthwhile under the new tax proposals. We have talked to a large number of companies... representative of the industry as a whole, and their calculations show that both new tax systems are too harsh.
>
> We are going to tell the government that the tax proposal will not work. The net impact when combined with corporate tax is too heavy on exploration. The RRT, as proposed, contains a threshold that is too low and a tax rate that is too high... and produces a very high effective tax rate. The [government's] paper shows a fundamental lack of understanding of the cost of risk capital to the industry. The totally inadequate exploration expenditure write-off provisions... take no account of the many dry hole ventures which must

occur if the industry is to explore the huge area of Australian sedimentary basins in an adequate manner.[15]

APEA also expressed concern over the allied government proposal to introduce a scheme of cash bidding for some exploration permits considered to be highly prospective for oil and gas discoveries. The idea was to award the areas to the highest bidder, provided financial and technical criteria were satisfied. But this was seen by the explorers as a straight out front-end tax which would diminish the amount of money available for exploration itself. In the end, after several more years and much political intrigue, both RRT and cash bidding were passed by Federal Parliament, while an equivalent to RRT onshore (Resource Rent Royalty) was put in place for production from Barrow Island field in Western Australia. The effect of the new legislation on exploration was difficult to judge, masked by the downturn in activity following the world oil price slump in 1985–86 and then the stockmarket/investment crash in 1987. Throughout this period APEA continued its calls for restoration of incentives, adding that it was in the national interest. On the eve of the 1986 conference in Adelaide, Association chairman and a director of Magellan Petroleum, Dennis Benbow, pointed out during an interview with the Melbourne *Age* that a bipartisan commitment was needed to a tax regime recognizing rewards that were appropriate to the risk involved in exploration.

> If the Australian public and the Federal Government sincerely wish to maximize self-sufficiency in crude oil in the interests of national security and public revenue, then attention must be paid to the tax issue — and very quickly. The two- to six-year lead time which occurs between discovery and production, and the fact that the government itself expects Bass Strait oil production to halve by 1991, highlights the importance of removing tax impediments to exploration and development as soon as possible.[16]

Benbow amplified these comments during his opening address to the conference.

> ... bearing in mind the government's stated objective of achieving '... the optimum economic level of liquid fuel self-sufficiency through... encouraging petroleum exploration and development', consideration should be given to investigating some form of incentive for explorers without production. There needs to be consultation between government and the industry on what would be most effective...
>
> No doubt government will object that APEA is calling for the reduction of taxes on one hand and for incentives on

> the other. My response is that if the government is going to continue its discrimination against the industry through secondary taxes, then it also should reinject a portion of that revenue gain into the industry by providing encouragement for Australian investors. We need them. We need the companies they support and we need the skills and the risk-taking commitment of the people those companies employ.[17]

In reply during the same conference, Senator Gareth Evans, who had taken over the Energy Portfolio from Senator Walsh, suggested that the companies which faced the greatest difficulties as a result of the oil price falls were the companies operating in the onshore basins of South Australia and Queensland. Those companies were faced with high levels of debt financing and were expected to cut back on high-risk exploration.

> ... it is not clear what, if any, short-term palliatives would reverse this situation, even if affordable by government... I believe that the difficulties currently being faced by these companies highlight the inherent problems of our production-based taxation regimes. From discussions with a number of them I believe that the resource rent royalty system does offer a real alternative which would, if quickly implemented, do something to ease the current pressures, and do a great deal to rationalize and guarantee the longer-term health of the industry.[18]

However, as the depression worsened during 1986, the Minister offered a concession to the beleaguered small producers by removing all excise from the first 30 million barrels of production from onshore fields and new offshore projects. This also doubled as a fairly significant incentive for onshore explorers in particular because, with the exception of the Jackson field, all the 1980s discoveries had been well below the 30-million barrel mark and the industry expected this trend to continue. Nevertheless it did not address the wider issue of incentives to explore the offshore regions which geologists still believed to be the most likely place to make substantial finds in Australia. By 1988 there was clear evidence that Bass Strait production was in decline and yet no replacements had been found. The explorers, through APEA, renewed their calls for a complete overhaul of the oil taxation system and were finally rewarded for their efforts in April 1989. New Minister of Resources, Senator Peter Cook, told delegates attending the Association's conference in Hobart that he would extend a planned Bass Strait excise review to include another look at the details within the resource rent tax system. Promising to consider any reasonable suggestions for improvement, he then invited the industry to make submissions and declared that a verdict would be given by the middle of 1990.

The tax review was very much a result of APEA's perseverence in what had been a long, hard and often repetitious debate on the issue. Although no one could yet claim the campaign was a resounding success, it was nevertheless a good illustration of the worth of the Association's continued high profile during the 1980s. Keith Orchison pointed out that, as much as anything else, a lobby group needed to call on the attributes of a marathon runner rather than those of a sprinter.

> Selling an argument over and over again — dressing it up differently to catch the audience's attention, keeping it running in the media so that the politicians cannot ignore it — is one of the least appreciated skills. But, without it, organizations like APEA cannot hope to succeed.[19]

The Association had come a long way since its stumbling and rather amateurish entry to the decade when even some member companies had begun to question its relevance as a representative body. In addition to lobbying via its own secretariat, during the 1980s it had succeeded in gathering allies who were working towards similar goals, albeit outside the immediate area of oil industry activity. Bodies like the Australian Institute of Petroleum (the oil refiners' association), the Australian Mining Industry Council, the Business Council of Australia, the two coal associations, the state chambers of mines plus the farmers' and foresters' groups were all brought into a 'sphere of cooperation' at various times. The number of forums for venting APEA's arguments had increased too with the advent of the council of state, Northern Territory and federal minerals and energy ministers (AMEC) which met twice a year, and the 1989 birth of the National Energy Consultative Council. However it would be wrong to suggest that the oil industry was constantly at loggerheads with government. The paradox was that the 1980s saw a great deal of cooperation between APEA committees and officers of the various governments' energy departments. A major part of it involved updating the regulatory regime for offshore work under the Petroleum (Submerged Lands) Act, but there was also close liaison over the difficult issues of safety (which arose from the Piper Alpha disaster in the North Sea), onshore drilling standards, and even the controversial negotiations over Australian content in the North West Shelf construction project.

Yet, clearly, there was still more to be done and as the 1980s came to a close, few explorers doubted that the decade ahead would be a vital prelude to the nation's economic wellbeing in the twenty-first century. Some believed the cause, in terms of maintaining a reasonable level of oil self-sufficiency, had already been lost. They felt that the impediments to exploration in Australia were too great and so they turned their attention to operations overseas. But the hard core of explorers, while acknowledging the difficulties and the lack of incentives, were prepared to try new initiatives and find a way through the web. Probably the best summation of this attitude came during 1986, the most adverse year of the past 10 years, when international energy consultant Peter Gaffney spoke, or rather sang, to delegates at the annual conference luncheon. Journalist Peter Gill recorded his performance for the Melbourne *Age*.

> After two days of sombre discussions on low oil prices, high taxes and exploration cuts, APEA delegates were probably not quite ready for the tack taken by Peter Gaffney . . . [who] kicked off his luncheon address by discarding his suit coat, pulling his tie askew and breaking into a ditty:
>
> 'I don't want to be a singer,
> I don't want to go to sea.
> I'd rather stay around
> Where oil and gas are found,
> Living off my earnings as a poor explorer.'
>
> APEA delegates started to wonder whether this expert was, in fact, admitting that the oil industry was bust and there was more future in song-and-dance routines. But Mr Gaffney was merely demonstrating a bit of the versatility that was now needed in the oil industry.
>
> His message to the oilmen was simple — they had to find opportunities in adversity . . . Don't do less of the same. Do more differently.[20]

PART SIX

THE 1990s

BOTH EYES ON THE HORIZON

Chapter 20

FEATHERS FOR THE PHOENIX

Australia's oil explorers enter the 1990s with the knowledge that their past efforts have found about 4.5 billion barrels of crude oil. They are equally aware that nearly 2.8 billion of those barrels have been produced and used. But less well appreciated is the fact that only about five percent of the total area underlain by the country's sedimentary basins has been searched to an even moderate degree. Obviously there are vast areas still to be looked at, and a much more widespread drilling programme is needed to fully assess Australia's potential reserves. The remaining 1.7 billion barrels of discovered oil are clearly an inadequate quantity for the nation's needs in the years ahead. Yet the sparseness of past exploration, coupled with the large areas involved, lie at the root of problems facing the nation's explorers as they contemplate the last decade of their second century and look at the horizon beyond. The feeling is that, compared with many countries, Australia has lower prospectivity, higher costs, less infrastructure and more difficult access to market. The explorers' association, APEA, quantified the perceived difficulties when it pointed out that the national ten-year average oil finding cost had risen from $1.50 to more than $8 a barrel (in 1989 dollars). Companies were simply not drilling enough wells in the country to bring the costs down.

> Despite the need for more exploration in Australia, activity has fallen sharply from levels of the early 1980s. Furthermore, 22 Australian-based companies have spent $2 195 million on petroleum acquisitions overseas and $1 029 million on overseas exploration since 1985.
>
> Australia's high costs reflect the remoteness of most prospective exploration areas, the lack of infrastructure, the relatively small scale of operations especially in terms of rig numbers, and Australian work practices. Similar depth wells onshore USA can be drilled for $A200 000 versus $A600 000 in Australia, while an offshore well costing $A1.5 million in the USA will cost $A6 million in Australia.[1]

While admitting that new discoveries made on the North West Shelf during the first half of 1990 (Woodside's Cossack and BHPP's Griffin/Chinook finds in particular) have prompted a momentary upturn in offshore drilling activity in spite of the high cost of Australian exploration, APEA does believe companies should be free to make their exploration decisions based on those costs alone. They should not have to include the impost of secondary taxation on a discovery in the risk-reward equation. The Association says that in the 1990s such taxes will be increasingly unwarranted as the country's indigenous oil declines, and it will continue the fight to have them withdrawn. At the same time, the industry generally welcomed the government's August 1990 Federal Budget announcement which allows companies to include exploration and development expenses incurred in all RRT-controlled offshore regions as deductions when calculating the tax, instead of the previous rule of allowing deductions for just those expenses incurred in the immediate production area. The change paralleled a government move to finally cut through the complicated levy/royalty system in Bass Strait and impose an RRT regime on all future oil and gas production from this region including production from the existing fields.

There is also a realization that Australian exploration companies themselves will have to incorporate a greater degree of commercial expertise than they have in the past. Geologist/economist Graeme Foley, formerly managing director of the BA Explorers Trust and now director of the Sydney-based firm Rimcorp Capital Ltd, suggests that the junior exploration companies in particular are 'short' on financial management skills.

> I regret to say that poor management is the primary reason for the inadequate performance of our [Australian oil] industry... In the past we blamed prospectivity, taxes, drilling costs, etc. for our problems. To me that is simply passing the buck. The responsibility for poor financial performance rests with management because it is management alone that controls the investment of a company's funds.
>
> There is a general preoccupation of 'listening to the rocks'. The time devoted to a farm-in decision is typically split 80 percent technical and 20 percent financial. Further, the wrong, or inadequately trained, people are often involved in the financial analysis of investment opportunities... Even though the level of technical expertise employed is generally high, the chances of an adequate rate of return being generated is low because poor investments are made.[2]

Executive director of Bridge Oil, Steven Koroknay, believes however that the industry has already swung from being a totally technically driven entity in the 1950s and 1960s to one in which financial management took substantial control in the 1980s.

> In the early days, shareholders' expectations were satisfied purely by capital appreciation of share price. Today, the capital appreciation of share price has to be accompanied by dividend to provide the total return on investment to satisfy shareholders... it is no longer enough to give the 'blue sky' story of the next well...
>
> ... successful oil companies need superior technical input to create value. But equally, they require superior commercial talent to maximize both short and long-term benefits of the added value.[3]

John Doran, managing director of Command Petroleum, agrees.

> Investors are impatient people these days. If there is no dividend, they interpret that as no growth. The important thing in the 1990s will be to have a cash flow, and the healthiest companies will be those which can attain cash flow without relying on an upturn in the world oil prices or any significant government tax concessions.[4]

Consultant Richard (Dick) Paten also has a formula for the successful oil company of the 1990s.

> Really it hasn't changed much over the years. If you look back, behind every surviving company there was a leading light — not necessarily a flamboyant character, but someone solid and competent in the industry. Today the emphasis on economics has probably increased, but the formula remains true. For the best chance at success the modern exploration company will have on board an entrepreneur, a good finance man and a sound explorationist. Between them they will have to gather together a small nucleus of people and then look after them.[5]

Despite having a potentially winning formula, the major problem facing all explorers during the 1990s will be the difficulty in raising finance, and there seems little doubt that the Australian exploration industry will contract physically as takeovers, mergers and purchase of the middle rank companies continue to take place. Some predictions go as far as to say that only the top 25 percent of Australian oil explorers on the lists in 1990 will still be there in the year 2000. On the other hand, such talk angers many of the country's geologists who still nurture a belief that the phoenix can rise given the opportunity and a little encouragement. Pioneer Reg Sprigg is one.

> Provided adequate price incentives remain and punitive taxes and levies are avoided, there is no doubt that many unsuspected

> oil and gas fields remain to be discovered and brought into production. Given continuing encouragement there is little reason for investor pessimism in the overall pattern.
>
> Despite this there are concerns. Petroleum exploration is risk enterprise. Healthy competition must be nurtured and maintained. Entrepreneurial activity is as important an ingredient now as ever. The discovery of oil or gas must first still be made 'in the minds of men and women'. A new generation of oil seekers must always be in the making.
>
> In this, Australia's tertiary educational systems must not be ignored. As more information is accumulated, so the idiosyncrasies of Australian sedimentary basins are broken down, and the new potential realized.[6]

Consultant Colin Glazebrook believes that much of the country's current predicament in terms of unreplenished reserves stems from the fact that, with few exceptions during the past 15 years, Australia's local companies have tended to cluster in the favoured basins.

> Eighty percent or so of the wells since the early 1970s have been drilled in seven basins — everyone clustering round some small honey pots. No one was exploring the areas where giants might be found.
>
> Mostly companies did not want to drill a dry hole because it would adversely affect their share price, and so they hedged a lot. What we have to do now is go back to the courage of the 1950s and 1960s. We have to go back out to the regions where God never trod and take more risks, instead of picking a drilling location next to the service station.
>
> Some people will say that all the big ones have been found. Maybe that's the case for easy-to-spot structural traps, but I still believe there is potential for large stratigraphic plays. They might be harder to define, but that doesn't mean they won't be there.[7]

Another pioneer of the 1950s and 1960s, Eric Webb, believes that more data should be extracted from the wells which are drilled.

> I worry even now that a lot of oil and gas has been passed by. Probably in the old days we should have tested wells more thoroughly where we found porosity. Maybe you can find an excuse for it back in the 1950s, but it still seems to be happening. Companies spend perhaps three or four million dollars doing initial surveys and siting a well, and then skimp on the costs of drilling it. I feel many wells should receive more attention so that the maximum data can be gleaned from them.[8]

During a talk to the New South Wales branch of PESA in 1985, geologist Ken Richards pursued this line of thinking even further.

> As an industry we have not properly analysed 'why our holes were dry'. We spend an inordinate amount of effort explaining away small oil discoveries, but tend to put our dry holes in a cupboard and forget them. There is a vast amount of data locked up in these dry holes for the enquiring mind.[9]

This view is shared by John Armstrong, executive general manager of exploration for SANTOS Ltd, who uses the Timor Sea as an example.

> To mid-1990 there were 73 wells drilled in the Timor Sea resulting in four producible fields and 14 other discoveries, some of which may be elevated to commercial status with further work. But the hidden figure is that as many as 32 of those 73 wells were invalid tests. They have been drilled on structures where there is no seal, or where the reservoir is absent, or even on structures which subsequently have been shown to be nonexistent.
>
> Part of the problem is in poor seismic data aquisition and poor interpretation of the data that is available. But there is also a lack of lateral geological thinking. I believe a number of potential traps in the region have fault-penetrated seals which cause them to 'leak'. We should be paying more attention to improving data-gathering techniques, trying to better understand the primary and secondary migration paths of the Timor Sea oil, and then drilling wells using accumulated logic.[10]

Adelaide-based consultant, Reg Thomas, believes the modern explorer should keep in mind the overall economics of a programme.

> There was a happy-go-lucky attitude among the early exploration geologists and few difficulties were treated seriously until a discovery was made. Nowadays technicians seem more level-headed. The potential problems are thought out ahead and brought up to company board level before they occur.
>
> That doesn't mean geologists now are inhibited, but they do have some idea of how much is needed for a commercial development before they go into the field and look. More and more in the future, exploration operations will need constant reappraisals of the economics involved.[11]

The 1980s, particularly during the early part of the decade, were characterized by a 'scatter-gun' approach to exploration programmes, where many Australian companies took up small percentage interests in a number of permits widely dispersed throughout the country. The philosophy was that the more varied play types a company was involved with, even in a minor way, the more chance there was of being successful and maybe being part of a big discovery. The 1990s, however, could see a revision of this policy in favour of an approach which tends to hinge on a thorough examination of one particular region. Perth-based explorer Arrow Petroleum NL (formerly Barrack Energy) has already put this idea into practice. David Maxwell is the company's managing director.

> Originally we began like a number of others with some percentage interests in permits here and there round Australia. But, looking at the exploration industry in the slump of 1986, we decided it would be best to consolidate our interests in one region.
>
> From that point we developed what you could call our four-point philosophy. First identify a basin with good hydrocarbon potential. In our case we decided the costs were too great offshore, so we opted for the onshore Perth Basin. The second step is to establish a dominant position in the chosen region in terms of permit interests. Although Perth Basin had both oil and gas discoveries, it was not perceived by many in the industry to have a great deal of commercial potential. So over a period of several years Arrow (Barrack) was able to take up substantial interests in the permits throughout the region — in some cases 100 percent interest.
>
> The point about this is that it enables you to move on to step three, and that is to study the geology of the whole region in great detail. In the old scatter-gun approach it was not often possible to correlate the data in one permit with that of another group across the boundary. In our regional approach we are involved with all the permits and so we can use all the data to draw up the total picture in the basin. Having done that we are in a better position to plan our exploratory drilling and site our wells, which is step four.[12]

Naturally enough this regional approach does not guarantee commercial discoveries, but it is indicative of the recent, more ordered and mature thinking within the Australian industry. Several other groups are following similar programmes and, given encouragement, they will make a valuable contribution to the determination of Australia's overall oil potential. Bevan Devine, general manager and director of SAGASCO Resources, based in Adelaide, believes that in the past there has been too much direct reliance on overseas 'look-alikes' in the country — particularly from the US and Europe — when really Australian sedimentary geology is different and Northern Hemisphere data is not applicable.

> The difference here is source rock and oil generation. In other countries source rock is widespread and the conditions for generation have been favourable over vast areas. Virtually all the geologist needs to do is look for a trap.
>
> In Australia, however, the source rock is poor, so identifying the hydrocarbon 'kitchens' is essential. Source rock and timing of generation are key elements and they will need to be looked at very carefully for the future programmes. At the beginning, explorers will tend to go for the basins with proven oil-generating capacity and subsequent migration pathways, although we still need some companies to move into less well-known regions and do grass-roots work.[13]

Jaap Poll, a former head of Ranger Oil in Australia and now managing director of Petroz based in Perth, is the first to agree. But he is quick to add that even grass-roots exploration should not return to the 'blind stabbing' of some early programmes. Writing about the Canning Basin during the early 1980s, his comments then are equally applicable now to the rest of the country as the oil industry looks towards the last years of the century.

> The genuine explorer will have an open mind as to where he or she may find oil. Any coincidence of potential reservoir and trap forms a legitimate initial target. Whether it has been recognized by others as such is irrelevant. Our expertise should be applied to the recognition, understanding and priority grading of all objectives.
>
> Because of the high cost of exploration in [regions like] the Canning Basin and the scarcity of risk money, explorers should more than ever resist drilling poorly defined targets. Too many wells are still being drilled on inadequate seismic control, thereby wasting funds and losing potential investors' confidence.[14]

There is no doubt that the technology surrounding oil exploration will continue to improve as computerization and laser techniques become more and more sophisticated. It is also considered likely that natural gas will 'have its day' — if not in the 1990s, then early in the next century. As gas becomes a more sought-after commodity, exploration and development within Australia's basins will be more widespread. The chance of new oil finds will increase on the back of this activity, while the economics of development will also be improved via the growth of infrastructure. Nevertheless advances will not come by themselves.

Geologist Murray Johnstone pointed out during an Australian lecture tour in 1986 that the industry cannot afford to become complacent about its achievements.

> Oil is becoming harder to find so we will have to use even more sophisticated techniques in our future search. We must not be blinded by current technology into thinking that we know everything about basin evaluation, or that current techniques cannot be bettered. We have only to look to our past 35 years to know that the next 35 will see equally dramatic changes in technology and geological concepts.[15]

At the beginning of 1990 there are about 80 companies engaged in Australian and Papua New Guinea petroleum exploration to a greater or lesser degree. In addition there are about 110 companies which provide services to the industry. How many will be able, or be inclined, to remain in the business and see the start of its third century is unclear. Certainly those that do stay will be part of a different environment. They will have endured ongoing oil price crises as well as debates on taxation, on the rights of land ownership, land access and land use, on pollution and the wide implications of the Greenhouse Effect, and on the need to have indigenous oil at all. They will also have undergone rigorous corporate purges and perhaps been favoured with new technical breakthroughs. Many will have gained valuable experience by increased exposure to the international oil industry through the Federal Government's deregulatory moves and their own overseas exploratory programmes during the 1980s. In the final analysis though, future success in the Australian and Papua New Guinea oil search will still depend on people, their original ideas and the willingness of their companies to try them out. The late Ken Richards, one of the country's best modern explorationists, saw the future in Australia's offshore regions. He also saw that, in the long run, success would depend almost entirely on the combined strength and efforts of geological minds and industry investment.

> I don't think industry can succeed by simply drilling holes using the same exploration philosophy as in the past. Over the next 15 years the capacity for a private oil exploration industry to survive in offshore Australia will depend directly on the amount of capital it is prepared to expend on scientifically analysing the issues fundamental to oil generation and migration in our continental margin basins.
>
> In this research effort I firmly believe we are largely on our own. It will need creative Australian thinking within an Australian context to resolve the problem. We have the brains, we have the basic data. We have in piecemeal fashion already researched some key elements.

> What we lack are the finances and the organization to coordinate the effort. Management, through APEA, and professionals, through the Petroleum Exploration Society of Australia, must combine to ensure industry addresses this issue as its number one priority for the immediate future.[16]

These thoughts can apply equally to the whole Australia/PNG industry on and offshore. In fact, they echo the sentiments expressed back in 1977 by the (then) Chancellor of the University of Sydney, Sir Hermann Black. In concluding his principal address to delegates at the APEA conference that year, he wished the conference luck, because he felt luck was needed during a difficult period of evolving resources policy. Then he added a rider.

> But most of all I wish it [the conference] fruitful and reasoned analysis, because in the end it is the power of reason in your industry at all levels — managerial, innovative and technical — which will make or break it.[17]

Perhaps the last word should come from Reg Sprigg, arguably one of the most active, enthusiastic and colourful of Australia's many explorers since the birth of the modern oil search in the 1950s.

> Australia in many ways is still a sleeping giant by way of energy minerals. Despite growing success ratios, only a fraction of the continent's potential has been put to the test. The future is as challenging and as exciting as ever.[18]

Appendix

CHRONOLOGY OF AUSTRALIA'S OIL EXPLORATION

1802	French world scientific expedition finds oil shale in New South Wales near Blue Mountains.
Early 1800s	Sealers and whalers use bitumen stranded along Australia's southern beaches to caulk boats and floor huts.
1820s	First settlers over Blue Mountains use oil shale blocks as fuel.
1839	John Lort Stokes of HMS *Beagle* finds bitumen in water wells sunk on banks of Victoria River, Northern Territory.
1840–45	Count Paul Strzelecki finds oil shale at Hartley Vale/Petrolea, New South Wales.
1849	Rev. W.B. Clarke finds oil shale at America Creek, Wollongong, New South Wales.
1852	Police Inspector C.W. Stuart notes presence of surface bituminous substance (Coorongite) at Salt Creek, South Australia.
1865	Rev. J.T. Woods publishes *History of Discovery and Exploration of Australia* in two volumes.
1865	J. Graham sets up Pioneer Kerosene Works at Port Kembla, New South Wales, using oil shale from America Creek.
1866	West Australian bituminous resins on display in Paris Universal Exhibition.
1866	Australia's first oil exploration well drilled at Salt Creek in the Coorong district of South Australia.
1869	One ton sample of Coorongite sent to Scotland for analysis shows petroleum content.
1860s	First naturally formed petroleum imported into Australia (from America).
1865–1905	Oil shale production at Hartley Vale, New South Wales.
1870–1907	Oil shale production at Joadja Creek, New South Wales.
1900	Natural gas found in water bore, Roma, Queensland.
1900	Natural gas found in water bore, Grafton racecourse, New South Wales.
1902	First exploration drilling for oil in Western Australia at Warren River in southwest of State.

1904	Prospector William Campbell samples oil from rocks near Victoria River, Northern Territory.
1906	Roma, Queensland, lit for 10 days by natural gas.
1906–23	Oil shale production at Newnes, New South Wales.
1907	First survey in Sepik region of German New Guinea.
1908	Blowout and fire at Roma No.1 gas well, Roma, Queensland.
1910–15	Oil shale production at Railton, Tasmania.
1911	Planters Thomas and Lett find gas seeps at Opa on Vailala River, Papua New Guinea.
1912	Joseph Carne finds oil seeps at Orevi and Upoia on Vailala River, Papua New Guinea. First exploration drilling in the country.
1912	Oil from seeps in Sepik region of German New Guinea analysed in Germany.
1913	Arthur Wade engaged by Australian Government to survey Papuan region of New Guinea.
1915	First recorded Tasmanian exploration drilling on Bruny Island.
1919	Stockman Okes finds pitch in basalt near Ord River, Western Australia.
1919	Water driller Price finds oil traces in bore near Fitzroy Crossing, Western Australia.
1920	Arthur Broughton proves Coorongite in South Australia to be vegetable origin.
1920	Australian and British Governments begin joint exploration of Papua and New Guinea using BP as exploration agent.
1920	Australian Government offers £50 000 reward for discovery of commercial oil.
1920–35	Oil shale production at Latrobe, Tasmania.
Early 1920s	'Johnson's' well reportedly finds gas/condensate on Bruny Island, Tasmania.
1922	British Government withdraws from Papua and New Guinea exploration. Australian Government and BP continue work.
1922–24	Freney Kimberley Oil finds oil traces during drilling in northwest Western Australia.
1924	Australia's first oil discovery (as opposed to condensate and oil traces) at Lake Bunga No.1 well near Lakes Entrance, Victoria.
1926	Australian Government passes Petroleum Prospecting Act providing pound for pound subsidies (up to £60 000) for oil exploration.
1927	Petroleum Prospecting Act extended to Papua New Guinea and increased to £120 000.

1927	Dr George Woolnough appointed Commonwealth Geological Advisor.
1927	Roma Oil Co's Roc No.1 gas/condensate discovery at Roma, Queensland, sets off stock exchange boom.
1928	First commercial sale of Australian oil (condensate) — 160 gallons [727 L] from Roc No.1, Roma, Queensland.
1928	Geophysical Survey Act passed by government to prompt surveys in Australia for oil, minerals and water.
1929	Roma absorption plant begins commercial production of condensate, Queensland.
1930–41	107 000 gallons [486 422 L] of oil produced from Lakes Entrance region, Victoria. Most sold to Melbourne and Metropolitan Tramways Board for lubricating suburban trams.
Early 1930s	Woolnough experiments with first applications of air photography to geology in northern Australia.
1931	Results of Imperial Geophysical Experimental Survey (IGES) lay foundations for commercial geophysics in Australia.
1932	Oil Search Ltd begins field work in Mandated Territory of New Guinea.
1932–34	Arthur Wade maps Kimberley region of Western Australia for Freney Kimberly Oil.
1934	Oil recovered from water bore at Wilkatana, near Port Augusta, South Australia.
1934–42	Government forms and supports Northern Australia Survey to investigate geology of areas north of 22nd parallel.
1935	Oil Search Ltd's Arcadia No.1 in Queensland finds gas in lower horizon than reservoirs at Roma.
1936	Petroleum Oil Search Act passed with £250 000 available to encourage drilling operations.
1936	Three-man Oil Advisory Committee formed of geologists Walter Woolnough, Arthur Wade and Keith Ward.
1937	First air survey of Papua and New Guinea.
1937–52	Oil shale production at Glen Davis, New South Wales.
1938	Australasian Petroleum Company (APC) formed by BP, Mobil and Oil Search.
1939–51	Shell conducts largest coordinated exploration effort in Australia to that time.
1939–54	War-interrupted drilling of Freney Kimberley's Nerrima Dome well in Western Australia on recommendation of Wade.
1940	Oil Advisory Committee disbanded. Harold Raggatt appointed Commonwealth Geological Advisor.
1941	American engineer Leo Ranney recommends oil shaft (Ranneywell) project at Lakes Entrance, Victoria, to Australian Government.

1941–48	APC's war-interrupted Kariava No.1 well drilled near Vailala River, Papua New Guinea.
1946	Australian Government forms Bureau of Mineral Resources, Geology and Geophysics (BMR). Harold Raggatt appointed first director.
1946	Australian Government withdraws from Ranneywell scheme. Austral Oil (Lakes Oil) takes over.
1947	Frome-Broken Hill conducts Australia's first commercial post-war oil surveys in Canning Basin of Western Australia.
1947	Ampol secures its first oil exploration permits, Exmouth region, Carnarvon Basin, Western Australia.
1948	First BMR field party begins geologically mapping Fitzroy Basin in Kimberley region of Western Australia.
1949	Australia's first reflection seismic survey shot by BMR in Roma area of Queensland.
Early 1950s	Introduction of helicopters to survey work, particularly in Papua and New Guinea.
1951	Seismograph Services of London, the first contract seismic crew in Australia, arrive at Rough Range area for WAPET.
1951	Ranneywell project abandoned after producing 172 590 gallons [784 594 L].
1952	Australian Associated Oilfields flows gas in its first well, at Hospital Hill, Roma, in Queensland.
1953	WAPET tests oil at Rough Range No.1 well, Carnarvon Basin, Western Australia.
1954	WAPET geologists make up first civilian group to visit Barrow Island after atomic tests in the region and conduct basic reconnaissance survey.
1956	Australasian Petroleum Company's Kuru No.1 well in Papua New Guinea blowout and uncontrolled gas flow for five months.
1957	First heli-rigs used in Papua and New Guinea by Australasian Petroleum Company.
1957	Petroleum Search Subsidy legislation passed.
1958	Introduction of system of graticular blocks as basis for drawing up permit boundaries.
1959	Australasian Petroleum Company's Puri No.1 wildcat in Papua New Guinea finds oil.
1959	Petroleum subsidy scheme broadened to include seismic and other survey work.
1959	Formation of Queensland Petroleum Exploration Association (QUPEX).
1959	Formation of Australian Petroleum Exploration Association (APEA).
1959	Geophysical Services International shallow-water survey for Frome-Broken Hill in Otway Basin, Victoria, first offshore seismic programme in Australia.

1959	Woodside (Lakes Entrance) Oil takes out first offshore exploration leases in Gippsland Basin, Victoria.
1959	Innamincka No.1, the first well in the Cooper Basin, is drilled by Delhi/SANTOS.
1959	WAPET's Meda No.1 well brings oil to surface for first time in Canning Basin, Western Australia. But noncommercial.
1959	Frome-Broken Hill tests significant gas flow in Port Campbell No.1 well, Victorian Otway basin.
1960	Institute Francaise du Petrole (IFP) contracted by Australian Government to survey the country's oil potential.
1960	Associated Group finds gas in Timbury Hills No.2 near Roma, Queensland.
1960	Gas blowout in Union Oil group's Cabawin No.1, Surat Basin, Queensland.
1960	Consultant Lewis Weeks recommends BHP explore Bass Strait.
1961	First APEA conference held in Melbourne.
1961	Cabawin No.1 tests gas and light oil.
1961	Union group's Moonie No.1 flows oil in Surat Basin, Queensland.
1962	Moonie declared Australia's first commercial oil field after 18 months appraisal work.
1963	Woodside awarded North West Shelf permits, Western Australia.
1963	Delhi/SANTOS find Gidgealpa gas field in Cooper Basin, South Australia.
1963	Ooraminna No.1 records hydrocarbon shows from Ordovician/Cambrian reservoirs in the Amadeus Basin of central Australia — the oldest such occurrence known in the world.
1963	IFP report is generally lukewarm about Australia's oil potential.
1963	APEA Petroleum Club formed in Sydney — the forerunner of the Association's technical division.
1964	Esso farms into BHP Bass Strait acreage, Victoria, Tasmania and South Australia.
1964	Moonie field comes on stream as Australia's first commercial oil project.
1964	Union Oil group finds Alton oil field in Surat Basin, Queensland.
1964	Phillips Petroleum group finds Gilmore gas field in central Queensland.
1964	Exoil finds Mereenie oil/gas field in Amadeus Basin, central Australia via spectacular blowout.
1964	WAPET finds oil at Barrow Island, Carnarvon Basin, Western Australia.

Mid-1960s	First use of satellite position fixing and digital recording instruments in seismic surveys.
1965	WAPET finds second, shallower reservoir in Barrow Island field.
1965	Esso/BHP finds Barracouta gas field, Gippsland Basin, Victoria.
1965	Magellan Petroleum finds Palm Valley field in Amadeus Basin, central Australia.
1966	WAPET finds Dongara gas field in Perth Basin, Western Australia.
1966	Esso/BHP finds Marlin gas field, Gippsland Basin, Victoria.
1966	Delhi/SANTOS finds Moomba gas field in Cooper Basin, South Australia.
1966	Alliance Oil Development's Caroline No.1 well in Otway Basin of South Australia finds commercial carbon dioxide field.
1967	Naturalist Harry Butler appointed conservation consultant to WAPET in early industry move to protect the environment.
1967	APEA Petroleum Club formalized into the APEA Professional Division (the forerunner of PESA) with a branch in each State.
1967	Esso/BHP finds Halibut, Kingfish and Dolphin oil fields, Gippsland Basin, Victoria.
1967	Woodside finds Golden Beach gas field, Gippsland Basin, Victoria.
1967	Woodside drills Ashmore Reef No.1 — first well on North West Shelf acreage.
1968	Esso/BHP finds Perch oil, Tuna oil/gas, Snapper gas and Flounder oil fields, Gippsland Basin, Victoria.
1968	Woodside's Legendre No.1 well finds first hydrocarbons (oil) on North West Shelf.
1968	Phillips' Pasca No.1 finds gas in Gulf of Papua.
1968	Esso/BHP strike fixed price arrangement with Australian Government for Bass Strait crude production.
1968	Joint Federal-State legislations for jurisdiction of continental shelf regions.
1969	Esso/BHP finds Mackerel oil and Bream oil/gas field, Gippsland Basin, Victoria.
1969	Aquitaine group's Petrel No.1 wildcat in Bonaparte Gulf finds gas with spectacular blowout.
1969	Delhi group finds Toolachee gas field in Cooper Basin, South Australia.
1970	AAR finds Rolleston gas field in Bowen Basin, Queensland.
1970	Hartogen drilling confirms Kincora as commercial gas field in Surat Basin, Queensland.

1970	Cooper Basin discoveries at Merrimelia (gas), Strzelecki (gas), Della (gas) and Tirrawarra (oil), all in South Australia.
Early 1970s	Introduction of LANDSAT photo imagery and seismic stratigraphy to geological survey work.
1971	Cooper Basin of South Australia discoveries Big Lake (gas), Moorari (oil) and Fly Lake (oil).
1971	Woodside North West Shelf gas/condensate discoveries — Scott Reef, North Rankin and Goodwyn, Western Australia.
1971	Aquitaine finds Tern gas field in Bonaparte Gulf, Western Australia.
1972	Five more Cooper Basin gas discoveries at Brumby, Brolga, Epsilon, Burke and Dullingari.
1972	Aquitaine finds Puffin oil field, Timor Sea, Northern Territory.
1972	Esso/BHP find Cobia field, Gippsland Basin, Victoria.
1972	Woodside find Angel gas/condensate field, North West Shelf, Western Australia.
1973	Petroleum Search Subsidy scheme terminated.
1974	APEA Professional Division becomes an independent, but still affiliated body and is renamed the Petroleum Exploration Association of Australia (PESA).
1974	Barrier Reef Royal Commission report bans further drilling on Great Barrier Reef, Queensland.
1974	Bridge Oil finds Silver Springs gas/condensate field, Surat Basin, Queensland.
1974	Drillship *Regional Endeavour* launched in New South Wales.
1975	Papua New Guinea gains Independence.
1975	Federal Government introduces world parity pricing for Australian crude oil.
1976	Semisubmersible *Ocean Endeavour* launched in Western Australia.
1976	Namur gas field in South Australia is first find in Eromanga Basin.
1976	Semisubmersible *Southern Cross* launched in Queensland.
1976	Bridge Oil finds Boxleigh gas/condensate field, Surat Basin, Queensland.
1977	Gazettal of deep-water Exmouth Plateau permits off Western Australia.
1977	APEA draws up pioneering Code of Environmental Practice.
1978	Esso/BHP finds Seahorse and Fortescue oil fields, Gippsland Basin, Victoria.
1978	Delhi/SANTOS group finds first commercial Eromanga oil at Strzelecki and Dullingari North fields, South Australia.

1978	Gulf Oil farms into APC's Papua New Guinea highland permits.
1979	Successful Palm Valley and Mereenie field development negotiations represent significant advance in Aboriginal relations.
1979	Beach Petroleum finds North Paaratte gas field in Otway Basin, Victoria.
1979	Surat Basin discoveries: Thomby Creek (oil) and Beldene (gas), Queensland.
1980	Western Australian Government and Amax Petroleum confront Aborigines in land rights claim at Noonkanbah Station in the Kimberleys over Fitzroy River No.1 well.
1980	Perth Basin discoveries Mt Horner (oil) and Woodada (gas), Western Australia.
1980	Esso/BHP finds Scarborough gas field, Exmouth Plateau, Western Australia.
1981	Denison Trough gas discoveries: Merivale, Punchbowl Gully, Yellowbank and Springvale, Queensland.
1981	Esso/BHP find Tarwhine oil field in Gippsland Basin, Victoria.
1981	Western Australian finds include Gorgon gas (offshore Carnarvon Basin), Blina oil (onshore Canning Basin) and Tubridgi gas (onshore Carnarvon Basin), plus Goodwyn No.6 well finds oil leg.
1981	Delhi/SANTOS group finds Jackson oil field, Eromanga Basin, Queensland.
1981	BHPP group finds Riverslea oil field, Surat Basin, Queensland.
1981-82	Queensland Eromanga Basin finds include Gunna (oil), Chookoo (gas), Nockatunga (oil), Tinpilla (oil), Thungo (oil).
1982	Queensland Surat Basin oil finds include Yellowbank Creek, Borah Creek, Sandy Creek and Waratah.
1982	Mesa finds South Pepper oil field, Carnarvon Basin, Western Australia.
1982	Home Energy group finds Sundown oil field, Canning Basin, Western Australia.
1983	Western Australian offshore finds include North Herald, Chervil and Harriet/Lenita (oil) and Wilcox (gas/condensate).
1983	Hartogen group finds Tintaburra oil field, Eromanga Basin, Queensland.
1983	BHPP group finds Jabiru oil field, Timor Sea, Northern Territory.
1983	Niugini Gulf group finds oil shows at Mananda in Papua New Guinea — first oil in the highlands region.

1983	Esso/BHP finds Whiting oil field, Gippsland Basin, Victoria.
1984	Niugini Gulf group finds Juha gas/condensate field, Papua New Guinea.
1984	BHPP group finds Challis oil field, Timor Sea, Northern Territory.
1984–85	Queensland Eromanga Basin oil discoveries include Bodalla South, Kenmore and Talgaberry.
1985	WAPET finds Saladin oil field, Carnarvon Basin, Western Australia.
1985	Resource Rent Tax introduced to offshore 'greenfield' areas.
1985–86	Queensland Surat Basin finds include Fairymount, Louise and Taylor oil fields.
1986	Esso/BHP/Shell group finds Kipper gas field, Gippsland Basin, Victoria.
1986	Niugini Gulf group finds Iagifu oil field, Papua New Guinea.
1986	World price of oil crashes to below $US10 per barrel.
1986	Bass Strait oil production begins its decline.
1987	Stock market crash affects all Australian listed explorers.
1988	Free market introduced for Australian crude oil marketing.
1988	Delhi/SANTOS group finds oil at Sturt and Tantanna, Eromanga Basin, South Australia.
1988	BHP Petroleum finds oil at Skua, Timor Sea.
1988	Ultramar group finds gas at Katnook, Otway Basin, South Australia.
1988	Chevron Niugini (Gulf) group finds Hedinia oil field in Papua New Guinea.
1988	International Petroleum group finds Pandora gas field in Papuan Gulf.
1989	Woodside finds Wanaea oil field, North West Shelf, Western Australia.
1989	Chevron Niugini group finds Agogo oil/gas field in Papua New Guinea.
1989	Government calls submissions for complete review of oil taxation.
1989	Australia and Indonesia sign Timor Gap Treaty to allow exploration in newly created joint development zone in Timor Sea.
1990	Woodside finds Cossack oil field, North West Shelf, Western Australia.
1990	SANTOS group finds oil at Talbot in Timor Sea.
1990	BHP Petroleum finds Griffin oil field, North West Shelf, Western Australia.
1990	Barrack Energy (now Arrow Petroleum) finds oil at North Yardonogo and gas at Beharra Springs, Perth Basin, Western Australia.

1990	Federal Government extends Resource Rent Tax regime to all Bass Strait oil and gas production with effect from July 1990. Government also extends RRT deductibility to all offshore areas within the regime.
1990	SANTOS group finds oil at Malgoona, Cooper/Eromanga Basin, South Australia.

Footnotes

CHAPTER BY CHAPTER

Frontispiece

1. David, M. E. *Professor David, The Life of Sir Edgeworth David,* Edward Arnold & Co., London, 1937.
2. Glazebrook, C. Interview with author, Melbourne, 1989.

Introduction

1. Haddock, J. Interview with author, Perth, 1989.

PART ONE

Chapter 1 — Early Glimpses

1. Sprigg, Reg. C. Towards 2000: Incentives versus Disincentives, Opening Address, APEA Conference 1986, Adelaide.
2. Millar, Ann. *'I See No End To Travelling', Journals of Australian Explorers 1813–1876,* Bay Books, Sydney, 1986.
3. Stokes, John Lort. Discoveries in Australia, an account of the coasts and rivers explored and surveyed during the voyage of the HMS Beagle in the years 1837–1843, by command of the Lords Commissioners of the Admiralty, London, 1846.
4. Rudd, E. A. & Sprigg, R. C. 'History of Oil Search in Australia and Papua New Guinea', *Economic Geology of Papua and New Guinea,* The Australian Institute of Mining and Metallurgy, Melbourne, 1976.
5. Carne, J.E. 'Kerosene Shale Deposits in New South Wales', *Memoirs of Geological Survey of New South Wales, Geology,* No.3, Sydney, 1903.
6. Author unknown. '150 Years of Oil Exploration in Victoria', *Australian Oil and Gas Directory,* Offshore Oil and Gas Directory Pty Ltd, Melbourne, 1984–85.
7. ibid.
8. ibid.
9. ibid.

10. ibid.
11. West Australian Petroleum Pty Ltd. *The Story of the Search for Oil in Western Australia,* Perth, undated, circa 1950s.
12. Carne, op. cit.
13. Ampol Ltd. *Oil For Tomorrow,* a company publication undated, but probably the 1950s.
14. Johns, R.K. (ed.). *History and Roles of Government Geological Surveys in Australia,* South Australian Government Publication, 1976.
15. Phipson, T.L. 'On the Composition of a Peculiar Substance from Wallabies' Holes, River Murray', *The Geologist,* London, April 1862.
16. Broomham, Rosemary. *First Light, 150 Years of Gas,* Hale & Iremonger, Sydney, 1987.
17. Grainger, Elena. *The Remarkable Reverend Clarke,* Oxford University Press, Melbourne, 1982.
18. Moyal, Ann. *'A Bright and Savage Land', Scientists in Colonial Australia,* William Collins, Sydney, 1986.
19. Curran, Rev. J.M. 'J.E.T Woods', *Centennial Magazine,* 1889-90, pp. 406-411.
20. O'Neill, Rev. George. *The Life of the Reverend Julian Edmund Tenison Woods (1832 — 1889),* a book dated 1929 possibly published by the Catholic Church, Melbourne.

Chapter 2 — An Alternative Route and a False Trail

1. Knapman, Leonie. *Joadja Creek, The shale oil town and its people 1870-1911,* Hale & Iremonger, Sydney, 1988.
2. Carne, J.E. 'Kerosene Shale Deposits of New South Wales', *Memoirs of Geological Survey of New South Wales, Geology,* No.3, 1903.
3. ibid.
4. Ferguson, Penelope. *History of Australia's Shale Oil Industry,* Southern Pacific Petroleum NL and Central Pacific Minerals NL. Sydney, undated (about 1981).
5. Reid, McIntosh. 'The Oil Shale Resources of Tasmania', *Geological Survey of Mineral Resources,* No.8, Vol.1, Tasmanian Department of Mines, Hobart, 1924.
6. Australian Institute of Petroleum. 'Australia's Shale Oil Industry', *Petroleum Gazette,* Vol.14, No.4, Melbourne, December 1966.
7. Carne, op. cit.
8. Australian Institute of Petroleum, op. cit.
9. Carne, op. cit.
10. Knapman, op. cit.
11. ibid.
12. Carne, op. cit.
13. ibid.
14. ibid.

15. ibid.
16. Knapman, op. cit.
17. Australian Institute of Petroleum, op. cit.
18. ibid.
19. Ferguson, op. cit.
20. Carne, op. cit.
21. Ferguson, op. cit.
22. Tasmanian Department of Mines. Report of Tasmanian Shale Oil Investigation Committee, *Geological Survey of Mineral Resources,* No.8, Vol.2, Hobart, 1932.
23. McCourt, T. & Mincham, H. *The Coorong and Lakes of the Lower Murray,* National Trust (Beachport Branch), Adelaide, July 1987.
24. Cane, R.F. 'Coorongite, Balkashote and Related Substances, an annotated bibliography', *Transactions and Proceedings of the Royal Society of South Australia,* Vol.101, August 1977.
25. McCourt & Mincham, op. cit.
26. Tyson, P. 'Persistence Pays off in Eccentric Search', *Bulletin,* Sydney, January 12, 1988.
27. Colyer, F. 'Fool's Oil', *Petroleum Gazette,* Australian Institute of Petroleum, Melbourne, June 1974.
28. Lewis, J.M. 'The Search for Oil', *From the Coorong to Port Bonython,* Proceedings of the Australian Institute of Petroleum, South Australian branch, Adelaide, 1983.
29. Cane, op. cit.
30. Colyer, op. cit.
31. Cane, op. cit.
32. Colyer, op. cit.
33. McCourt & Mincham, op. cit.
34. Lewis, op. cit.
35. McCourt & Mincham, op. cit.
36. Lewis, op. cit.
37. Ward, L.K. 'A Review of Mining Operations in the State of South Australia', *South Australian Mines Department,* Issue No.24, 1916.
38. Colyer, op. cit.
39. Author unknown. 'Captain De Hautpick Back In Australia', *Smith's Weekly,* Sydney, December 10, 1927.
40. De Hautpick, Captain E. 'Coorongite, a petroleum product', *Mining Journal,* London, July 28, 1923.
41. *Smith's Weekly,* op. cit.
42. McCourt & Mincham, op. cit.
43. Broughton, A.C. 'Notes on the Formation of Coorongite', *Transactions and Proceedings of the Royal Society of South Australia,* Vol.44, 1920.
44. McCourt & Mincham, op. cit..
45. Basedow, H. *Adelaide Observer,* August 14, 1925.
46. Cane, op. cit.

47. ibid.

Chapter 3 — Promoters and Explorers Take to the Field

1. Carne, J.E. 'Kerosene Shale Deposits in New South Wales', *Memoirs of Geological Survey of New South Wales, Geology,* No.3, 1903.
2. Campbell, I.R. & Parry J.C. 'Oil Drilling Onshore and Offshore', *Mining in Western Australia* (ed. Prider, R.T.), University of Western Australia Press, Perth, 1979.
3. Author unknown. *Oil For Tomorrow,* Ampol company publication, circa 1955.
4. Durack, M. Family papers and news clippings, Perth.
5. Durack, M. *Sons in the Saddle,* Constable and Co. Ltd., London, 1983.
6. Raggatt, H.G. & Crespin I. 'Summary of Drilling Activities in Australia and New Guinea', Report No.1943/63, *Mineral Resources Survey,* Dept. Supply and Shipping, Canberra, 1943.
7. *Oil For Tomorrow,* op. cit.
8. Durack, Family papers, op. cit.
9. Purcell, P. 'The Canning Basin, Western Australia — an introduction', *Proceedings of Canning Basin Symposium,* Geological Society of Australia/Petroleum Exploration Society of Australia, Perth, 1984.
10. Durack, Family papers, op. cit.
11. Rudd, E. Interview with author, Adelaide, 1987.
12. Durack, Family papers, op. cit.
13. O'Driscoll, E. Interview with author, Perth, 1987.
14. ibid.
15. ibid.
16. Durack, Family papers, op. cit.
17. Poll, J. 'Onshore Canning Basin — An Historical Perspective', *Journal of Petroleum Exploration Society of Australia,* No.3, 1983, pp. 11-20.
18. Allen, R.J. 'Petroleum Resources of Queensland', *APEA Journal,* 1975.
19. Watson, L. & E. *The Roma Story,* Roma, circa 1965.
20. West, B.G. Notes compiled from BMR and Queensland Government records.
21. ibid.
22. Watson, op. cit.
23. ibid.
24. ibid.
25. Byrne, C. Interview with author, Caloundra, 1987.
26. Smith, C.W. Interview with author, Melbourne, 1989.
27. ibid.
28. ibid.

29. Watson, op. cit.
30. Byrne, op. cit.
31. ibid.
32. Rudd, op. cit.
33. Wade, A. 'Fifty Years Searching For Oil in Queensland', *Queensland Government Mining Journal,* June 20, 1950.
34. ibid.
35. Thomas, B. M. 'In For The Long Haul: 50 Years Of Shell Exploration In Australia', *APEA Journal,* 1990.
36. Wade, op. cit.
37. ibid.
38. White, J. Interview with author, Canberra, 1987.
39. Carne, J.E. 'Occurrence of Coal, Petroleum and Copper in Papua', *Bulletin of Territory of Papua,* No.1, 1913.
40. ibid.
41. ibid.
42. ibid.
43. ibid.
44. ibid.
45. Wade, A. *Report on Petroleum in Papua,* Commonwealth Government Report, No.14, Canberra, undated.
46. Author unknown. 'The Search For Australian Oil', *Petroleum Gazette,* Vol. 3, No.1, June 1954.
47. ibid.
48. Author unknown. 'The Oil Exploration Work in Papua and New Guinea conducted by Anglo Persian Oil Co on Behalf of the Government of the Commonwealth of Australia, 1920–1929' (4 volumes plus 2 atlas volumes), unpublished report, Canberra.
49. Raggatt & Crespin, op. cit.
50. Purcell, P.G. 'Marienberg-1, Sepik Basin', in Carman, G.J. & Z. (eds), *Petroleum Exploration in Papua New Guinea,* Proceedings of the First PNG Petroleum Convention, Port Moresby, February 1990.
51. Rudd, op. cit.
52. Carey, S.W. Interview with author, Melbourne, 1987.
53. ibid.
54. ibid.
55. ibid.
56. ibid.
57. Carey, S.W. 'Preliminary Notes of the Recent Earthquake in New Guinea', *Australian Geographer,* Vol.11, No.8, December 1935.
58. Stanley, G.A.V. *Review of Geology and Petroleum Prospects of Permit 22, Papua,* Volume 1, report to the Papuan Apinaipi Petroleum Company Ltd, Brisbane, September 1960.
59. Stach, L. Interview with author, Melbourne, 1988.
60. ibid.
61. ibid.

62. O'Driscoll, op. cit.
63. Condon, A. Interview with author, Sydney, 1987.
64. Carey, interview, op. cit.
65. Stach, op. cit.
66. Richter, R. Interview with author, Brisbane, 1987.
67. Stach, op. cit.
68. Campbell, W.A. Sworn statement, quoted in an unpublished review of the petroleum prospects of the Victoria River Basin written for Chillagoe Minerals, Brisbane, circa 1985.
69. Durack, Family papers, op. cit.
70. ibid.
71. ibid.
72. Meudell, G.D. 'History of Petroleum in Australia', *Victorian Historical Magazine,* Vol.11, 1926, pp. 128–137.
73. Twelvetrees, W.H. *The Search for Petroleum in Tasmania,* Tasmanian Mines Department, Circular No.2, 1917.
74. McIntosh Reid, A. 'North Bruny Oil Prospects', unpublished report, Dept. Mines, Hobart, 1929.
75. Cannon, M. *The Landboomers,* Melbourne University Press, Melbourne, 1966.
76. Meudell, G.D. *The Pleasant Career of a Spendthrift,* George Routledge & Sons, London, 1929.
77. Meudell, G.D. Collection of maps on petroleum, Mitchell Library, Sydney.
78. Raggatt & Crespin, op. cit.
79. Carne, J.E. 'Kerosene Deposits of New South Wales', op. cit.
80. ibid.
81. Jones, L.T. 'Notes on Petroleum and Natural Gas and the Possibilities of Their Occurrence in New South Wales', *Mineral Resources,* No.31, New South Wales Geological Survey, 1921.
82. Raggatt & Crespin, op. cit.
83. Meudell, G.D. 'The History of Petroleum in Australia', op. cit.
84. Raggatt & Crespin, op. cit.
85. ibid.
86. Raggatt & Crespin, op. cit.
87. Wopfner, H. 'From Coorongite to Cooper Basin and Beyond', an address to APEA Professional Division, Adelaide, October 1972.
88. Colyer, F. 'Fool's Oil', *Petroleum Gazette,* Australian Institute of Petroleum, June 1974.
89. Unknown author. 'Captain de Hautpick Back in Australia to Trace Oil By Smell', *Smith's Weekly*, Sydney, December 10, 1927.
90. Lewis, J. 'The History of the Oil Industry in South Australia', *From the Coorong to Port Bonython,* Proceedings of the Australian Institute of Petroleum, South Australian branch, Adelaide, October 1983.
91. ibid.

92. Meudell, G.D. *The Pleasant Career of a Spendthrift,* op. cit.
93. Wopfner, op. cit.
94. Lewis, op. cit.
95. ibid.
96. Rudd, op. cit.
97. O'Neil, B. *In Search of Mineral Wealth — The South Australian Geological Survey and Department of Mines to 1944,* special publication, South Australian Dept. Mines and Energy, Adelaide, 1982.
98. Wade, A. 'The Supposed Oil-Bearing Areas of South Australia', *Bulletin* No.4, Geological Survey of South Australia, Adelaide, 1915.
99. Lewis, op. cit.
100. ibid.
101. ibid.
102. Sprigg, R.C. 'The Search for Commercial Oil and Gas in South Australia', *From the Coorong to Port Bonython,* Proceedings of the Australian Institute of Petroleum, South Australian branch, Adelaide, October 1983.
103. Rudd, op. cit.
104. Lewis, op. cit.
105. Sprigg, R.C. 'The Long Lead-up to Commercial Oil Discovery in the Mesozoic Eromanga Basin', *Proceedings of Eromanga Basin Symposium,* Geological Society of Australia/Petroleum Exploration Society of Australia, Adelaide, 1982.
106. Author unknown. '150 Years of Oil Exploration in Victoria', *Australian Oil and Gas Directory,* Offshore Oil and Gas Directory Pty Ltd, Melbourne, 1984–85.
107. Boutakoff, N. 'Oil in Victoria', *Mining and Geological Journal,* Victorian Dept. Mines, Vol.4, No.4, September 1951.
108. Author unknown. ' "Oil" About Oil In Victoria', *Smith's Weekly,* Sydney September 17, 1927.
109. Author unknown. *Gippsland East — Its Geology and Mining Development (including Lakes Entrance Oil Field),* Victorian Mines Dept., Melbourne, 1936.
110. ' "Oil" About Oil in Victoria', op. cit.
111. Meudell, G.D. *The Pleasant Career of a Spendthrift,* op. cit.
112. Meudell, G.D. 'History of Petroleum in Australia', op. cit.
113. ' "Oil" About Oil in Victoria', op. cit.
114. ibid.
115. Smith, op. cit.
116. Baragwanath, W. 'Boring For Oil, Lakes Entrance', *Records of Geological Survey of Victoria,* Vol.5, No.4, 1952.
117. Scarce, K. Interview with author, Bairnsdale, 1988.
118. Withers, R. Interview with author, Melbourne, 1987.
119. Raggatt & Crespin, op. cit.
120. Clarke, A.B. Interview with author, Lakes Entrance, 1989.

121. Boutakoff, op. cit.
122. Clarke, op. cit.
123. Author unknown. 'How Oil Is Being Won at Lakes Entrance', *Age*, Melbourne, February 22, 1949.

Chapter 4 — The Government Impetus

1. Sprigg, R. Letter to author, 1989.
2. Crespin, I. *Memoirs of a Micropalaeontologist — Dr Irene Crespin*, BMR Paper, Canberra, 1980.
3. David, M.E. *Professor David — The Life of Sir Edgeworth David*, Edward Arnold & Co., London, 1937.
4. Woolnough, W.G. 'Professor David — an Appreciation', *Lone Hand*, June 1969, pp. 201-205.
5. David, op. cit.
6. Johns, R.K. (ed.). *History and Role of Government Geological Surveys in Australia*, South Australian Department of Mines and Energy, Adelaide, 1976.
7. Author unknown. 'The Search For Petroleum', *Times*, Imperial and Foreign Trade Supplement, London, 19 February 1921, p.16.
8. Meudell, G.D. *The Pleasant Career of a Spendthrift*, George Routledge & Sons, London, 1929.
9. Raggatt, H.G. *Early Attempts to Form a Commonwealth Geological Survey*, BMR Record 1956/150, Canberra, 1956.
10. Lewis, J. 'The History of the Oil Industry in South Australia', *From The Coorong to Port Bonython*, Proceedings of the Australian Institute of Petroleum, South Australian branch, Adelaide, October 1983.
11. Wade, A. 'Appendix to Geological Survey of South Australia', *Bulletin*, No.4, South Australian Department of Mines, 1915.
12. Robertson, C.S. 'Australia's Petroleum Prospects: Changing Perceptions Since the Beginning of the Century', *APEA Journal*, Vol.1, 1988.
13. Meudell, op. cit.
14. Andrews, E.C. 'Prospecting for Petroleum in Australia', *Economic Geology*, Vol.19, No.2, 1924.
15. Robertson, op. cit.
16. Condit, D.D. 'Geological Factors in Oil Prospecting with Reference to Australia', *Australian Geographer*, Vol.2, No.6, 1935.
17. Jensen, H.I. 'The Probable Oil Formations of North East Australia', *Pan Pacific Science Congress Proceedings*, Vol.2, 1923, pp. 1254-76.
18. Cambage, R.H. 'Australian Resources of Liquid Fuels', *Royal Society of New South Wales Journal and Proceedings*, Vol.58, 1924, pp. 15-60.
19. Raggatt, H.G. & Montgomery, J.H. 'Arthur Wade (1878-1951)', *Bulletin of the American Association of Petroleum Geologists*, Vol.35, 1951, pp. 2643-45.

20. Robertson, op. sit.
21. Raggatt, 'Early Attempts...', op. cit.
22. Robertson, op. cit.
23. Amos, D.J. *The Story of the Commonwealth Oil Refineries and the Search for Oil,* booklet published by E.J. McAlister & Co., Adelaide, 1935.
24. Woolnough, W.G. *Report on Tour of Inspection of the Oil Fields of the United States of America and Argentina and on Oil Prospects in Australia,* Report to Parliament of Commonwealth of Australia, Canberra, March 1931.
25. Woolnough, W.G. 'Prospecting For Oil', in Ion Idriess, *Prospecting For Gold, Other Minerals and Precious Stones,* Angus & Robertson, Australia, 1932.
26. Raggatt, H.G. 'Walter George Woolnough', *Bulletin of the American Association of Petroleum Geologists,* No.43, 1959, pp. 2035–40.
27. Author unknown. 'He Sought The Good Oil', an article on Woolnough in *People Magazine,* July 18, 1951.
28. Laws, Wing Commander. Report of an address to Perth Legacy Club, *West Australian,* Perth, 8 November 1933.
29. Woolnough, W.G. 'The Aeroplane in the Service of Geology', Clarke Memorial Lecture, *Royal Society of New South Wales Journal and Proceedings,* Vol.70, 1936, pp. 39–62.
30. Raggatt, 'Walter George Woolnough', op. cit.
31. Thyer, R.F. 'Georoots — Early Geophysical Prospecting in Australia', *Bulletin of the Australian Society of Exploration Geophysicists,* Vol.10, No.4, December 1979.
32. ibid.
33. Crespin, I. *Recollections on Growth of Commonwealth Interest in Geological Sciences,* BMR Record No.1967/157, Canberra, 1957.
34. ibid.
35. Fisher, N.H. 'Sir Harold Raggatt 1900–1968', *Journal of Geological Society of Australia,* Vol.16, Part 1, Sydney, 1969.
36. Raggatt, 'Early Attempts...', op. cit.
37. Croll, I.H.C. 'Prospecting For Oil in Australia', *Mining and Geological Journal of Victoria,* Vol.1, No.2, 1938.

PART TWO

Chapter 5 — BMR Leads the Way

1. Fisher, N. 'Sir Harold Raggatt (1900–1968)', *Journal of Geological Society of Australia,* Vol.16, Part 1, Sydney, 1969.
2. Johns, R.K. (ed.). *History and Role of Government Geological Surveys in Australia,* South Australian Department of Mines, Adelaide, 1976.

3. Ćrespin, I. *Memoirs of a Micropalaeontologist,* BMR publication, Canberra, 1980.
4. ibid.
5. Fisher, N. Interview with author, Sydney, 1987.
6. ibid.
7. Traves, D.M. Interview with author, Surfers Paradise, 1987.
8. Guppy, D. Interview with author, Canberra, 1986.
9. ibid.
10. ibid.
11. Fisher, Interview, op. cit.
12. Guppy, op. cit.
13. ibid.
14. Lindner, A. Interview with author, Sydney, 1989.
15. ibid.
16. Guppy, op. cit.
17. Johnstone, D. Interview with author, Perth, 1988.
18. Traves, D.M. & Casey, J.N. 'Exploring For Oil in the Canning Basin Desert (Western Australia)', *Walkabout,* Sydney, January 1, 1956.
19. Reynolds, M. Interview with author, Melbourne, 1987.
20. Smith, R. Interview with author, Melbourne, 1987.
21. Haddock, J. Interview with author, Perth, 1988.
22. Smith, op. cit.
23. ibid.
24. ibid.
25. ibid.
26. Taylor-Rogers, H. Interview with author, Canberra, 1986.
27. ibid.
28. ibid.

Chapter 6 — Walking with Walkley

1. Polkinghorne, A. Interview with author, Melbourne, 1986.
2. Guppy, D. Interview with author, Canberra, 1986.
3. Polkinghorne, op. cit.
4. ibid.
5. Reeves, F. 'Australian Oil Possibilities', *Bulletin of American Association of Petroleum Geologists,* Vol.35, No.12, 1951.
6. Raggatt, H. 'Search for Oil in Australia and New Guinea', *Royal Society of New South Wales Proceedings,* pp. S5–S21, 1955.
7. Reeves, op. cit.
8. Knife, C. Interview with author, Sydney, 1987.
9. Johnstone, M. Interview with author, Melbourne, 1987.
10. Author unknown. 'Oil — Lifestream of Progress', Caltex house magazine, supplementary issue, New York, December 1954.
11. Author unknown. *Oil For Tomorrow,* Ampol publication, Sydney, circa 1955.

12. 'Oil — Lifestream of progress', op. cit.
13. Haddock, J. Interview with author, Perth, 1989.
14. Evans, O. Interview with author, Perth, 1989.
15. Haddock, op. sit.
16. Johnstone, op. sit.
17. Knife, op. sit.
18. Author unknown. *WAPET News Digest,* Perth, April 1955.
19. Haddock, op. cit.
20. Parry, J. Interview with author, Perth, 1988.
21. Johnstone, op. sit.
22. Elliot, R. Interview with author, Perth, 1988.
23. Parry, op. sit.
24. ibid.
25. ibid.
26. Haddock, op. sit.
27. Cook, B. Interview with author, Perth, 1988.
28. ibid.
29. Elliot, op. cit.
30. Author unknown. *WAPET News Digest,* Perth, April 1956.
31. Haddock, op. cit.
32. Author unknown. *WAPET News Digest,* Perth, May 1959.
33. Johnstone, op. cit.
34. Elliot, op. cit.

Chapter 7 — Impetuous Disciples

1. Sprigg, R. 'The Search For Commercial Oil and Gas in South Australia', *From the Coorong to Port Bonython,* Proceedings of the Australian Institute of Petroleum, South Australian branch, Adelaide, October 1983.
2. Llewellyn, K. Interview with author, Ballarat, 1987.
3. ibid.
4. Zehnder, J. Interview with author, Sydney, 1987.
5. ibid.
6. Llewellyn, op. cit.
7. Zehnder, op. cit.
8. Laing, C. 'Early years in Queensland/PNG Oil Exploration', an address to Petroleum Exploration Society of Australia, Queensland branch, Brisbane, July 1985.
9. Prince, A. Interview with author, Sydney, 1988.
10. Smith, C.W. Interview with author, Melbourne, 1989.
11. Author unknown. 'Search For Oil Continues', *Petroleum Gazette,* Vol.4, No.3, p. 5.
12. Withers, R. Interview with author, Melbourne, 1987.
13. ibid.
14. ibid.

15. ibid.
16. ibid.
17. ibid.
18. Richards, K. Interview with author, Sydney, 1988.
19. Byrne, C. Interview with author, Caloundra, 1987.
20. Richter, R. Interview with author, Brisbane, 1987.
21. Laing, op. cit.
22. Dumbrell, R. Interview with author, Sydney, 1987.
23. Siller, C.W. Interview with author, Brisbane, 1987.
24. ibid.
25. Author unknown. *The ODE Story*, company publication, Sydney, October 1981.
26. Laing, op. cit.
27. Madden, L. Interview with author, Sydney, 1987.
28. Siller, op. cit.
29. Paten, R. Interview with author, Brisbane, 1987.
30. McPhee, I. Interview with author, Melbourne, 1987.
31. Calderwood, M. Interview with author, Brisbane, 1988.
32. ibid.
33. ibid.
34. Derrington, S. Interview with author, Brisbane, 1987.
35. Clow, D. Interview with author, Brisbane, 1987.
36. Laing, op. cit.
37. Derrington, op. sit
38. Author unknown. 'AOG Spuds in at Kurrajong Heights — NSW', *Australasian Oil & Gas Journal*, February 1955.
39. McGarry, D. Interview with author, Sydney, 1987.
40. Sprigg, 'The Search for Commercial Oil and Gas ...', op. cit.
41. Sprigg, R. Correspondence with author, 1989.
42. Wopfner, H. 'From Coorongite to Cooper Basin and Beyond', an address to APEA Professional Division, Adelaide, October 1972.
43. ibid.
44. Fitzpatrick, B. Correspondence with author, 1987.
45. ibid.
46. Sprigg, R. Interview with author, Melbourne, 1987.
47. Richter, op. cit.
48. Derrington, op. cit.
49. Madden, op. cit.

Chapter 8 — A Technological Awakening

1. Johnstone, M. 'Years of Dramatic Change — The Growth in Expertise in Australian Oil Exploration 1950–1986', PESA Australian Distinguished Lecture, 1986.
2. Robertson, C.S. 'Australia's Petroleum Prospects: Changing Perceptions Since the Beginning of the Century', *APEA Journal*, Sydney, 1988.

3. ibid.
4. Playford, P.E. & Johnstone, M.H. 'Oil Exploration in Australia', *Bulletin of American Association of Petroleum Geologists,* Vol.43, No.2, February 1959.
5. Johnstone, 'Years of Dramatic Change...', op. cit.
6. Hill, Professor D. Interview with author, Brisbane, 1988.
7. Johnstone, D. Interview with author, Perth, 1989.
8. ibid.
9. Johnstone, 'Years of Dramatic Change...', op. cit.
10. Doyle, H.A. 'Geophysics in Australia', submitted for publication in *Earth Sciences History,* 1987.
11. Author unknown. 'Drilling Rigs', *The WAPET Story,* a company publication, circa 1958.
12. ibid.
13. Campbell, I.R. & Parry, J.C. 'Oil Drilling Onshore and Offshore', in *Mining in Western Australia* (ed. Rex Prider), University of Western Australia Press, Perth, 1979.
14. Johnstone, 'Years of Dramatic Change...', op. cit.
15. Johnstone, M. Interview with author, Melbourne, 1986.
16. West, B. 'Wellsite Services in Australia', *Petroleum in Australia: The First Century,* APEA, Sydney, 1988.
17. ibid.
18. Author unknown. 'Papuan Venture', *The Accelerator* (house magazine of Commonwealth Oil Refineries), August 1954.
19. Author unknown. 'Cavalcade of Transport in Oil Exploration in New Guinea', *Australasian Oil & Gas Journal,* May 1957.
20. Author unknown. 'Oil Search Moves On Wheels', *WAPET News Digest,* Perth, June 1957.
21. Author unknown. 'Bulldozers Of The Sea', *WAPET News Digest,* Perth, May 1957.
22. ibid.

Chapter 9 — Independence or Bust

1. Robertson, C.S. 'Australia's Petroleum Prospects: Changing Perceptions Since the Beginning of the Century', *APEA Journal,* 1988.
2. Sprigg, R. Interview with author, Melbourne, 1987.
3. Spooner, Senator W. *Hansard,* Vol.S.11, November 25, 1957, p. 1572.
4. Taylor-Rogers, H. Interview with author, Canberra, 1986.
5. Traves, D. 'Commonwealth Subsidy', *APEA Journal,* 1962.
6. Williams, L.W. 'A Review of the Petroleum Search Subsidy Acts', *APEA Journal,* 1974.
7. White, J. Interview with author, Canberra, 1987.
8. Raynor, J.M. 'Governmental Activities in Petroleum Exploration', *APEA Journal,* 1969.

9. Prince, A. Interview with author, Sydney, 1987.
10. Withers, R. Interview with author, Melbourne, 1987.
11. Siller, C.W. Interview with author, Brisbane, 1987.
12. Fisher, N. Interview with author, Sydney, 1986.
13. Siller, Interview, op. cit.
14. Author unknown. *QUPEX, Twenty-Fifth Anniversary 1959–1984,* QUPEX Steering Committee, Brisbane, November 1984.
15. Siller, C.W. 'QUPEX During 30 Years', an address to QUPEX, Brisbane, March, 1989.

PART THREE

Chapter 10: The Onshore Comeback

1. Author unknown. 'The Bounty of Natural Gas', *Petroleum Gazette,* Melbourne, Vol.9, No.1, February 1960.
2. Stach, L. Interview with author, Melbourne, 1988.
3. Cundill, J. Interview with author, Perth, 1987.
4. McCulloch, R. Interview with author, Brisbane, 1987.
5. Rudd, E. Interview with author, Adelaide, 1987.
6. McGarry, D. Interview with author, Sydney, 1987.
7. Dumbrell, R. Interview with author, Melbourne, 1987.
8. Graves, D. *Cabawin-Moonie-Alton, Australia's First Commercial Oil, A Personal Album of Events,* Brisbane, November 1986.
9. McGarry, op. cit.
10. Author unknown. Notes on Moonie oil discovery, BMR notes and papers, Canberra 1986.
11. Graves, op. cit.
12. Author unknown. 'Queensland Scene of Renewed Activity, with Offshore Leases Opened', *Petroleum Gazette,* Vol.22, No.3, September 1980.
13. Graves, op. cit.
14. Clow, D. Interview with author, Brisbane, 1987.
15. Dumbrell, op. cit.
16. Wright, T. Interview with author, Adelaide, 1988.
17. Watts, T. Interview with author, Adelaide, 1988.
18. Author unknown. Cameo note on H.A. Hackathorn, *Bulletin,* Sydney, 13 April, 1960.
19. Twist, R. Interview with author, Perth, 1987.
20. Fuller, J. Interviews with author, Sydney, 1980 and 1987.
21. Spinks, R. Correspondence with author, April 1987.
22. McCulloch, op. cit.
23. Prackert, E. Interview with author, Brisbane, 1987.
24. Fitzpatrick, B. Correspondence with author, June, 1987.
25. Wopfner, H. 'From Coorongite to Cooper Basin and Beyond', an address to APEA Professional Division, Adelaide, October 1972.

26. Stephens, G. Interview with author, Perth, 1987.
27. Considine, J. Interview with author, Adelaide, 1987.
28. Johns, N. Interview with author, Alice Springs, 1987.
29. Siller, C.W. Interview with author, Brisbane, 1987.
30. Dumbrell, op. cit.
31. Hopkins, R. Interview with author, Brisbane, 1987.
32. Taylor-Rogers, H. Interview with author, Canberra, 1986.
33. Hopkins, op. cit.
34. Hollingsworth, R. Interviews with author, Adelaide, 1987 and 1988.
35. Author unknown. 'Search In Unusual Places', *WAPET News Digest*, August 1967.

Chapter 11: The Island Programmes

1. Stach, L. Interview with author, Melbourne, 1988.
2. McCulloch, R. Interview with author, Brisbane, 1987.
3. Author unknown. 'The Basin and The Brecha', *Delhi Magazine*, January-February 1962.
4. Brown, B. Interview with author, Melbourne, 1988.
5. Haddock, J. Interview with author, Perth, 1989.
6. Cook, B. Interview with author, Perth, 1988.
7. Nasr, S. Interview with author, Sydney, 1989.
8. Elliot, R. Interview with author, Perth, 1988.
9. Clow, D. Interview with author, Brisbane, 1987.
10. Author unknown. 'Barrow Island — Where Oil and Wildlife Co-exist In Harmony', *Australian Oil and Gas Directory*, Oil and Gas Directory Pty. Ltd., Melbourne 1984-85.
11. Cook, op. cit.
12. Haddock, op. cit.

Chapter 12: The Offshore Revelation

1. Lonie, M. Interview with author, Melbourne, 1988.
2. Richards, K. Interview with author, Sydney, 1988.
3. Sprigg, R. Address on receiving inaugural Weeks Medal, Sydney, May 1982.
4. Reynolds, M. Interview with author, Melbourne, 1987.
5. Lonie, op. cit.
6. Hopkins, B.M. Interview with author, Melbourne, 1989.
7. Hopkins, B.M. 'Oil Under The Sea', *BHP Review*, Melbourne, June 1964.
8. Lonie, op. cit.
9. Richards, op. cit.
10. Evans, J. Interview with author, Melbourne, 1989.
11. Richards, op. cit.
12. Evans, op. cit.

13. Adams, A. Interviews with author, Longford, 1987 and 1988.
14. Author unknown. *Barracouta, The First 20 Years: 1967–1987*, Esso publication, Sydney, 1987.
15. Adams, op. cit.
16. Washington, P. Interview with author, Sale, 1987.
17. Morgans, J. Interview with author, Melbourne, 1987.
18. Thomas, R. Interview with author, Adelaide, 1986.
19. Durkee, E. Correspondence with author, November, 1986.
20. Boutakoff, N. Personal papers, March 1973.
21. Withers, R. Interview with author, Melbourne, 1987.
22. Boutakoff, op. cit.
23. Garbin, V. Interview with author, Perth, 1987.
24. Edmond, G. Interview with author, Perth, 1987.
25. Morgans, op. cit.
26. Langton, D. Interview with author, Melbourne, 1987.
27. Johnstone, D. Interview with author, Perth, 1988.
28. Haddock, J. Interview with author, Perth, 1989.
29. Washington, op. cit.
30. White, J. Interview with author, Canberra, 1987.
31. Spinks, R. 'Offshore Drilling Operations in the Gulf of Papua', *APEA Journal*, 1970.

Chapter 13: A Question Of Unity

1. Guillemot, J. & Tissot, B. *Petroleum Prospects in Australia, General Review of Main Sedimentary Basins*, Institute Francais du Petrole, August 1960.
2. Trumpy, D. *Mission in Australia, Recent Developments of Petroleum Prospects in Australia*, Institute Francaise du Petrole, Bureau des Etudes Geologiques, September 1963.
3. Robertson, C.S. 'Australia's Petroleum Prospects: Changing Perceptions Since the Beginning of the Century', *APEA Journal*, 1988.
4. Sprigg, R. Interview with author, Melbourne, 1987.
5. Prince, A. Interview with author, Sydney, 1987.
6. Sprigg, R. 'The Search for Commercial Oil and Gas in South Australia', *From the Coorong to Port Bonython*, Proceedings of the Australian Institute of Petroleum, South Australian branch, Adelaide, October 1983.
7. Sprigg, Interview, op. cit.
8. Withers, R. Interview with author, Melbourne, 1987.
9. Williams, L.W. 'A Review of the Petroleum Search Subsidy Acts', *APEA Journal*, 1974.
10. Author unknown. 'Oil Drilling Plans Upset by Cut in Subsidy', *Australian Financial Review*, 9 August 1962.
11. Sprigg, R. 'APEA Seeks Subsidy Increase', press release issued by APEA, 19 July, 1965.

12. Williams, op. cit.
13. Blumer, J.M. 'The Petroleum Tenement Map of Australia', *APEA Journal,* 1982.
14. Reynolds, M.A. *The Sedimentary Basins of Australia and the Stratigraphic Occurrences of Hydrocarbons,* BMR Record 1965/196, Canberra, 1965.
15. Thompson, J.E. *The Sedimentary Basins of the Territory of Papua and New Guinea and the Stratigraphic Occurrence of Hydrocarbons,* BMR Record 1965/176, Canberra, 1965.
16. Johnstone, M. 'Years of Dramatic Change — The Growth of Expertise in Australian Oil Exploration 1950-1986', PESA Distinguished Lecture, 1986.
17. Sprigg, R. 'Towards 2000: Incentives Versus Disincentives', Opening Address, APEA Conference, Adelaide, 1986.
18. Taylor-Rogers, H. Interview with author, Canberra, 1986.
19. Mott, R.H. 'Australian Offshore Petroleum Law', Research Paper, University of Melbourne, August 1971.
20. Wilkinson, R. 'APEA: The First 28 Years', *Petroleum in Australia, The First Century,* Australian Petroleum Exploration Association, March 1988.
21. Watson, S.J. 'Recent Advances in Geophysical Techniques', *APEA Journal,* 1968.
22. McGarry, D. 'Education and the Needs of the Petroleum Industry in Australia', *APEA Journal,* 1967.

PART FOUR

Chapter 14 — Both Feet on the Treadmill

1. Author unknown. 'How Much Is Enough ?', *Petroleum Gazette,* Vol.16, No.2, June 1970.
2. Wells, W. Interview with author, Melbourne, 1988.
3. Rudd, E. 'Thomas William Hamilton Dee — An Appreciation', *Australasian Oil & Gas Review,* December 1970.
4. Wells, op. cit.
5. Campbell, I. Interview with author, Melbourne, 1989.
6. Author unknown. 'Cooper's Country Comes Good', *Petroleum Gazette,* Vol.16, No.8, December 1971.
7. Imbert, M. Interview with author, Melbourne, 1989.
8. Cadart, M. Interview with author, Melbourne, 1988.
9. St John, P. Interview with author, Melbourne, 1989.
10. Brown, B. Interviews with author, Melbourne, 1988-1989.
11. Wilkinson, R. 'Henri Flirted With Music, But Married The Sea', Worlds Apart, supplement in *Offshore Engineer,* London, May 1976.
12. Wilkinson, R. *A Thirst For Burning,* David Ell Press, Sydney, 1988.
13. Curnow, C. Interview with author, Sydney, 1989.

14. Wright, C. 'Geoff Donaldson — last of the Pioneers', *Australian Business,* 20 March 1985.
15. Reynolds, M. Interview with author, Melbourne, 1987.
16. Edmond, G. Interview with author, Perth, 1987.
17. Garbin, V. Interview with author, Perth, 1987.
18. Agostini, D. Interview with author, Perth, 1989.
19. Whitley, W.R. 'Effect of Tropical Cyclones on Offshore Drilling Operations on North West Shelf, Australia', *Proceedings of the Symposium on the Impact of Tropical Cyclones on Oil and Mineral Development in North-west Australia,* Australian Department of Science, Perth, 1975.

Chapter 15 — Labor Deals Its Hand

1. Author unknown. 'Oil Subsidy Hits Locals', Melbourne *Herald,* 22 October 1970.
2. Murray, R. 'Oil Explorers Want A Higher Australian Price; Users Might Object', *Australian Financial Review,* 17 December 1971.
3. Author unknown. 'Oiling The Oil Exploration Effort', *Finance Week,* 14 April 1972.
4. Author unknown. 'Burmah Director Attacks Oil Search Policies', *Australian Financial Review,* 11 April 1972.
5. Wright, C. 'Geoff Donaldson: Last of the Pioneers', *Australian Business,* 30 March 1985.
6. Author unknown. 'Disaster, Says APEA Director', Hobart *Mercury,* 9 May 1973.
7. Webb, T. Address to annual meeting of Alliance Oil Development Australia NL, 24 October 1973.
8. Webb, E. Interview with author, Melbourne, 1989.
9. Parkin, L. Interview with author, Adelaide, 1987.
10. Cheesewright, A. 'Government Must End Restraints On Oil Search', Melbourne *Herald,* 24 July 1975.
11. Easley, C. 'Vice-Chairman's Concluding Address', *APEA Journal,* 1970.
12. Sprigg, R. 'Conservation of Australia's Arid Land in Relation to Continuing Exploration and Developmental Needs', *APEA Journal,* 1973.

Chapter 16 — Unexpected Success and a Deep-sea Gamble

1. Brackenridge, A.H. 'Concluding Conference Address', *APEA Journal,* 1976.
2. Charlton, R.M. 'Petroleum 1978: Focus On A Vital Resource', *APEA Journal,* 1978.
3. Author unknown. 'Bass Strait', *Petroleum Gazette,* Vol.22, No.3, September 1980.

4. Brown, B. Interview with author, Melbourne, 1989.
5. Hollingsworth, R. Interview with author, Adelaide, 1986.
6. Curnow, C. Interview with author, Sydney, 1989.
7. Cameron, P. Interview with author, Melbourne, 1989.
8. Author unknown. 'Oil Play On The Plateau', *Petroleum Gazette*, June 1978.
9. St John, P. Interview with author, Melbourne, 1989.

Chapter 17 — The Changing Tools of Trade

1. Coffman, J.B. 'Technology, Risk and Cost', *APEA Journal*, 1979.
2. Author unknown. 'Bright Spot In Oil Search', *Petroleum Gazette*, September 1974.
3. Poll, J.J.K. 'Onshore Gippsland Geochemical Survey: A Test Case For Australia', *APEA Journal*, 1975.
4. Mulready, J. Interview with author, Melbourne, 1989.
5. Author unknown. 'Dry Welds And Saturated Men', *Petroleum Gazette* Vol.16, No.2, June 1970.
6. Author unknown. 'JIMiny ! A Marine Monster', *Petroleum Gazette* Vol.19, No.5, July 1977.
7. Author unknown. 'Float Off !' *Petroleum Gazette*, Vol.18, No.5, April 1975.
8. McKinney, J.M. 'Deep Water Drilling And Production Systems Overview', *APEA Journal*, 1979.
9. Blumer, J.M. 'The Petroleum Tenement Map Of Australia', *APEA Journal*, 1982.
10. Robertson, C.S. 'Australia's Petroleum Perspectives: Changing Perceptions Since The Beginning Of The Century', *APEA Journal*, 1988.

PART FIVE

Chapter 18 — In Search of a Saviour

1. Garland, E.H.C. 'New Decade — New Challenges', *APEA Journal*, 1980.
2. Sykes, I. Interview with author, Melbourne, 1989.
3. Glazebrook, C. Interview with author, Melbourne, 1989.
4. Ell, R. Interview with author, Moomba, 1988.
5. McCulloch, R. Interview with author, Brisbane, 1987.
6. Author unknown. 'Is Blina A Fluttering Eyelid Of A Sleeping Giant?' *Petroleum Gazette*, spring 1984.
7. Author unknown. 'Lonely Christmas trees Bereft Of Tinsel Still Promise Joy', *Petroleum Gazette*, Vol.23, No.3, September 1982.
8. Angove, R. 'The Point Torment Seismic Survey: A Semi-Portable Seismic Operation', *APEA Journal*, 1985.

9. Doran, J. Interview with author, Sydney, 1989.
10. Cameron, P. Interview with author, Melbourne, 1989.
11. Imbert, J.M. Interview with author, Melbourne, 1989.
12. Author unknown. *Jabiru: A Billion Dollar Baby!*, Research report, Peter Haines & Company, Sydney, December 1983.
13. Author unknown. 'Despite Let-Down, Timor Sea Holds Prospects', *Petroleum Gazette*, spring 1984.
14. Mackenzie, A. Interview with author, Troughton Island, 1988.
15. MacErlich, J. Interview with author, Timor Sea, 1988.
16. Elliot, C. Interview with author, Melbourne, 1989.
17. Durkee, E. Correspondence with author, 1988.
18. Author unknown. 'Iagifu', *Petroleum Gazette*, autumn 1987.
19. Imbert, op. cit.
20. ibid.

Chapter 19 — A Matter of Incentives

1. Richards, K.A. 'Concluding conference address', *APEA Journal*, 1980.
2. Orchison, K. Interviews with author, Brisbane/Sydney, 1988–1990.
3. Hopkins, R. Interview with author, Brisbane, 1987.
4. Reynolds, M. Interview with author, Melbourne, 1987.
5. Orchison, Interview, op. cit.
6. Toyne, P. Interview with author, Melbourne, 1981.
7. Orchison, Interview, op. cit.
8. Orchison, K. *Land Use in Australia: A Petroleum Exploration and Production Industry Perspective*, APEA, Sydney, May 1988.
9. Morozow, O. 'Access to Land for Exploration: The Adoption of Multiple Land-Use Principles in South Australia', *APEA Journal*, 1988.
10. Orchison, Interview, op. cit.
11. Robertson, C.S. 'Australia's Petroleum Prospects: Changing Perceptions Since the Beginning of the Century', *APEA Journal*, 1988.
12. West, B. 'Wellsite Services in Australia', *Petroleum in Australia: The First Century*, APEA, Sydney, 1988.
13. Mulready, J. Interview with author, Melbourne, 1986.
14. Richards, K.A. 'Chairman's Address', *APEA Journal*, 1983.
15. Erskine, D. 'APEA Warns Extra Tax Will Trigger Cuts In Oil Exploration', *Australian*, 18 May 1984.
16. Gill, P. 'APEA Chief Asks For Oil Tax Relief', Melbourne *Age*, 7 April 1986.
17. Benbow, D. 'Chairman's Address', *APEA Journal*, 1986.
18. Evans, Senator G. 'Australian Government Policies and the 1986 Oil Shock', *APEA Journal*, 1986.
19. Orchison, Interview, op. cit.
20. Gill, P. 'When Oil Sells For A Song, Expert Shows He's In Tune', Melbourne *Age*, 10 April 1986.

PART SIX

Chapter 20 — Feathers for the Phoenix

1. APEA, *Submission to Federal Government concerning oil excise review*, October 1989.
2. Foley, G. 'Raising Risk Capital', Australian Petroleum Exploration Association, Canberra seminar, 26 September 1990.
3. Koroknay, S. 'Oil Exploration and Development in Australia: Is Blue Sky Enough ?', 1989 Conference On Energy, Sydney, August 1989.
4. Doran, J. Interview with author, Sydney, 1989.
5. Paten, R. Interview with author, Brisbane, 1988.
6. Sprigg, R. Interview with author, Melbourne, 1987.
7. Glazebrook, C. Interview with author, Melbourne, 1989.
8. Webb, E. Interview with author, Melbourne, 1989.
9. Richards, K. 'Unconventional Wisdom', address to PESA New South Wales branch annual dinner, 1985.
10. Armstrong, J. Interview with author, Melbourne, 1990.
11. Thomas, R. Interview with author, Adelaide, 1987.
12. Maxwell, D. Interview with author, Perth, 1989.
13. Devine, B. Interview with author, Adelaide, 1989.
14. Poll, J. 'Onshore Canning Basin: An Historic Perspective', *PESA Journal*, No.3, 1983.
15. Johnstone, M. 'Years of Dramatic Change: The Growth in Expertise in Australian Oil Exploration 1950-1986', PESA Australian Distinguished Lecture, 1986.
16. Richards, op. cit.
17. Black, Sir Hermann. 'Principal Address', *APEA Journal*, 1977.
18. Sprigg, op. cit.

Index

CHARACTERS

Index

PLACES AND EVENTS